China Water Forum 2022

China Water Forum 2022

Editors

Qiting Zuo
Fuqiang Wang
Jiaqi Zhai
Xiuyu Zhang
Dunxian She
Lei Zou
Rong Gan
Zengliang Luo

MDPI

Basel • Beijing • Wuhan • Barcelona • Belgrade • Novi Sad • Cluj • Manchester

Editors
Qiting Zuo
Zhengzhou University
Zhengzhou
China

Fuqiang Wang
North China University of Water Resources and Electric Power
Zhengzhou
China

Jiaqi Zhai
China Institute of Water Resources and Hydropower Research
Beijing
China

Xiuyu Zhang
North China University of Water Resources and Electric Power
Zhengzhou
China

Dunxian She
Wuhan University
Wuhan
China

Lei Zou
Chinese Academy of Sciences
Beijing
China

Rong Gan
Zhengzhou University
Zhengzhou
China

Zengliang Luo
China University of Geosciences
Wuhan
China

Editorial Office
MDPI
St. Alban-Anlage 66
4052 Basel, Switzerland

This is a reprint of articles from the Special Issue published online in the open access journal *Water* (ISSN 2073-4441) (available at: www.mdpi.com/journal/water/special_issues/china_water_forum_2022).

For citation purposes, cite each article independently as indicated on the article page online and as indicated below:

Lastname, A.A.; Lastname, B.B. Article Title. *Journal Name* **Year**, *Volume Number*, Page Range.

ISBN 978-3-0365-8639-7 (Hbk)
ISBN 978-3-0365-8638-0 (PDF)
doi.org/10.3390/books978-3-0365-8638-0

Contents

water

Editorial

A Summary of China's Water Security Status and Issues from a Special Issue

Qiting Zuo [1,2], Fuqiang Wang [3], Jiaqi Zhai [4], Xiuyu Zhang [3], Dunxian She [5], Lei Zou [6], Rong Gan [1] and Zengliang Luo [7,*]

1 School of Water Conservancy Engineering, Zhengzhou University, Zhengzhou 450001, China
2 Yellow River Institute for Ecological Protection & Regional Coordinated Development, Zhengzhou University, Zhengzhou 450001, China
3 College of Water Resources, North China University of Water Resources and Electric Power, Zhengzhou 450045, China
4 China Institute of Water Resources and Hydropower Research (IWHR), Beijing 100038, China
5 School of Water Resources and Hydropower Engineering, Wuhan University, Wuhan 430072, China
6 Institute of Geographic Sciences and Natural Resources Research, Chinese Academy of Sciences, Beijing 100101, China
7 School of Geography and Information Engineering, China University of Geosciences, Wuhan 430074, China
* Correspondence: luozengliang@cug.edu.cn

Affected by global climate change and rapid socio-economic development, China faces serious water security issues, especially in terms of water shortages, flood disasters, and water-related ecological and environmental problems. These challenges associated with China's water issues have restricted the country's social and economic development, and attracted global attention. Due to the complexity of the water-related issues in China, the solution to these requires comprehensive research across different disciplines.

A series of academic exchanges have been carried out to discuss China's water issues, including the well-known "China Water Forum", which focuses on discussing solutions to China's water issues, the "Water Science Development Forum", which focuses on promoting the multidisciplinary integration of water science and other relevant disciplines to discuss water issues in China, and the "Water Science Lectures", which is an open, shared and non-profit public welfare lecture.

In order to better discuss China's water issues, the Editorial Department of *Water* cooperated with "China Water Forum", "Water Science Development Forum", and "Water Science Lectures" to set up a series of columns. At least one column is planned each year currently. The column established in 2022, "China Water Forum 2022", has ended and the new column set up in 2023, namely "China Water Forum 2023", has begun. A total of 11 papers were accepted in the column "China Water Forum 2022". According to the papers published in the column, we summarize China's water security status and issues to help readers understand the relevant research progress.

Four themes are designed in "China Water Forum 2022", as follows:

(1) Climate change and hydrology;
(2) Water problems and human–water relationship control;
(3) Water environment and ecology;
(4) Water information technology and modeling.

The eleven papers published in this Special Issue discuss China's water issues from different aspects. They were divided into three categories based on their research themes, including Category A: "the groundwater issues in parts of China" [1,2], Category B: "the water ecological and environmental issues of rivers and lakes" [3–6], and Category C: "the sustainable utilization of water resources and human–water sustainable development" [7–11].

For Category A, "The groundwater issues in parts of China", Wang et al. [1] set up a numerical model to simulate the saltwater upconing and recovery process in a small coral

Citation: Zuo, Q.; Wang, F.; Zhai, J.; Zhang, X.; She, D.; Zou, L.; Gan, R.; Luo, Z. A Summary of China's Water Security Status and Issues from a Special Issue. *Water* **2023**, *15*, 1628. https://doi.org/10.3390/w15081628

Received: 28 March 2023
Accepted: 5 April 2023
Published: 21 April 2023

island located on the Xisha Islands in the South China Sea to determine the exploitable coefficient of the freshwater lens in the island. They concluded that the exploitable coefficient of the freshwater lens in the small coral island is smaller than that in inland areas. Wu et al. [2] compared the performance of data-driven models to predict groundwater level with Hebei Plain as a case study. Their results showed that the gated recurrent unit model performed better than other data-driven models such as the support vector machine, long-short term memory, and multi-layer perceptron. Therefore, the determination of ground water level is worth exploring further in the future. The combination of data-driven and physics-based models may help to achieve better results.

For Category B, "The water ecological and environmental issues of rivers and lakes", Lu et al. [3] calculated the minimum lake water demand in East Juyan Lake, an inland desert terminal lake of the Heihe River in northwest China. They found that the area of the lake has increased over the past 15 years due to artificially ecological water diversion. A minimum annual water demand of 54×10^6 m^3/year was suggested for the lake; however, it is not currently satisfied. The results of Gao et al. [4] showed that silt contributes the most to lake sediment in the Hulun Lake, China, followed by sand. Zhang et al. [5] investigated the characterization of dissolved organic matter of sediments in streams of the Wuhan City. They showed a considerable impact of heavy metals on the characterization of dissolved organic matter concentrations and components. Liu et al. [6] proved an annual production flow reduction of about −28.2% in the Songhua River Basin under climate change and water use. The contribution of climate change and water use to annual runoff reduction is 77.0% and 23.0%, respectively. They also found that groundwater recharge increased by 9.2% and 4.1% during the freezing and thawing periods in the Songhua River Basin. According to the published papers, water ecological and environmental issues in rivers and lakes of China are still urgent problems to be solved. In detail, the most popular topics of discussion are ecological water demand, water pollutions, and streamflow alteration induced by climate change and human activities.

With regard to Category C, "The sustainable utilization of water resources and human-water sustainable development", Zuo et al. [7] proposed a carbon dioxide emission equivalent analysis method to quantify the carbon dioxide emission equivalent in 31 provinces in China. The results showed that reservoir storage, tap water allocation, and wastewater treatment are the main contributors to carbon dioxide emissions equivalent for water resource development, allocation, and protection behaviors, respectively. They suggest that increasing the proportion of hydroelectric power generation, improving ecological water security capacity, and strengthening the level of wastewater treatment and reclaimed water reuse are effective measures to promote carbon neutrality. Liang et al. [8] established a green development-level evaluation system to assess the green development of the Chengdu–Chongqing City Group. They showed that the overall green development level of the Chengdu–Chongqing City Group is on an upward trend. Li et al. [9] assessed the harmony degree and adaptive utilization capacity of water resources for sub-systems of water resources, economy, society, and ecology in the Tarim River Basin. They showed that the main factors affecting the harmony development and the adaptive utilization of water resources are per capita GDP, the proportion of non-agricultural output value in GDP, and the per capita net income of rural residents. Zhang et al. [10] analyzed the water-saving level in 31 provinces in China using support vector machine model and found that the water-saving level is high in Beijing City, Henan Province and Zhejiang Province. Overall, the water-saving level in North China, Central China and Southeast China was higher than that in Northwest China, Southwest China and Northeast China. Ju et al. [11] discussed the concept of a happy river. They posit that a happy river should be able to maintain its own health, support high-quality economic and social development in the river basin and the region, reflect harmony between humans and water, and provide people in the river basin with a high sense of security and the ability to gain satisfaction. They further analyzed the meaning of happy rivers from five levels, including water security, water resources, water environment, water ecology, and water culture.

Author Contributions: Conceptualization and Writing—Original Draft, Q.Z. and Z.L.; Writing—Review & Editing, F.W., J.Z., X.Z., D.S., L.Z. and R.G. All authors have read and agreed to the published version of the manuscript.

Funding: This research was supported by the Natural Sciences Foundation of China (No. 52279027), and the National Key Research and Development Program of China (No. 2021YFC3200201); Henan International Joint Laboratory of Water Cycle Simulation and Environmental Protection, and Zhengzhou Key Laboratory of Water Resource and Environment.

Acknowledgments: The authors acknowledge the contributions of all authors of the eleven papers in this Special Issue.

Conflicts of Interest: The authors declare no known conflict of interest.

References

1. Wang, R.; Shu, L.; Zhang, R.; Ling, Z. Determination of Exploitable Coefficient of Coral Island Freshwater Lens Considering the Integrated Effects of Lens Growth and Contraction. *Water* **2023**, *15*, 890. [CrossRef]
2. Wu, Z.; Lu, C.; Sun, Q.; Lu, W.; He, X.; Qin, T.; Yan, L.; Wu, C. Predicting Groundwater Level Based on Machine Learning: A Case Study of the Hebei Plain. *Water* **2023**, *15*, 823. [CrossRef]
3. Lu, J.; Li, L.; Jiang, E.; Gan, R.; Liu, C.; Deng, Y. Ecological Water Demand Estimations for Desert Terminal Lake Survival under Inland River Water Diversion Regulation. *Water* **2023**, *15*, 66. [CrossRef]
4. Gao, H.; Zhang, R.; Wang, G.; Fan, Y.; Zhu, X.; Wu, J.; Wu, L. A Dam Construction Event Recorded by High-Resolution Sedimentary Grain Size in an Outflow-Controlled Lake (Hulun Lake, China). *Water* **2022**, *14*, 3878. [CrossRef]
5. Zhang, H.; Liu, Z.; Xu, J.; Yang, J.; Zhang, X.; Tao, S. Characterization of Dissolved Organic Matter of Sediments in Urban Streams Using EEMs–PARAFAC and Absorption Spectroscopy: A Case Study in Wuhan, China. *Water* **2022**, *14*, 3181. [CrossRef]
6. Liu, S.; Zhou, Z.; Liu, J.; Li, J.; Wang, P.; Li, C.; Xie, X.; Jia, Y.; Wang, H. Analysis of the Runoff Component Variation Mechanisms in the Cold Region of Northeastern China under Climate Change. *Water* **2022**, *14*, 3170. [CrossRef]
7. Zuo, Q.; Zhang, Z.; Ma, J.; Zhao, C.; Qin, X. Carbon Dioxide Emission Equivalent Analysis of Water Resource Behaviors: Determination and Application of CEEA Function Table. *Water* **2023**, *15*, 431. [CrossRef]
8. Liang, Y.; Zhang, L.; Leng, M.; Xiao, Y.; Xia, J. System Simulation and Prediction of the Green Development Level of the Chengdu-Chongqing City Group. *Water* **2022**, *14*, 3947. [CrossRef]
9. Li, X.; Zuo, Q.; Zhai, J.; Zhao, Y.; Wang, Y.; Han, S. Evaluation of Adaptive Utilization Capacity of Water Resources and Analysis of Driving Element: A Case Study of Tarim River Basin. *Water* **2022**, *14*, 3820. [CrossRef]
10. Zhang, W.; Hou, S.; Yin, H.; Li, L.; Wu, K. Evaluation of Regional Water-Saving Level Based on Support Vector Machine Optimized by Genetic Algorithm. *Water* **2022**, *14*, 2615. [CrossRef]
11. Ju, Q.; Liu, C.; Jiang, S. Integrated Evaluation of Rivers Based upon the River Happiness Index (RHI): Happy Rivers in China. *Water* **2022**, *14*, 2568. [CrossRef]

water

MDPI

Article

Determination of Exploitable Coefficient of Coral Island Freshwater Lens Considering the Integrated Effects of Lens Growth and Contraction

Ran Wang, Longcang Shu *, Rongrong Zhang * and Zihan Ling

College of Hydrology and Water Resources, State Key Laboratory of Hydrology-Water Resources and Hydraulic Engineering, Hohai University, Nanjing 210098, China
* Correspondence: lcshu@hhu.edu.cn (L.S.); 20200609@hhu.edu.cn (R.Z.);
Tel.: +86-138-5194-1641 (L.S.); +86-139-1384-0412 (R.Z.)

Abstract: Groundwater on small coral islands (so-called freshwater lens) is an important water resource for residents and local ecosystems. However, an overexploitation of it may induce a contamination by saltwater. In this paper, we strive to determine the exploitable coefficient of the freshwater lens considering the integrated effects of lens growth and contraction and examine the impacts of well layout schemes on the evolution of the freshwater lens. For this purpose, a numerical model is setup to simulate the saltwater upconing and recovery process under pumping conditions during different evolution stages. Our results show that long-term and higher intensity pumping activities are suggested to be conducted at the latter stage of the lens evolution. Meanwhile, the seasonal contraction of the freshwater lens caused by the seasonal variation in rainfall is characterized by a quicker response of center thickness than maximum thickness of the lens, which further impacts the pumping intensity. The results also indicate that the exploitable coefficient (ρ) of the freshwater lens in small coral island is generally smaller than that in inland areas, ranging from 0.09 to 0.37 under different well layout schemes. Additionally, it is also affected by the uncertainty of hydrogeological parameters. Finally, a safe exploitable coefficient is proposed under the most unfavorable parameter combination for the studied island. The study has important implications for the protection and sustainable exploitation of subsurface freshwater resources on island.

Keywords: freshwater lens; small coral island; numerical simulation; sustainable exploitation

Citation: Wang, R.; Shu, L.; Zhang, R.; Ling, Z. Determination of Exploitable Coefficient of Coral Island Freshwater Lens Considering the Integrated Effects of Lens Growth and Contraction. *Water* **2023**, *15*, 890. https://doi.org/10.3390/w15050890

Academic Editor: Peiyue Li

Received: 14 January 2023
Revised: 21 February 2023
Accepted: 22 February 2023
Published: 25 February 2023

1. Introduction

Coral islands are distributed in the tropical and subtropical oceans of the Pacific and Indian oceans where corals are easy to grow [1]. Because of their extreme geographical isolation, unique geological structure, and vulnerable water resources, they attracted the attention of researchers in various fields [2,3]. The freshwater lens is a limited renewable underground freshwater resource on the coral island, which exists in the form of a floating "lens" [4–6]. Unlike most marine islands, coral islands have a double-layer aquifer system, which is composed of Holocene aquifer and Pleistocene aquifer [7]. The Holocene aquifer is unconformably covered by the Pleistocene limestone [5], and the freshwater lens is stored in the Holocene aquifer. Due to the high permeability of the soil and weathering layer of the small coral island, it is difficult to form surface runoff, and the freshwater lens became the only natural water supply source on the island [5]. Therefore, whether it is the early development of uninhabited natural islands or the daily life of inhabited islands, the reasonable development and utilization of the freshwater lens can play an important role in alleviating the water supply difficulties [1–3], with great economic, military, and social benefits. The amount of available groundwater is usually obtained by the exploitable coefficient method, while the exploitable coefficient of small coral islands is closely related to many factors, such as the hydrogeological conditions, the development of the freshwater lens, etc. Therefore, it is necessary to discuss the determination method of the exploitable coefficient in combination with the formation and evolution of the freshwater lens, the seasonal variation in precipitation, and the specific pumping well layout, so as to provide help for the sustainable development and utilization of the freshwater lens in the small coral island.

In order to scientifically and rationally develop and utilize the fresh groundwater resources of coral islands, scholars conducted in-depth research on the formation and internal fluid dynamics of freshwater lenses [8–13]. Zhou et al. [14] and Yao [15] simulated the formation and evolution of the freshwater lens on natural and artificial islands through numerical models; Li et al. [16] simulated the dynamic process of the freshwater lens under the conditions of recharge and pumping through physical experiments. Dose et al. [17] explored the influence of hydrogeological parameters on the flow movement of the freshwater lens through a physical sand tank experiment and numerical simulation; Post et al. [18] studied the long-term impact of abstraction on the freshwater lens through field observation and numerical simulation. These studies show that freshwater lenses are highly sensitive to climate factors and human activities [19,20], which must be considered in the sustainable development and utilization of freshwater lenses [21,22].

Although the fresh groundwater in the small coral island is very limited due to the limitations of natural conditions and technology, it is necessary to exploit freshwater lenses to meet the water demand at the early stage of island development. Unlike inland aquifers, the quality and quantity of freshwater available to small island communities depend on the mixing of seawater and freshwater, seawater intrusion, and human activities [5]. In many studies, the definition of drinking water in the freshwater lens is 2.5% relative to the salinity of seawater, or the chloride concentration is about 500 mg/L [11,13,23,24]. Underwood et al. [11] believed that the exploitable freshwater lens appeared at the place where the recharge was greater than or equal to 2000 mm/y and the island width is larger than 250 m. Traditionally, the sustainable exploitation of atolls ranges from 25% to 50% based on the percentage of recharge [5,25]. Ibrahim et al. [26] used 30% of the rainfall as an approximate estimate of sustainable exploitation. Pauw et al. [27] and Post et al. [18] suggested that the numerical simulation for evaluating sustainable exploitation should be conducted at a decadal scale, even for small islands.

The above-mentioned research provides a rich knowledge background for the research on sustainable exploitation of freshwater lenses in small islands. However, more factors should be considered in the estimation of ideal sustainable exploitation, such as the acceptable salinity limit, the growth stage of freshwater lenses, the seasonal contraction of freshwater lenses, the well layout plan, and the heterogeneity of media.

This paper discusses a method to determine the exploitable coefficient of freshwater lenses in small coral islands, and considers the comprehensive effect of lens growth and contraction and the impact of well layout on the evolution of the freshwater lens. By simulating the growth stage of the freshwater lens and the seasonal contraction of the freshwater lens caused by the uneven distribution of rainfall in a year, the optimum pumping stage of the freshwater lens in the coral island was studied. In addition, considering the uncertainty of hydrogeological parameters, the safe exploitable coefficient under the given well layout scheme is calculated, which is supposed to provide a reference for the exploitable coefficient range of freshwater lenses in small coral islands.

2. Study Area

The coral island is located on the Xisha Islands in the South China Sea. It is 1.98 km long from east to west and covers an area of 2.1 km^2. Due to the tropical monsoon and humid marine climate, the annual average temperature is 26.5 °C, and the average annual rainfall is 1505 mm, most of which falls from June to November. The average elevation of the island is about 4.4 m, and the depression in the middle is 2–3 m lower than the surrounding sandbank, which is developed by the lagoon. The sand mat with flat terrain is the main part of the island, which is slightly higher than the depression. The sandbank formed by clastic sand from coral shells surrounds the islands. The narrow beach slopes towards the sea. Reefs much larger than islands are hidden under the sea. The island is covered with tropical plants, such as pittosporum fortunei. The coverage of the vegetation on the islands and reefs shows that the freshwater is sufficient for the growth of plants.

The coral island is formed by the accumulation or cementation diagenesis of coral reef fragments, shells, gravel, algae, and other biological debris on the coral reef. Geological profiles based on 10 boreholes show that the subsurface can be divided into a Holocene and an underlying Pleistocene unit. Coral islands and reefs have no surface runoff due to loosened surface structures and strong permeability. Rainfall is the only natural recharge to the groundwater system. The depth of groundwater in the coral island is about 0.3–2.9 m. The indigenous people on the island are fishermen. Before the completion of the desalination plant and the sewage treatment plant, the water supply mainly depends on the groundwater with a withdrawal of about 400 m^3/d.

3. Materials and Methods

3.1. Conceptual Model

The main natural recharge source of groundwater is atmospheric precipitation infiltration. The underground can be divided into Holocene and underlying Pleistocene units. The contact between Holocene and underlying Pleistocene sediments, called 'Thurber discontinuity" or "Holocene Pleistocene unconformity" (HPU) [24], occurs at a depth of 15–20 m below sea level. Pleistocene coral reef limestone is characterized by developed pores and dissolution pores, strong permeability, and easy seawater circulation [14]. Holocene granular sediments are mainly composed of coral clastic medium sand and coral clastic gravelly sand. The hydraulic properties of the Holocene and Pleistocene units of the coral island are obviously different. The hydraulic conductivity (K) of Holocene limestone is obviously higher than that of Holocene sediments. The conceptual sketch of the coral island is shown in Figure 1. The map includes a simplified geological setting and shows the dual nature of the water system formed by Holocene sediments overlying Pleistocene karst limestone.

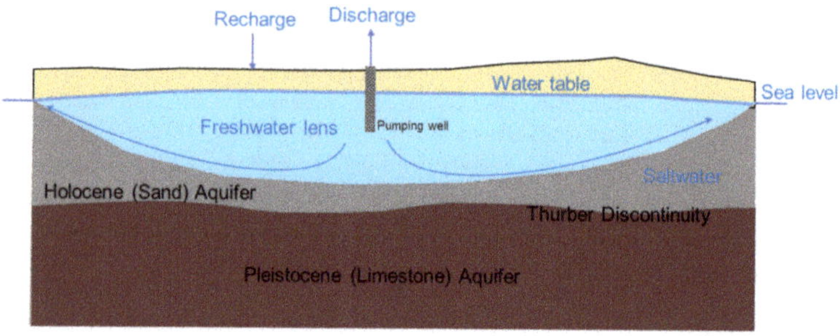

Figure 1. Conceptual model of small coral island.

3.2. Parameter Setting of Numerical Model

A two-dimensional numerical density-dependent groundwater flow and solute transport model is developed to represent the profile of a coral island using SEAWAT [28,29]. This interpretation model is used to simulate the dynamics of a coral island freshwater lens under similar hydrogeological conditions; as a result, calibration is not required [30]. The variable density groundwater flow equation is as follows:

$$K\left(\frac{\partial^2 H}{\partial x^2}+\frac{\partial^2 H}{\partial z^2}\right)+K\eta\frac{\partial c}{\partial z}=S_s\frac{\partial H}{\partial t}+n\eta\frac{\partial c}{\partial t}-\frac{\rho}{\rho_0}q_s \quad (1)$$

where K is the hydraulic conductivity (m/d), H is the piezometric head (m), c is solute concentration (kg·m^{-3}), S_s is the water storage rate (m^{-1}), t is time (d), n is the medium's porosity; η is the density coupling coefficient (kg·m^{-3}), ρ is the mixed fluid density (kg·m^{-3}), ρ_0 is the freshwater density (kg·m^{-3}), and q_s is the volume of source and sink items entering unit volume aquifer in unit time (d^{-1}).

The solute transport equation is as follows:

$$\frac{\partial c}{\partial t} + u_i(\frac{\partial c}{\partial x_i}) = \frac{\partial}{\partial x_i}(D_{ii}\frac{\partial c}{\partial x_i}) + \frac{(c^* - c)}{n}q_s \quad (2)$$

where x_i is the infiltration direction (m), D_{ii} is the hydrodynamic dispersion coefficient tensor (m^2·d^{-1}), u_i is the flow velocity (m·d^{-1}), and c^* is the mass concentration of the source or sink fluid (kg·m^3).

The model was running for 100 years, during which the freshwater lens developed. The model reached a stable state within 50 years (i.e., the shape of the lens remained relatively stable). The model includes 22 layers with a depth of 50 m to improve the vertical resolution in the model simulation, and is discretized into 41 columns in the horizontal direction and refined at the boundary. A constant head boundary is defined along both sides of the area to simulate sea level. The density is specified as 1025 kg/m^3, representing typical sea water composition. The constant concentration boundary is assigned to the same grid cell as the constant head boundary, and the chloride concentration is 19,000 mg/L. Assuming that the loose sediment of the coral island is initially saturated with seawater, the initial chloride concentration in the whole model area is specified as 19,000 mg/L. The ground of the model is based on boreholes. The model grid is shown in Figure 2.

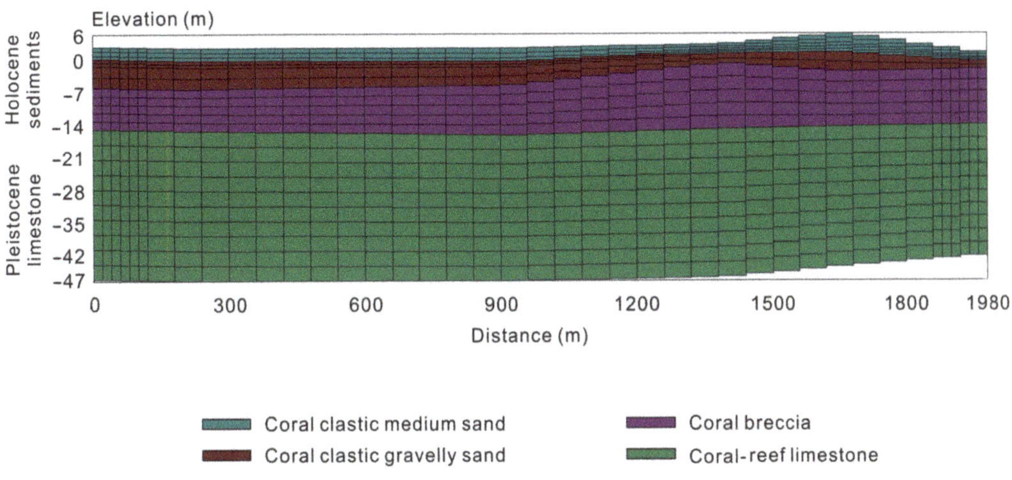

Figure 2. The grid of the developed model.

The hydraulic conductivity of main aquifers is determined according to petrology and previous studies [14,31,32]. The upper model layer represents coral sand, and the hydraulic conductivity (K) ranges from 60 m/d to 150 m/d. The lower model layer represents coral reef limestone. Because there are many pores and fractures of different sizes, it has good permeability, and the hydraulic conductivity (K) is 1000 m/d [33]. The effective porosity (ne) is 0.25–0.45, and the specific yield (Sy) is 0.1–0.2.

Precipitation is the only recharge of freshwater to the hydrogeological system, so it is the main mechanism to simulate the development of the freshwater lens in the model [34]. The coefficient of the precipitation recharge (α) is used to calculate the monthly replenishment. Production wells will not be activated until the model reaches a stable state. After the model reaches the steady state, the production wells are started for 20 years to simulate the saltwater coning. Table 1 summarizes the hydrogeological parameters assigned to the SEAWAT model.

Table 1. Parameter values used in the model for the coral island groundwater system.

Settings	Parameter	Units	Value
Basic setup	Island width	m	1980
	Thickness	m	50
	Grid	\	41 × 22
	Simulated time step	d	36,525
Flow model	Recharge	mm/y	Monthly average recharge
	Effective porosity	\	0.25–0.45
	Holocene *K*	m/d	60–150
	Pleistocene *K*	m/d	1000
	Specific yield	\	0.1–0.2
Transport model	Longitudinal dispersivity	m	5
Density-dependent model	Reference fluid density	kg/m^3	1000
	Seawater density	kg/m^3	1025

4. Results and Discussion

4.1. Growth and Contraction of Freshwater Lens under Natural Conditions

4.1.1. Growth Characteristics and Stage Division of Freshwater Lens

According to the drinking water standard formulated by the World Health Organization (WHO), the chloride (Cl^-) ion concentration should not exceed 250 mg/L [35]. In this study, a 250 mg/L concentration contour was selected as the division between freshwater and saltwater. The growth of the freshwater lens is a long and slow dynamic process. The simulation results show that it takes 45 years for the freshwater lens to grow to its maximum thickness and for the reserve to be basically stable, and the central thickness will also reach its maximum in the 60th year. The water head distribution shows the basic characteristics of "high in the middle, low around". The maximum freshwater head (H_m) is 0.32 m, forming a hydraulic gradient from the center to the edge, which can ensure that freshwater is continuously discharged into the sea from the thickest part of the lens. The shape of the stabilized freshwater lens and the groundwater flow rate are shown in Figure 3. After being recharged, the water within the freshwater lens flows from top to the bottom and discharges water around the island. Meanwhile, the seawater moves upward, "lifting" the freshwater lens so that it floats on the land and the entire groundwater system is in a dynamic equilibrium. The center of the freshwater lens produces a low velocity area, forming a stagnation zone (Figure 3). The flow velocity increases from the center to the boundary of the lens. Due to the difference in seawater density and freshwater density, which produces larger buoyancy in the center of the lens, the center thickness (T_c) is smaller than the maximum thickness (T_m) (Figure 3). However, the difference will become smaller if the lower interface salinity standard increases.

The growth of the freshwater lens can be divided into three stages according to the temporal variation in the thickness and freshwater storage of the freshwater lens (Figure 4):

- Stage I: 0–20 years, two extremely thin freshwater areas are formed underground on both sides near the middle of the island. The freshwater lens develops rapidly in the form of a "doughnut", but the central thickness increases slowly. During this stage, the main role of precipitation recharge is to dilute and flush the salinity of the original groundwater under the island. The groundwater velocity near the middle of the island is the largest;
- Stage II: 20–40 years, this stage is the main stage for the stable growth of the freshwater lens. The freshwater on both sides near the middle of the island are combined into one. The maximum thickness of the freshwater lens increases synchronously with the

central thickness, and the central thickness increases significantly faster than Stage I. during this stage, a large great deal of freshwater seeps into the freshwater lens through the surface as the precipitation continues to recharge the groundwater, the thickness of the upper freshwater of the lens becomes larger, and the interface between freshwater and saltwater becomes deeper. Meanwhile, the salinity gradient becomes larger, the corresponding vertical mixing weakens, and the horizontal range of the freshwater lens gradually expands;

- Stage III: After 40 years, the thickness of the freshwater lens increases slowly. After 45 years, the maximum thickness of the freshwater lens is stabilized. Later, the central thickness increases slowly, and reaches the maximum in 60 years. During this stage, the supply and discharge reach a dynamic balance, the thickness and scope of the freshwater lens will not increase, and the freshwater lens will enter a relatively stable status.

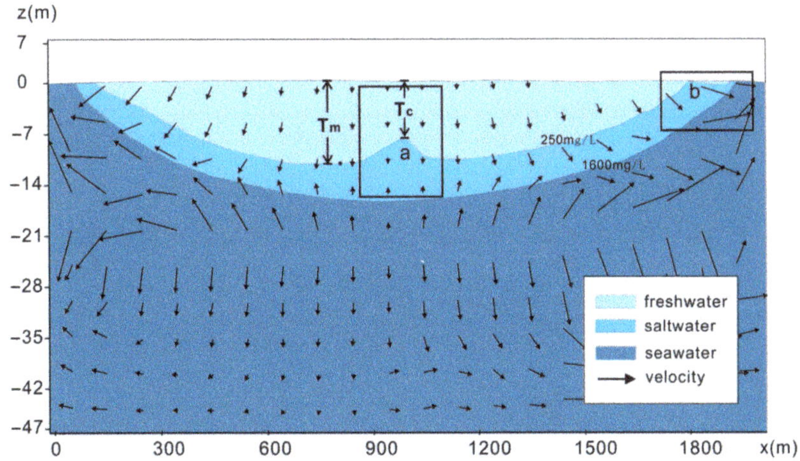

Figure 3. Morphology and velocity distribution of the freshwater lens (a: stagnation zone).

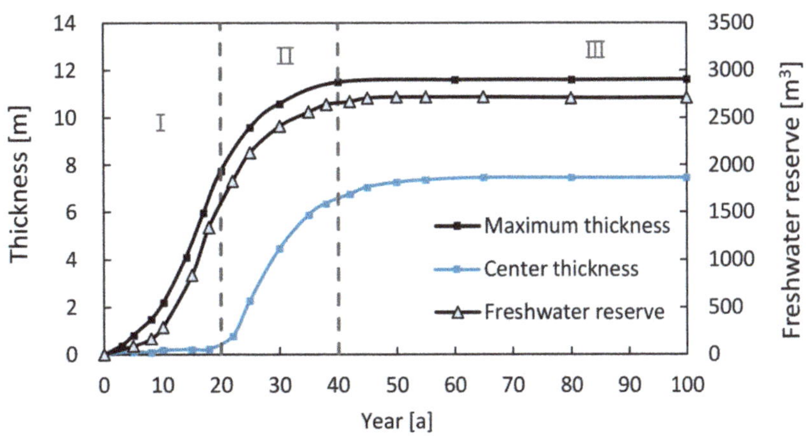

Figure 4. Division of growth stages of freshwater lens.

4.1.2. Seasonal Variation in Freshwater Lens

Precipitation impacts the difference of density and concentration between saltwater and freshwater, and further regulates the dynamic changes of freshwater lens. H_m increases continuously from June when the rainy season starts and usually reaches the maximum value in autumn when the rainy season ends. Although the annual variation in H_m is small, it is a key control factor for the freshwater lens to reach the maximum depth and plays an important role in maintaining the thickness of the freshwater lens.

The shape change of the freshwater lens is mainly reflected in its geometric thickness (Figure 5). The simulation results show that the change in T_m obviously lags behind T_c (about half a year), and T_m presents a smaller annual variation than T_c. Larger T_m indicates more recoverable groundwater, and due to the slower response of T_m to precipitation, the annual maximum T_m usually appears in the winter when the island is most vulnerable to water shortage in the year, which guarantees the more desired pumping amount during the dry season.

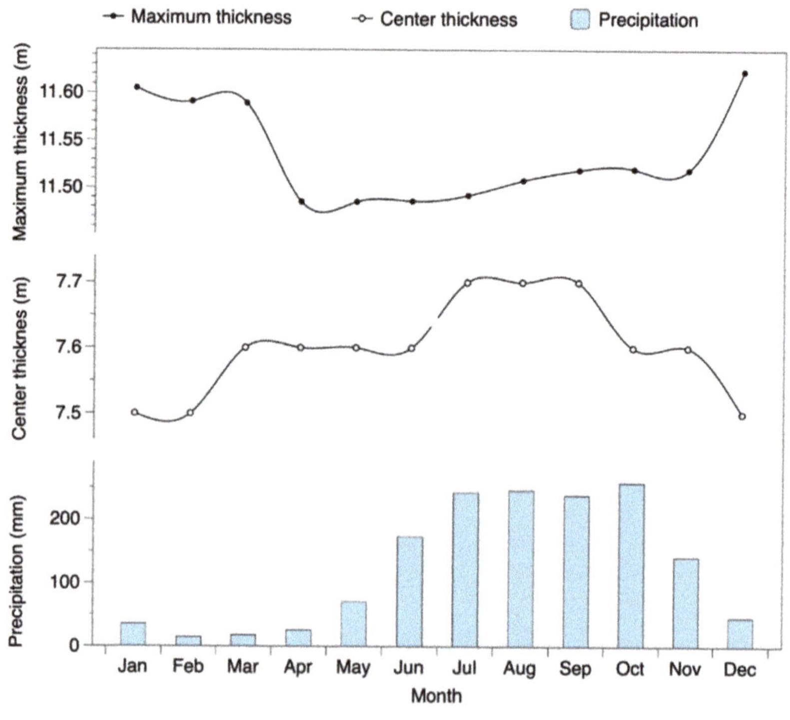

Figure 5. Seasonal variation in freshwater lens thickness.

The freshwater storage (Wa) of the profile is determined by the specific yield (μ) and the shape of the freshwater lens. It reaches the maximum value in July and minimum value in November. The annual variation Wa is only 24.6 m^3, accounting for 0.9% of the average Wa, and the mean precipitation has little effects on Wa (Figure 6).

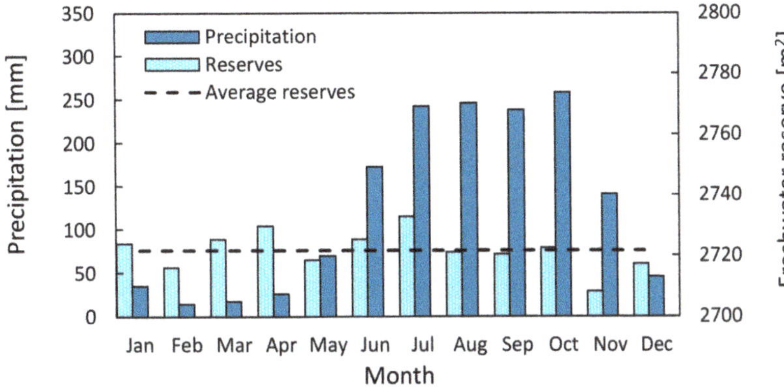

Figure 6. Seasonal change of freshwater lens reserves.

The seasonal change of the spatial distribution of the freshwater lens is mainly reflected in the edge area. Figure 7 is a local enlarged view of region b in Figure 3. In winter and spring, the edge of the freshwater lens becomes thinner and even discontinuous, leading to a narrower freshwater lens. In summer, the thickness of the freshwater lens begins to recover gradually and reaches the maximum in autumn. The seasonal cycle may be related to the obvious sand dike topography in the east. Therefore, the edge of the freshwater lens is very vulnerable and could be impacted by droughts, seawater flooding events, and human activities, which results in a reduction in the freshwater lens.

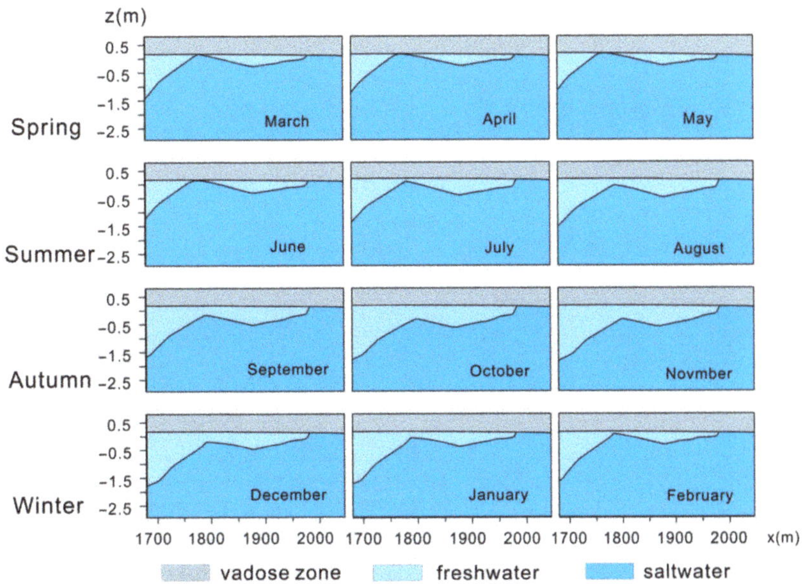

Figure 7. Seasonal change in the edge of the freshwater lens.

The uneven seasonal distribution of precipitation can also cause the natural contraction of the freshwater lens in terms of its thickness, storage, and distribution. In general, the maximum freshwater head is most sensitive to the seasonal change of precipitation, but because the value is very small, this change is not easy to notice; the second is the central thickness and the maximum thickness of the freshwater lens, but their changes are not synchronous. In contrast, the seasonal change of precipitation has a greater impact on the central thickness; and the seasonal variation in reserves is the least obvious. The thinning of the lens edge in spring leads to the reduction in the lens range. These changes mean that the natural contraction of the freshwater lens should be considered when determining the exploitable amount of groundwater.

4.2. Degradation and Recovery of Freshwater Lens under Pumping Conditions

In the three different growth stages, the adaptability of the freshwater lens to pumping is different. To explore the response of the freshwater lens to the pumping process in the three stages, the pumping will be carried out for 5 years at Stage I, Stage II, and Stage III, with the pumping volume of 0.2, 0.3, and 0.5 m^3/d. Because the freshwater lens in Stage I is very thin and has a limited distribution range, the pumping well will be quickly broken down by saltwater, so this stage is not suitable for exploitation.

The change process of the thickness of the freshwater lens in the pumping process of Stage II is shown in Figure 8. Compared with Stage I, the freshwater lens in Stage II has developed to a larger volume, and the effect of short-term exploitation at a low pumping rate on the freshwater lens thickness can be offset by the natural increase of lens. However, in the early stage of Stage II, during which T_c is very small, there is a high risk of the production well being broken down by saltwater. The recovery time is much shorter (10 years shorter) if the pumping starts in early Stage II rather than late State II (Table 2).

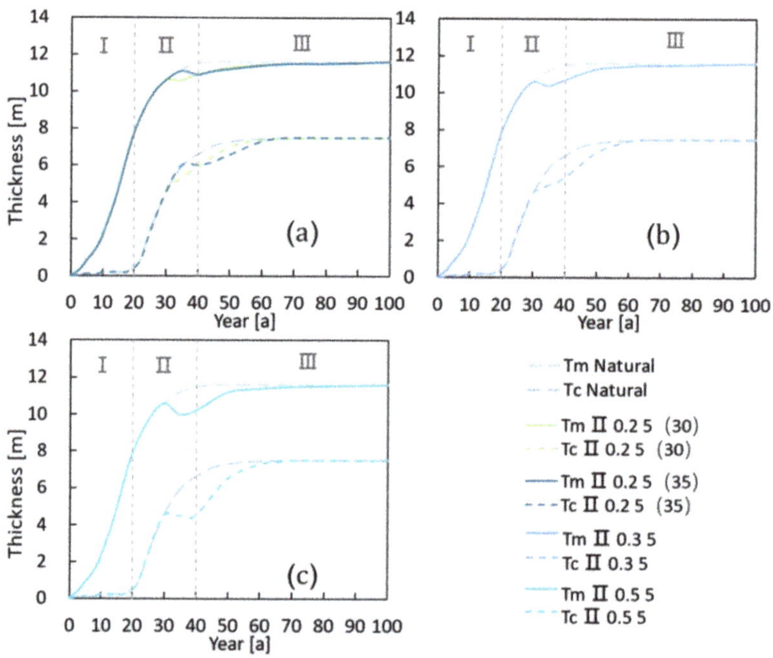

Figure 8. Thickness changes of the freshwater lens in Stage II under pumping conditions (**a**) Q = 0.2 m^3/d, pumping starts in the middle of Stage II (30), and pumping starts in the late stage of Stage II (35), (**b**) Q = 0.3 m^3/d, and (**c**) Q = 0.5 m^3/d.

Table 2. Changes and recovery time of freshwater lens thickness at different stages and pumping volumes.

Stages	Pumping Duration (y)	Pumping Rates (m^3/d)	Minimum Value (m)		Thickness Variation (m)		Recovery Time (y)	
			T_m	T_c	T_m	T_c	T_m	T_c
II	5(30)	0.2	10.6	5.3	0	0.8	35	35
	5(35)	0.2	10.9	6	−0.2	0	25	25
	5	0.3	10.4	5	−0.2	0.4	30	30
	5	0.5	10	4.5	−0.6	0	35	35
III	5	0.2	11.1	6.8	−0.45	−0.6	35	25
	5	0.3	11	6.4	−0.55	−1	35	30
	5	0.5	10.5	5.7	−1.05	−1.7	35	35

In Stage III (Figure 9), the freshwater lens is developed and maintains a relatively stable shape, with large T_m and T_c. Compared with the pumping at the end of Stage II, the reduction in the thickness of the freshwater lens in Stage III is larger under the same pumping rates (Table 2). However, because the thickness base of the freshwater lens in this stage is larger, even if the thickness is reduced, it is still thicker than that in Stage II, which makes the freshwater lens in Stage III able to withstand greater pumping intensity. The duration is very long, which means there is a possibility of long-term pumping. In general, the freshwater lens is more adaptable to pumping in Stage III.

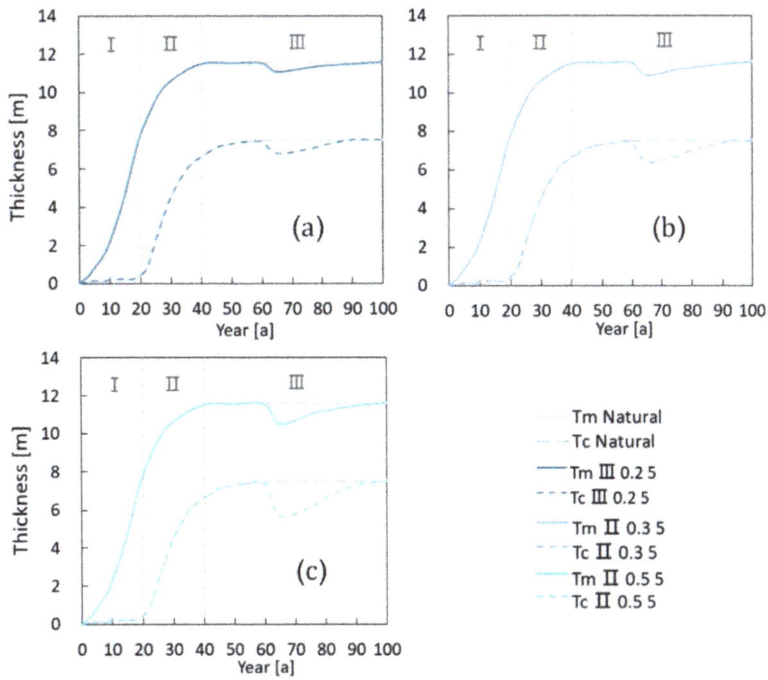

Figure 9. Thickness changes of the freshwater lens in Stage III under pumping conditions (**a**) Q = 0.2 m^3/d, (**b**) Q = 0.3 m^3/d, and (**c**) Q = 0.5 m^3/d.

If other conditions do not change, the time required for the freshwater lens to recover to its natural state after pumping suspension is mainly affected by the pumping intensity and the growth stage of the freshwater lens when pumping starts (Figure 10). In the early stage of Stage II, the pumping volume is 0.2 m^3/d for five years, and the recovery time to the natural state after pumping is 30 years. In the later stage of Stage II, the pumping will be carried out for five years with the same pumping amount, and the impact on the thickness can last for 25 years. Compared with the pumping in the earlier stage of Stage II, the recovery time will be shortened by five years. In Stage III, the pumping volume is 0.2, 0.3, and 0.5 m^3/d, respectively, and the recovery time after pumping suspension increases with the increase in pumping intensity, which is 25, 30, and 35 years, respectively.

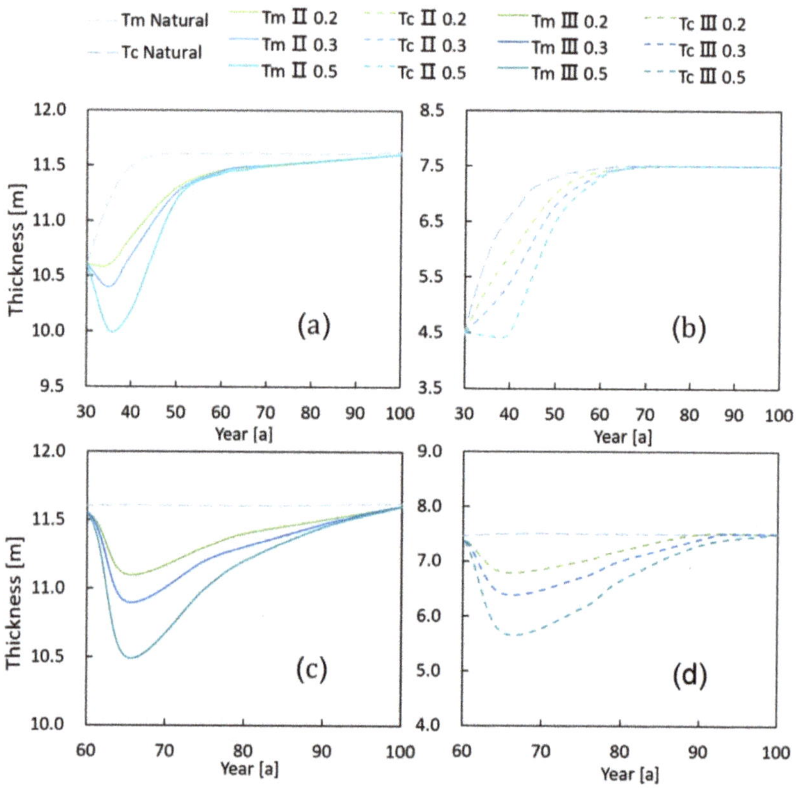

Figure 10. Recovery process of freshwater lens thickness after stopping pumping ((**a**): changes in T_m in Stage II; (**b**): changes in T_c in Stage II; (**c**): changes in T_m in Stage III; and (**d**): changes in T_c in Stage III).

To sum up, pumping is not recommended in the early stage of Stage II; in the middle and late stage of Stage II, relatively short and low intensity pumping can be accepted; and in Stage III, long-term pumping with relatively higher intensity is possible, but it should be noted that the pumping intensity should not be too large, and the basic requirement that the wells should not be broken down by saltwater during the pumping period should be met.

4.3. Calculation of the Safe Exploitable Coefficient

The sustainable exploitation amount should not exceed the exploitable amount of groundwater. The exploitable coefficient (ρ) refers to the ratio of the exploitable amount of groundwater (Q) in an area to the total recharge amount of groundwater (W) in the same area:

$$\rho = Q/W \quad (3)$$

where ρ is the exploitable coefficient; Q is the exploitable amount of groundwater in the area, W is the total recharge of groundwater in the same area. Furthermore, ρ should not be greater than 1. The closer the ρ value is to 1, the better the exploitation condition of aquifer is, and the smaller the ρ value is, the worse the exploitation condition is.

The assessment of sustainable exploitation must be based on decades of time scale [18]. This paper adopts the concept of total critical pumping rate (Q_T) (the maximum pumping rate of all pumping wells without salinization under pumping conditions) for an inverse calculation of maximum exploitation through numerical simulation [22]. Nine well layout schemes along a straight line are designed in this study, which include the combination of different well numbers, well spacing, and the length of screen. Under the premise of continuous pumping for 20 years without salinization, the maximum total pumping rates obtained by each scheme have a large difference, reflected by the large variation in exploitation coefficients ranging from 0.09 to 0.37 (Table 3). This means that on a small island, the specific layout plan of pumping wells has a huge impact on the exploitable amount of groundwater.

Table 3. Well layout plan and results of Q_T, T_c, T_m, and ρ.

Test Number	Well Layout Plan			Results			
	Screen Length (m)	Number of Wells	Distance between Wells (m)	Q_T (m^3/d)	T_c (m)	T_m (m)	ρ
1	4	4	100	0.407	4.3	9.2	0.17
2	2	4	200	0.73	1.1	7.4	0.30
3	2	6	150	0.907	0.5	6.1	0.37
4	3	4	150	0.555	2.3	8.2	0.23
5	2	2	100	0.33	2.2	9.8	0.13
6	4	6	200	0.735	1.7	7	0.30
7	3	6	100	0.562	2.8	7.5	0.23
8	4	2	150	0.215	3.3	10	0.09
9	3	2	200	0.33	4.6	9.8	0.13
Max ρ	2	6	150	0.907	0.5	6.1	0.37
Min ρ	4	2	150	0.215	3.3	10	0.09

Due to the uncertainty of hydrogeological parameters, even under the determined well layout scheme, Q_T is also uncertain. In this study, the impact of hydrogeological parameter uncertainty on Q_T is investigated by conducting simulations under three combinations of net recharge (R), hydraulic conductivity (*K*), and porosity (n) (Table 4) in a linear well layout plan (two wells with a length of 3 m and a spacing of 200 m). The exploitable amount of groundwater obtained under this scheme is not the largest, but the seasonal contraction of T_c and T_m of the freshwater lens is comprehensively considered to ensure that the reduction in T_c and T_m will maximize the benefit of Q_T increase when the thickness of the freshwater lens is not less than 2 m. Therefore, the Q_T and ρ are safe. The pumping period is 20 years in the growth Stage III of the freshwater lens.

Table 4. Parameter combinations and calculation results.

	Parameter Combinations			Results			
	ΔR*	ΔK*	Δn*	T_m (m)	Q_T (m^3/d)	T_r	ρ
Scenario 1:	0	0	0	9.3	0.330	35	0.13
Scenario 2:	−30%	30%	−30%	6.3	0.174	31	0.1
Scenario 3:	30%	−30%	30%	12.6	0.573	37.1	0.18

Notes: ΔR* is the percentage of net recharge change, ΔK* is the percentage of hydraulic conductivity change, and Δn* is the percentage of porosity change.

The simulation results of the three scenarios are shown in Figure 11. Table 4 indicates that if the extraction volume is used for continuous pumping, as long as the hydrogeological conditions do not change significantly, the pumping well will not be damaged by saltwater. Our result reflects that the difference of hydrogeological conditions has a significant impact on the exploitable amount of groundwater, and the uncertainty of hydrogeological parameters must be considered when determining the safe exploitable quantity.

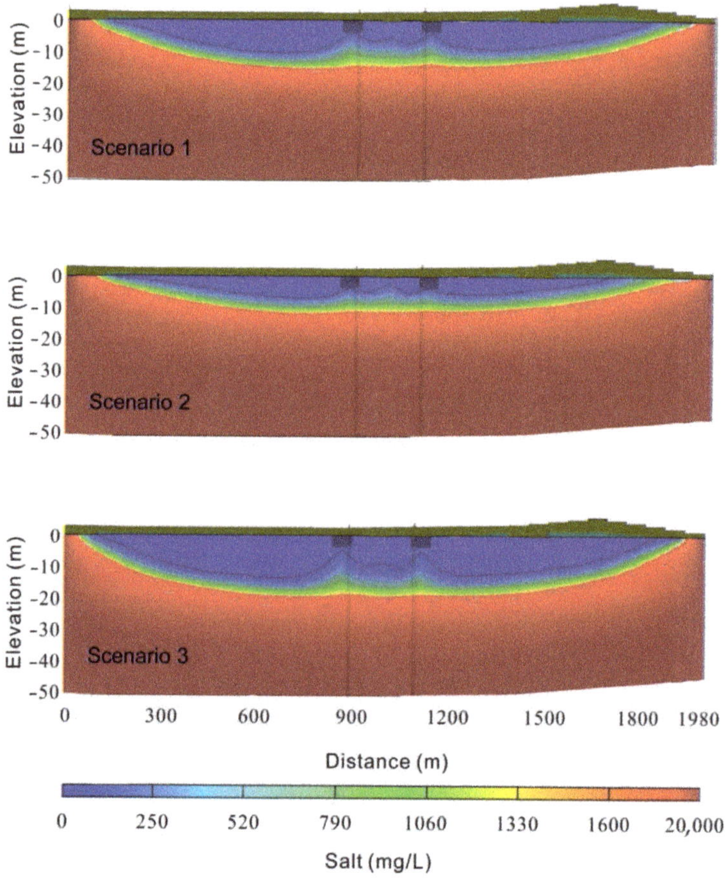

Figure 11. Pumping simulation results.

The recovery time, exploitable volume, and exploitable coefficient of freshwater lens under three scenarios are recorded in Table 4, and ρ ranges from 0.1 to 0.18 under different combinations of hydrogeological parameters. Therefore, under the determined well layout plan, the safe exploitable coefficient is only 0.1, which is very small, but it can protect the freshwater lens as much as possible from damage. The smaller exploitable coefficient corresponds to larger recovery time, which shows that the combination of parameters conducive to pumping is not conducive to the recovery of freshwater lenses.

5. Conclusions

In this paper, the characteristics of each stage in the formation process of a coral island freshwater lens is studied, and the most suitable exploitation stage of freshwater lens is analyzed considering the seasonal contraction of freshwater lens caused by rainfall, different well layout schemes, and hydrogeological parameter combinations. Furthermore, the exploitable coefficient and the partial safe exploitable coefficient under a certain well layout scheme are calculated.

Water resources assessment, development, and management in small islands should be conducted considering the specific characteristics of the study island. Our research provides a reference to determine the recoverable coefficient of other similar small islands around the world. Previous studies on the sustainable exploitation of islands were conducted based on the experience of managers [25]. In this paper, Q_T was used to deduce the exploitable amount and exploitable coefficient of the freshwater lens on the basis of considering the exploitation stage, seasonal variation in rainfall, and parameter uncertainty, which provides numerical solutions for the study of the sustainable exploitation of the freshwater lens. The recoverable coefficient calculated in this paper varies with different well layout schemes, generally ranging from 0.09 to 0.37, which is consistent with the results from previous studies [5,25,26]. However, under a determined pumping well layout scheme, the calculated exploitable coefficient is determined to be 0.1, which is small but safe, by considering the lens contraction caused by rainfall and the most unfavorable combination of hydrogeological parameters.

The exploitability coefficient of small coral islands is significantly lower than that of inland areas (0.6–0.95). The reason is that the freshwater on the island is constantly mixed with saltwater and partly discharges into the sea, while the other part maintains a certain head difference to resist seawater intrusion. In this situation, a large part of the recharge cannot be exploited. In addition, the pumping conditions on the island are worse than those in inland areas, resulting in a generally small exploitable coefficient on the island.

The freshwater lens is a valuable freshwater resource in island areas and exhibits seasonal responses to precipitation under natural conditions. In terms of water supply for the small coral island, managers should pay attention to the sustainability of the pumping process and the protection of pumping wells, which requires a deeper understanding of the island's hydrogeological conditions. It is necessary to select appropriate pumping methods according to the characteristics of the island itself, determine the exploitable coefficient scientifically, pertinently, and timely to adjust the pumping plan and pumping plan according to the development of the island, and finally achieve "one island, one policy".

Author Contributions: Data curation, Z.L.; funding acquisition, R.Z.; methodology, R.W.; project administration, L.S.; supervision, L.S.; validation, Z.L.; writing—original draft, R.W.; writing—review and editing, R.Z. All authors have read and agreed to the published version of the manuscript.

Funding: This research was made possible by the Key Technologies and Application Demonstration of Groundwater Over-Extraction Control and Protection in Huang-Huai-Hai Region (No. 2021YFC3200502) to the second author and the Belt and Road Special Foundation of the State Key Laboratory of Hydrology-Water Resources and Hydraulic Engineering (2021490911) to the third author.

Institutional Review Board Statement: Not applicable.

Informed Consent Statement: Not applicable.

Data Availability Statement: Data is contained within the article.

Acknowledgments: All authors are very grateful to the editor and the anonymous reviewers for their valuable comments: which have greatly improved the paper.

Conflicts of Interest: The authors declare no conflict of interest.

References

1. Chui, T.F.M.; Terry, J.P. Influence of sea-level rise on freshwater lenses of different atoll island sizes and lens resilience to storm-induced salinization. *J. Hydrol.* **2013**, *502*, 18–26. [CrossRef]
2. Gohar, A.A.; Cashman, A.; Ward, F.A. Managing food and water security in Small Island States: New evidence from economic modelling of climate stressed groundwater resources. *J. Hydrol.* **2019**, *569*, 239–251. [CrossRef]
3. UNESCO. *Small Island Developing States—UNESCO's Action Plan*; Ikhlef, K., Nakashima, D., Eds.; United Nations Educational, Scientific and Cultural Organization: Paris, France, 2016; 32p.

4. Gingerich, S.B.; Voss, C.I.; Johnson, A.G. Seawater-flooding events and impact on freshwater lenses of low-lying islands: Controlling factors, basic management and mitigation. *J. Hydrol.* **2017**, *551*, 676–688. [CrossRef]
5. White, I.; Falkland, T. Management of freshwater lenses on small Pacific islands. *Hydrogeol. J.* **2010**, *18*, 227–246. [CrossRef]
6. Shuhei, Y.; Satoshi, I.; Tsutomu, K.; Kazuhisa, K.; Takeo, T.; Katsushi, S. Using hydrogeochemical indicators to interpret groundwater flow and geochemical evolution of a freshwater lens on Majuro Atoll, Republic of the Marshall Islands. *Hydrogeol. J.* **2020**, *28*, 1053–1075.
7. Fetter, C. Position of the saline water interface beneath oceanic islands. *Water Resour. Res.* **1972**, *8*, 1307–1315. [CrossRef]
8. Bokuniewicz, H.; Pavlik, B. Groundwater seepage along a barrier island. *Biogeochemistry* **1990**, *10*, 257–276. [CrossRef]
9. Bryan, E.; Meredith, K.T.; Baker, A.; Post, V.E.; Andersen, M.S. Island groundwater resources, impacts of abstraction and a drying climate: Rottnest Island, Western Australia. *J. Hydrol.* **2016**, *542*, 704–718. [CrossRef]
10. Schneider, J.C.; Kruse, S.E. A comparison of controls on freshwater lens morphology of small carbonate and siliciclastic islands: Examples from barrier islands in Florida, USA. *J. Hydrol.* **2003**, *284*, 253–269. [CrossRef]
11. Underwood, M.R.; Peterson, F.L.; Voss, C.I. Groundwater lens dynamics of atoll islands. *Water Resour. Res.* **1992**, *28*, 2889–2902. [CrossRef]
12. Vacher, H.L.; Wallis, T. Comparative hydrogeology of fresh-water lenses of Bermuda and Great Exuma Island, Bahamas. *Groundwater* **1992**, *30*, 15–20. [CrossRef]
13. Wallace, C.D.; Bailey, R.T. Geohydrologic factors governing atoll island groundwater resources. *J. Hydrol. Eng.* **2017**, *22*, 05017004. [CrossRef]
14. Congzhi, Z.; Li, H.; Qin, Y.; Zhendong, F. Three-dimensional numerical simulation of freshwater lens in coral islands. *J. Hydraul. Eng.* **2010**, *41*, 560–566.
15. Yao, Y.; Andrews, C.; Zheng, Y.; He, X.; Babovic, V.; Zheng, C. Development of fresh groundwater lens in coastal reclaimed islands. *J. Hydrol.* **2019**, *573*, 365–375. [CrossRef]
16. Li, Y.; Shu, L.; Zhen, L.; Li, H.; Wang, R.; Opoku, P. Laboratory Physical Experiments on the Saltwater Upconing and Recovery of Island Freshwater Lenses: Case Study of a Coral Island, China. *Water* **2021**, *13*, 1137. [CrossRef]
17. Dose, E.J.; Stoeckl, L.; Houben, G.J.; Vacher, H.L.; Vassolo, S.; Dietrich, J.; Himmelsbach, T. Experiments and modeling of freshwater lenses in layered aquifers: Steady state interface geometry. *J. Hydrol.* **2014**, *509*, 621–630. [CrossRef]
18. Post, V.E.; Bosserelle, A.L.; Galvis, S.C.; Sinclair, P.J.; Werner, A.D. On the resilience of small-island freshwater lenses: Evidence of the long-term impacts of groundwater abstraction on Bonriki Island, Kiribati. *J. Hydrol.* **2018**, *564*, 133–148. [CrossRef]
19. Fu, W.; Moore, J.K.; Primeau, F.W.; Lindsay, K.; Randerson, J.T. A growing freshwater lens in the Arctic Ocean with sustained climate warming disrupts marine ecosystem function. *J. Geophys. Res. Biogeosci.* **2020**, *125*, e2020JG005693. [CrossRef]
20. Sulzbacher, H.; Wiederhold, H.; Siemon, B.; Grinat, M.; Igel, J.; Burschil, T.; Günther, T.; Hinsby, K. Numerical modelling of climate change impacts on freshwater lenses on the North Sea Island of Borkum using hydrological and geophysical methods. *Hydrol. Earth Syst. Sci.* **2012**, *16*, 3621–3643. [CrossRef]
21. Ling, Z.; Shu, L.; Sun, Y.; Wang, R.; Li, Y. Impact of island urbanization on freshwater lenses: A case study on a small coral island. *Water* **2021**, *13*, 3272. [CrossRef]
22. Wang, R.; Shu, L.; Li, Y.; Opoku, P.A. Pumping Well Layout Scheme Design and Sensitivity Analysis of Total Critical Pumping Rates in Coral Island Based on Numerical Model. *Water* **2021**, *13*, 3215. [CrossRef]
23. Lloyd, J.; Miles, J.; Chessman, G.; Bugg, S. A ground water resources study of a pacific ocean atoll-tarawa, gilbert islands 1. Jawra. *J. Am. Water Resour. Assoc.* **1980**, *16*, 646–653. [CrossRef]
24. Bailey, R.T.; Jenson, J.W.; Olsen, A.E. Estimating the ground water resources of atoll islands. *Water* **2010**, *2*, 1–27. [CrossRef]
25. Peterson, F.L. Hydrogeology of the Marshall Islands. *Dev. Sedimentol.* **2004**, *54*, 611–636. [CrossRef]
26. Ibrahim, S.A.; Bari, M.R.; Miles, L. Water Management in Maldives with Special Emphasis on Desalination. In *Proceedings of the Pacific Regional Consultation on Water in Small Island Countries, July 29th–August 3rd 2002;* Carpenter, C., Stubbs, J., Overmars, M., Eds.; Asian Development Bank and South Pacific Applied Geoscience Commission: Sigatoka, Fiji; Available online: http://citeseerx.ist.psu.edu/viewdoc/download?doi=10.1.1.113.913&rep=rep1&type=pdf (accessed on 17 November 2016).
27. Pauw, P.S.; van der Zee, S.E.; Leijnse, A.; Oude Essink, G.H. Saltwater upconing due to cyclic pumping by horizontal wells in freshwater lenses. *Groundwater* **2016**, *54*, 521–531. [CrossRef]
28. Langevin, C.D.; Shoemaker, W.B.; Guo, W. *MODFLOW-2000. The US Geological Survey Modular Ground-Water Model–Documentation of the SEAWAT-2000 Version with the Variable-Density Flow Process (VDF) and the Integrated MT3DMS Transport Process (IMT);* Center for Integrated Data Analytics Wisconsin Science Center Wisconsin: Madison, WI, USA, 2000; pp. 1258–2331.
29. Harbaugh, A.W.; Banta, E.R.; Hill, M.C.; McDonald, M.G. *Modflow-2000, the U. S. Geological Survey Modular Ground-Water Model-user Guide to Modularization Concepts and the Ground-Water Flow Process;* United States Department of the Interior: Washington, DC, USA, 2000.
30. Anderson, M.P.; Woessner, W.W.; Hunt, R.J. *Applied Groundwater Modeling: Simulation of Flow and Advective Transport*, 2nd ed.; Academic Press: Cambridge, MA, USA, 2015.
31. Zhou, C.; Qiao, H.; Du, R. Simulation and exploitation of the freshwater lens in Coral Island. *J. Logist. Eng. Univ.* **2016**, *32*, 1–10.
32. Sheng, C.; Xu, H.; Zhang, Y.; Zhang, W.; Rem, Z. Hydrological properties of calcareous sands and its influence on formation of underground freshwater lens on islands. *J. Jilin Univ.* **2020**, *50*, 1127–1138. (In Chinese)

33. Oberdorfer, J.A.; Hogan, P.J.; Buddemeier, R.W. Atoll Island Hydrogeology: Flow and Freshwater Occurrence in a Tidally Dominated System. *J. Hydrol.* **1990**, *120*, 327–340. [CrossRef]
34. Holding, S.; Allen, D. From days to decades: Numerical modelling of freshwater lens response to climate change stressors on small low-lying islands. *Hydrol. Earth Syst. Sci.* **2015**, *19*, 933–949. [CrossRef]
35. World Health Organization. *Guidelines for Drinking-Water Quality*, 3rd ed.; World Health Organization: Geneva, Switzerland, 2004; Volume 1.

water

MDPI

Article

Predicting Groundwater Level Based on Machine Learning: A Case Study of the Hebei Plain

Zhenjiang Wu, Chuiyu Lu, Qingyan Sun, Wen Lu, Xin He, Tao Qin, Lingjia Yan and Chu Wu *

State Key Laboratory of Simulation and Regulation of Water Cycle in River Basin, China Institute of Water Resources and Hydropower Research, Beijing 100038, China
* Correspondence: wuchu@iwhr.com

Abstract: In recent years, the groundwater level (GWL) and its dynamic changes in the Hebei Plain have gained increasing interest. The GWL serves as a crucial indicator of the health of groundwater resources, and accurately predicting the GWL is vital to prevent its overexploitation and the loss of water quality and land subsidence. Here, we utilized data-driven models, such as the support vector machine, long-short term memory, multi-layer perceptron, and gated recurrent unit models, to predict GWL. Additionally, data from six GWL monitoring stations from 2018 to 2020, covering dynamical fluctuations, increases, and decreases in GWL, were used. Further, the first 70% and remaining 30% of the time-series data were used to train and test the model, respectively. Each model was quantitatively evaluated using the root mean square error (RMSE), coefficient of determination (R^2), and Nash–Sutcliffe efficiency (NSE), and they were qualitatively evaluated using time-series line plots, scatter plots, and Taylor diagrams. A comparison of the models revealed that the RMSE, R^2, and NSE of the GRU model in the training and testing periods were better than those of the other models at most groundwater monitoring stations. In conclusion, the GRU model performed best and could support dynamic predictions of GWL in the Hebei Plain.

Keywords: groundwater level prediction; data-driven models; gated recurrent units; model performance; Hebei Plain

Citation: Wu, Z.; Lu, C.; Sun, Q.; Lu, W.; He, X.; Qin, T.; Yan, L.; Wu, C. Predicting Groundwater Level Based on Machine Learning: A Case Study of the Hebei Plain. *Water* **2023**, *15*, 823. https://doi.org/10.3390/w15040823

Academic Editors: Qiting Zuo, Fuqiang Wang, Jiaqi Zhai, Xiuyu Zhang, Dunxian She, Lei Zou, Rong Gan and Zengliang Luo

Received: 1 February 2023
Revised: 16 February 2023
Accepted: 16 February 2023
Published: 20 February 2023

1. Introduction

The Hebei Plain is one of the most water-sensitive areas of China. Its per capita water resources amount to less than 12.5% of the national total [1], and 70% of the water consumption depends on groundwater [2]. Consequently, the decline in GWL has precipitated various ecological and environmental issues in the Hebei Plain, such as land subsidence, soil salinization, the expansion of cones of depression, and aquifer dewatering [3].

Physical and statistical models are the main tools used to predict GWL. Physical models can describe the groundwater system and reflect changes in groundwater, but their practical applications are hindered by heavy computational loads and the need for large volumes of hydrogeological data [4,5]. Conversely, statistical models, such as machine learning and deep learning models, are an effective alternative that do not require the specific characterization of physical properties, accurate physical parameters, or the modeling of the physical processes of a groundwater system [6–9]. Widely used statistical models include the support vector machine (SVM), long-short term memory (LSTM), multi-layer perceptron (MLP), and gated recurrent unit (GRU) models. Moreover, the use of the Gravity Recovery and Climate Experiment (GRACE) gravity satellite and global hydrological model has been identified as a promising alternative method for predicting groundwater levels. By utilizing remote sensing data and numerical models, this method provides valuable insights into the distribution of groundwater resources, allowing for more informed decision-making and effective management of these vital resources [9–11].

SVMs are a type of generalized nonlinear model for classification and regression analysis [12], and their solutions adopt a macro-perspective to solve quadratic constraint optimization [13]. Asefa et al. [14] proposed a solution for groundwater monitoring and prediction networks using the SVM method. Yoon et al. [15] compared time-series models based on artificial neural networks (ANNs) and SVM to predict GWL. Moreover, Tapak et al. [16] used SVM to predict the GWL of the Hamadan–Bahar Plain in West Iran. SVMs have also been used to predict hydrological factors, such as river flow [17,18].

LSTM is an improved recurrent neural network (RNN) that was developed to address the exploding gradient problem using forget and update gates to regulate gradient [19], with GWL predictions extensively tested. Vu et al. [20] used LSTM to reconstruct, fill gaps, and extend existing time-series of GWL data in Normandy, France. Further, Wunsch et al. [21] compared the LSTM and nonlinear autoregressive networks with exogenous input and proved the proficiency and accuracy of LSTM for GWL predictions.

An MLP is a type of feedforward ANN consisting of an input layer, single or multiple hidden layers, and an output layer, and each node (neuron) in a layer is connected to every node in the following layer [22]. MLPs have been widely used in hydrological models [23,24] and agricultural pollution models [25–27]. Sahoo et al. [7] combined an MLP and genetic algorithm to predict GWL changes in agricultural areas of the United States.

GRUs are an optimized version of an LSTM that shorten the model training time and simplify the gated structure [28]. Jeong et al. [29] predicted GWL sequences using the LSTM, GRU, and autoregressive with exogenous input models. Zhang et al. [30] used various models to compare the simulations of the water level of the middle route of the South-to-North Water Transfer Project and proved that the performance of GRU and LSTM is similar but GRU has a comparatively faster learning curve. Additionally, Chen et al. [31] automatically calibrated groundwater parameters by combining the GRU model with particle swarm optimization.

The GWL in the Hebei Plain exhibits highly nonlinear variability due to various factors such as precipitation, evapotranspiration, and human activities. This variability may result in poor model prediction. Additionally, some nonlinear machine learning models may not accurately process the noise and features that are present in the real situation of the study area. Therefore, this paper aims to explore the mathematical relationships of the GWL time-scale data themselves and to perform dynamic prediction of the GWL in the Hebei Plain by comparing support vector machine (SVM), long-short term memory (LSTM), multilayer perceptron (MLP), and gated recurrent unit (GRU) models. We evaluate each model qualitatively and quantitatively using dynamic fluctuation, dynamic rise, and dynamic fall types of sites, respectively, in order to obtain a dynamic prediction model applicable to the GWL in the Hebei Plain. The remaining paper is organized as follows: Section 2 introduces the research area, data sources, model methods, and evaluation indicators, along with the technical approach of this study. Section 3 introduces the main findings and discusses the performance indicators of each model. Section 4 sets out the main conclusions of this study.

2. Materials and Methods

2.1. Study Area

The Hebei Plain is located in North China (114°33′–119°42′ N 36°05′–39°93′ E) (Figure 1). Per capita water resources in the region are approximately 386 m^3, which is approximately one eighth of China's national average, thus making it an extremely water-scarce region. It lies in the semi-humid, semi-arid climate zone and experiences a temperate continental monsoon climate. This region has four distinct seasons with rainy and hot periods coinciding. Annual precipitation is unevenly distributed but mainly falls in summer, with an annual average of 450–550 mm. Precipitation from June to July accounts for approximately 75% of the annual precipitation, and surface water resources are relatively scarce. The main water source for irrigation is groundwater, and agricultural groundwater consumption accounts for 74.5–76.6% of groundwater exploitation.

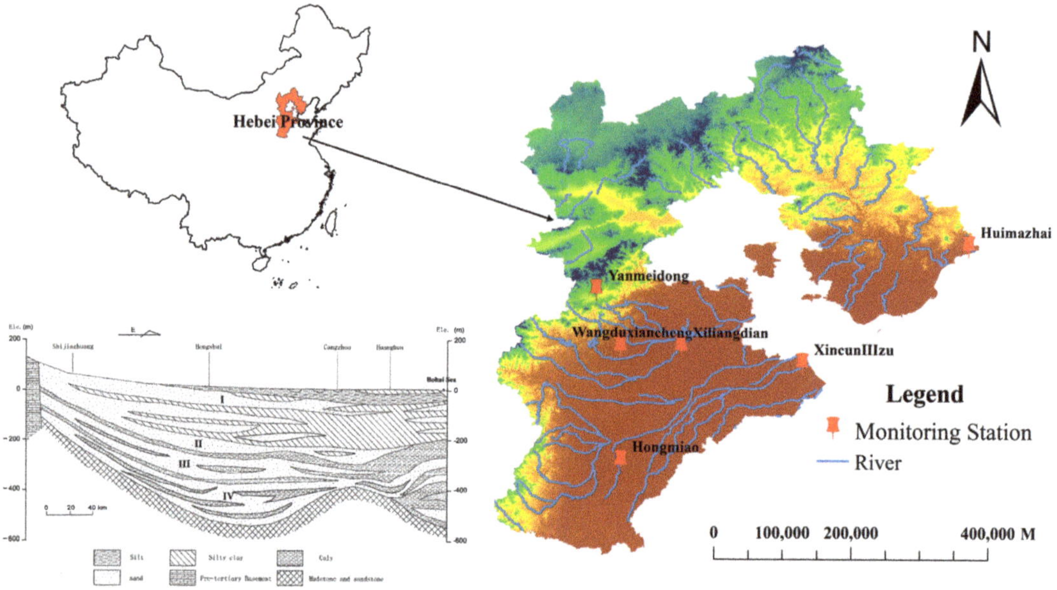

Figure 1. Location and hydrogeological profile of the study area in the Hebei Plain.

The North China Plain has complex hydrogeological conditions, but it mainly comprises basins of Quaternary (Cenozoic) loose sediment. Quaternary aquifer media are divided into four groups (from top to bottom): Quaternary Holocene Q4, Upper Pleistocene Q3, Middle Pleistocene Q2, and Lower Pleistocene Q1. The buried depths of the bottom boundaries of the aquifer group are 40–60 m, 120–170 m, 250–350 m, and 350–550 m, respectively. Overexploitation of groundwater for many years has resulted in many groundwater cones of depression in the first, second, and third aquifer groups in the Hebei Plain, causing nonlinear changes in the groundwater flow direction and GWL height in the region.

2.2. Data Sources

GWL data collected during 2018–2020, with a time interval of 4 h, were obtained from China's National Groundwater Monitoring Project. The trends in GWL in the Hebei Plain during the study period consisted of dynamic fluctuations, increases, and decreases. In total, six monitoring stations covering these three change types were selected as the study objects, which provided 32,880 datapoints. The time series of each station was divided, and any missing series was added so that the data could be converted into a format recognized by each model. The station data sample formats are shown in Table 1.

Table 1. Data samples in the study area.

Number	Type	Station	City	GWL	Sequence Length (Day)
1	dynamic	Huimazhai	Qinhuangdao	33.83	5480
2	fluctuations	Hongmiao	Xingtai	17.74	5480
3	dynamic	Xiliangdian	Baoding	−20.23	5480
4	increase	Yanmeidong	Baoding	1236.14	5480
5	dynamic	Wangduxiancheng	Baoding	−42.33	5480
6	decrease	XincunIIIzu	Huanghua	−44.21	5480

To intuitively reflect the daily changes in the GWL of the three types of selected monitoring stations, line graphs were drawn showing the time scale of GWL for 2018–2020. Figure 2a shows a station with dynamic fluctuations, where the GWL was the same at the beginning and end of the research period. Figure 2b shows a station with a dynamic increase. Although there were fluctuations during the study period, the GWL at the end of the study period was higher than at the beginning. Further, Figure 2c shows a station with a dynamic decrease, wherein the GWL at the end of the study period was lower than at the beginning.

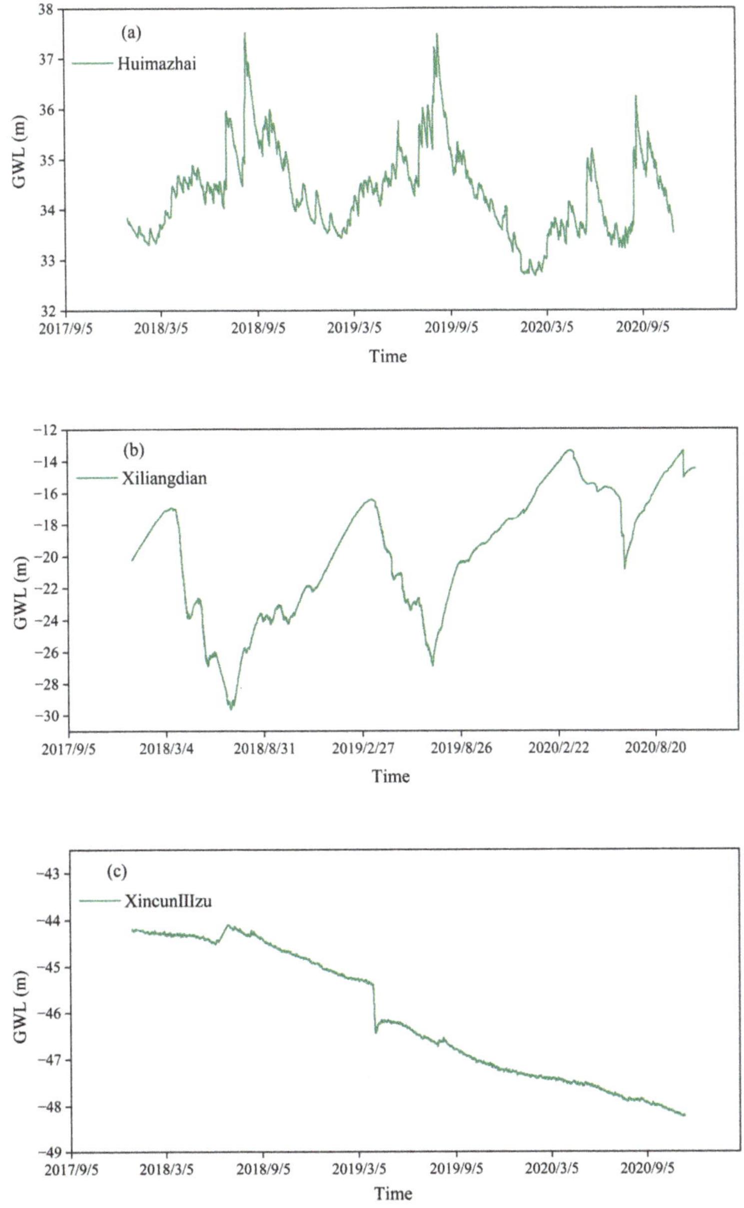

Figure 2. Three types of GWL data samples: (**a**) dynamic fluctuations; (**b**) dynamic increase; (c) dynamic decrease.

2.3. Methods

Data-driven models, such as ANNs, can easily approximate the complex behavior and responses of physical systems; additionally, they can quickly optimize many case scenarios with different constraints. Compared with the multiple assumptions, complex input variables, and parameter calibration of physical models, the input variables of data-driven models are easier to measure and quantify. In particular, machine learning can help to predict the GWL in areas that lack hydrogeological survey data. In this study, SVM, LSTM, GRU, and MLP models were used to predict GWL at the selected monitoring stations. Each of these models has been described below.

2.3.1. Support Vector Machine

SVM is a linear discriminant classification method based on the maximum margin. SVMs are the most widely used machine learning models for predicting GWL, as they can maximize prediction accuracy. They use the linear kernel function, polynomial kernel function, Gaussian kernel radial basis function (RBF), and sigmoid kernel function, which greatly optimize the nonlinear prediction capability of the model. Moreover, it is considered the best theory for small-sample statistical estimations and predictive learning, and can precisely predict GWL.

RBF has a strong nonlinear mapping capability and is suitable for predicting moment-to-moment changes in GWL as follows:

$$k(xi, xj) = \exp(-\gamma(xi - xj)^2) \tag{1}$$

where γ is an artificially determined positive real number parameter and (x_i, x_j) is the training sample.

2.3.2. Long-Short Term Memory

The LSTM model processes and analyzes time-series data by selectively extracting saved information and combining the selected information with subsequently input time-series data. The network can locally predict each fragmented sequence of GWL data, and the prediction deviations are passed back, to dynamically predict a GWL sequence.

LSTM is an improved recurrent neural network (RNN). The cycle of an ordinary RNN passes through the hidden state (H), but the LSTM output has two states: H and memory state (C). Further, three "gates" are added to the LSTM to process the input information differently. Figure 3 shows the internal structure of an LSTM cell.

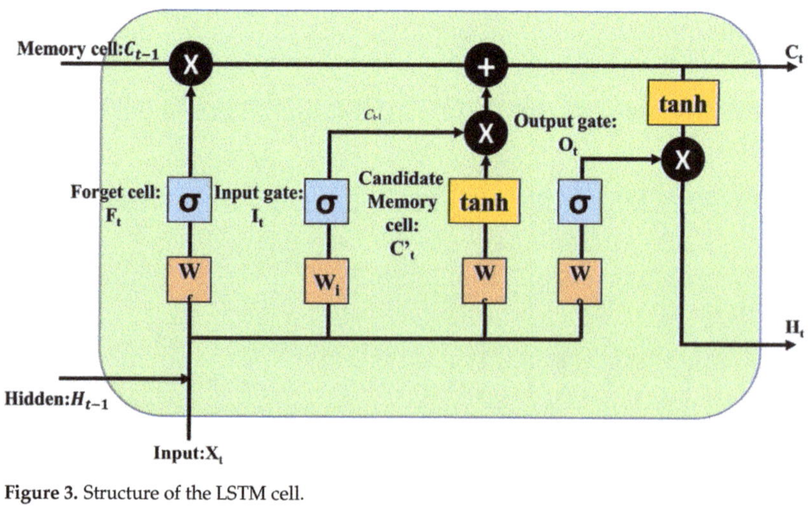

Figure 3. Structure of the LSTM cell.

At the core of LSTM is the objective of controlling the internal state to allow it to retain and filter information from previous moments. LSTM has three gates (forget gate F_t, input gate I_t, and output gate O_t) in the hidden layer to control the signal input and output. The forget gate uses H_{t-1} and X_t as inputs to control how much information needs to be forgotten in the internal state of the previous moment C_{t-1}; further, the input gate selectively memorizes input information and determines the quantity saved to the cell state C_t in the current moment X_t. The output gate controls how much information the internal state C_t needs to output to the external state in the current moment. The entire network equation of the LSTM cell can be given as follows:

$$\begin{bmatrix} \tilde{C}_t \\ O_t \\ I_t \\ F_t \end{bmatrix} = \begin{bmatrix} \tanh \\ \sigma \\ \sigma \\ \sigma \end{bmatrix} \left(W \begin{bmatrix} X_t \\ H_{t-1} \end{bmatrix} + b \right) \quad (2)$$

$$C_t = F_t \times C_{t-1} + I_t \times \tilde{C}_t \quad (3)$$

$$H_t = O_t \times \tanh(C_t) \quad (4)$$

where σ is the sigmoid activation function, which limits the output to between 0 and 1. Since each element of the output matrix (or vector) after the sigmoid layer is a real number between 0 and 1, which is then dot-multiplied with other information, it effectively controls the passage of information. In this range, "0" indicates that the information does not pass at all, and "1" indicates that the information passes entirely. This allows the network to regulate the flow of information through the "gate".

2.3.3. Gated Recurrent Units

GRU optimizes the gated structure of LSTM (Figure 4), and its training process is easier to converge. Compared with LSTM, GRU has only two gates: the update and reset gates. The update gate is constructed from the forget and input gates in the LSTM. The reset gate is recomposed from memory cells and hidden layer states. The GRU network controls the change in state of hidden units over time through its special gated structure. This avoids inaccurate parameter training due to the vanishing gradient problem or exploding gradient problem during long-term propagation.

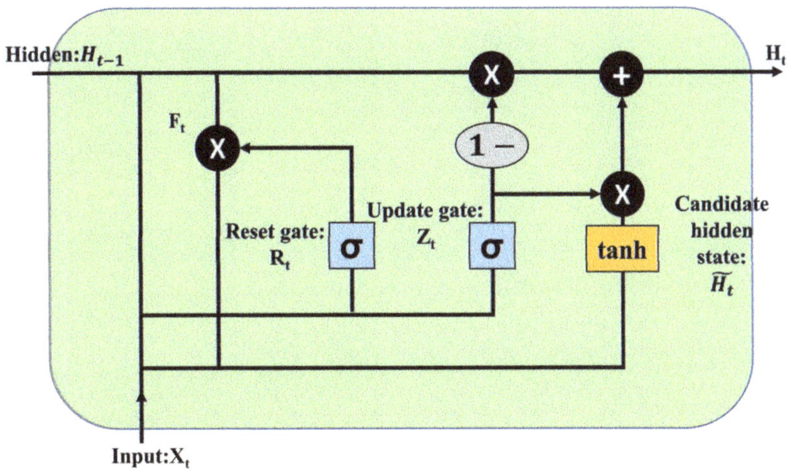

Figure 4. Structure of the GRU cell.

In Figure 4, the update and reset gates are denoted by Z_t and R_t, respectively, where Z_t is used to control the extent to which H_{t-1} is retained in H_t, and R_t is used to control the extent to which the current candidate set $\widetilde{H}_t$ will be written into H_{t-1}.

First, the gating state is obtained from the state in the previous moment H_{t-1} and the input X_t in the current moment. Subsequently, using sigmoid nonlinear transformation, the data are mapped to [0,1], which acts as the gating signal. Once the gating signal has been obtained, the reset gate is "reset" as a coefficient before the previous moment H_{t-1} to obtain the candidate hidden state $\widetilde{H}_t$. The equations are as follows:

$$R_t = \sigma(W_r \cdot [H_{t-1}, X_t]) \tag{5}$$

$$Z_t = \sigma(W_z \cdot [H_{t-1}, X_t]) \tag{6}$$

$$\widetilde{H}_t = \tanh\left(W_{\widetilde{h}} \cdot [R_t \times H_{t-1}, X_t]\right) \tag{7}$$

where σ is the sigmoid activation function. While calculating a gate's hidden state, the sigmoid function is used to obtain a result between 0 and 1, and while calculating a candidate hidden state, the activation function uses the tanh function. W_r, W_z, and $W_{\widetilde{h}}$ are the reset gate, update gate, and the weight matrix, respectively, for calculating the candidate hidden state.

2.3.4. Multi-Layer Perceptron

MLP is an ANN used for predictions. It learns the relationships between inputs and outputs using a large volume of data and can be used for nonlinear modeling. An MLP is composed of multiple layers of neurons, and each node (neuron) in a layer is connected with a certain weight to every node in the following layer (Figure 5). The main disadvantage of this method is that when layers or the nodes in each layer increase, overfitting and model training issues arise.

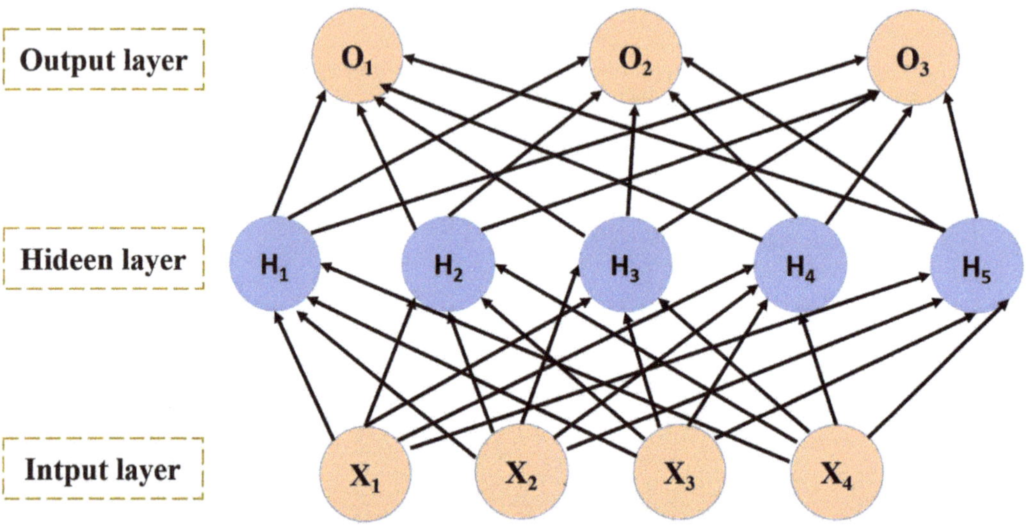

Figure 5. Structure of MLP model.

In Figure 5, the output of the MLP hidden layer H node is given as

$$H_j = g(\sum_{i=1}^{n} \omega_{ij} x_i + b_j) \tag{8}$$

where ω and b are the weight and deviation, respectively, and g is the activation function, commonly the sigmoid, tanh, or rectifier linear unit (ReLU) activation function. Because the ReLU function avoids the vanishing gradient problem and its convergence speed is faster than that of the sigmoid and tanh activation functions, ReLU was used as the activation function in this study.

2.4. Performance Evaluation of Models

The performance of the SVM, LSTM, GRU, and MLP models can be evaluated using the root mean square error (RMSE), coefficient of determination (R^2), and Nash–Sutcliffe efficiency (NSE), which are calculated as follows:

$$R^2 = \left[\frac{\left[\sum Q_o Q_p\right] - \left[\frac{\sum Q_o \sum Q_p}{N}\right]}{\sqrt{\left[\sum Q_P{}^2 - \frac{(\sum Q_P)^2}{N}\right]\left[\sum Q_o{}^2 - \frac{(\sum Q_o)^2}{N}\right]}}\right]^2 \tag{9}$$

$$RMSE = \sqrt{\frac{1}{N}\sum_{i=1}^{N}(Q_o - Q_p)^2} \tag{10}$$

$$NSE = 1 - \left[\frac{\sum_{i=1}^{N}(Q_o - Q_p)^2}{\sum_{i=1}^{N}(Q_o - \overline{Q}_o)^2}\right] \tag{11}$$

where Q_o is the observed GWL, Q_p is the predicted GWL, N is the length of the groundwater series, and $\overline{Q}o$ is the mean value of the observed GWL.

2.5. Groundwater Level Prediction Methodology

As shown in Figure 6, we first input missing values and processed outliers in the original dataset and converted the format into a data type recognized by the model. Subsequently, we split the dataset, with the first 70% of the time-scale GWL measured sequence used as the training set. We converted the one-dimensional training data into a two-dimensional matrix containing input data and testing data, with a window of 10, i.e., GWL_0–GWL_9 was used to predict GWL_{10}, GWL_1–GWL_{10} was used to predict GWL_{11}, etc. GWL data of each station were converted into this format. Later, we used the SVM, LSTM, GRU, and MLP models to build and train the dataset, calculated the loss functions of each model, observed the consistency between predicted and actual values, and judged the robustness of each model based on its dynamic changes. The degree of convergence of the loss function was used as the basis for the training results of each model to continuously train the model. After training was completed, 30% of the time-scale GWL measured sequence for each station was used as the testing set. The testing set data were also converted into a two-dimensional matrix as described above and input into the trained model. The R^2, RMSE, and NSE evaluation indicators were used to evaluate the performance of the models, to select the model that was most suitable for dynamically predicting the GWL in the Hebei Plain.

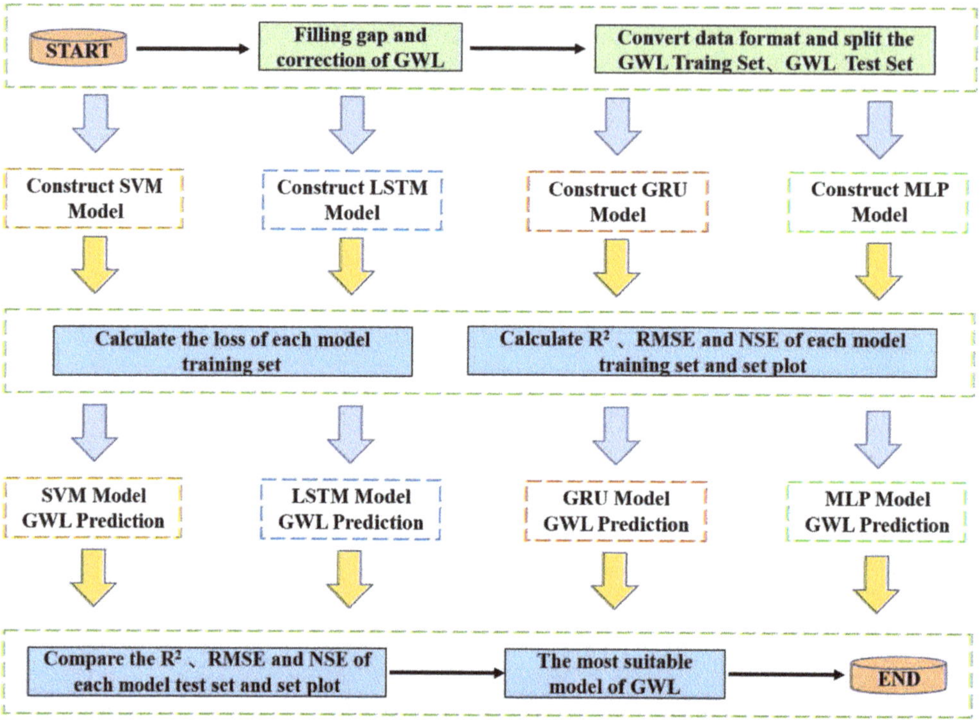

Figure 6. Running process of the SVM, LSTM, GRU, and MLP models.

3. Results and Discussion

The results of the GWL predictions for each station in Hebei Province acquired through the machine learning techniques are presented in this section. The Huimazhai and Hongmiao stations were the dynamically fluctuating stations. The Huimazhai station is located in the urban area of Qinhuangdao, adjacent to the Shi River and Liaodong Bay, and it showed an average GWL of 34.36 m. During the study period, its GWL in summer was mostly higher than in spring and winter, resulting in dynamic fluctuations. This may be due to a reduced need to irrigate crops in the rainy and summer season across the entire Qinhuangdao area. Moreover, changes in GWL were consistent with changes in precipitation. The Hongmiao station is located in Xingtai City, where groundwater is exploited for crop irrigation. However, Xingtai has conducted a groundwater restoration pilot project and promoted the cultivation of drought-resistant winter wheat; additionally, groundwater replenishment work has also been conducted. The GWL in this area fluctuated due to the influence of changes in human activities, but it had a mean GWL of −65.09 m. The dynamically increasing stations were Xiliangdian and Yanmeidong. The Xiliangdian station is located in Gaoyang County, Baoding, and had a mean GWL of −19.41 m. The GWL in this station was lower in spring and summer than in other seasons, which was likely due to the expansive area of water-intensive crops in spring and summer. Nevertheless, because this area is located in the middle route of the South-to-North Water Diversion Project, its GWL has been increasing. Further, the Yanmeidong station is located in Laiyuan County, Baoding, and showed an average GWL of 1235.67 m, and its GWL increased overall during the study period. This was likely due to the station being located in a mountainous area that receives relatively high levels of precipitation; moreover, because the area under irrigated farmland in the region was small, agricultural water consumption was low. The dynamically decreasing stations were Wangduxiancheng and XincunIIIzu. The Wangduxiancheng station is located in Wangdu County, Baoding, and had a mean GWL of 16.43 m, and XincunIIIzu is located in Huanghua and had a mean GWL of

−43.13 m. Despite the implementation of groundwater extraction projects at the two stations, their water levels decreased gradually during the study period due to increases in the area under winter wheat.

In this study, GWL data were divided into training and validation sets, with the GWL of the validation set predicted using mathematical relationships discovered in the training set data. The SVM, LSTM, GRU, and MLP models were used to develop a GWL prediction model, and time-series scatter plots, relative error, and Taylor diagrams were used to qualitatively evaluate the performance of the models, while statistical and hydrological model indicators, such as RMSE, R^2, and NSE, were used to quantitatively evaluate their performance.

3.1. GWL Prediction Using SVM Model

Gaussian and linear kernel functions were used for SVM-based runoff modeling. The former was more effective than the latter; therefore, we established an SVM model with the Gaussian function as the kernel function to predict the GWL of the six monitoring stations with the three types of dynamic changes (fluctuations, increases, and decreases). As shown in Figure 7, the SVM model overestimated the GWL of the dynamically fluctuating stations and slightly underestimated the GWL of the stations with dynamic increases and decreases.

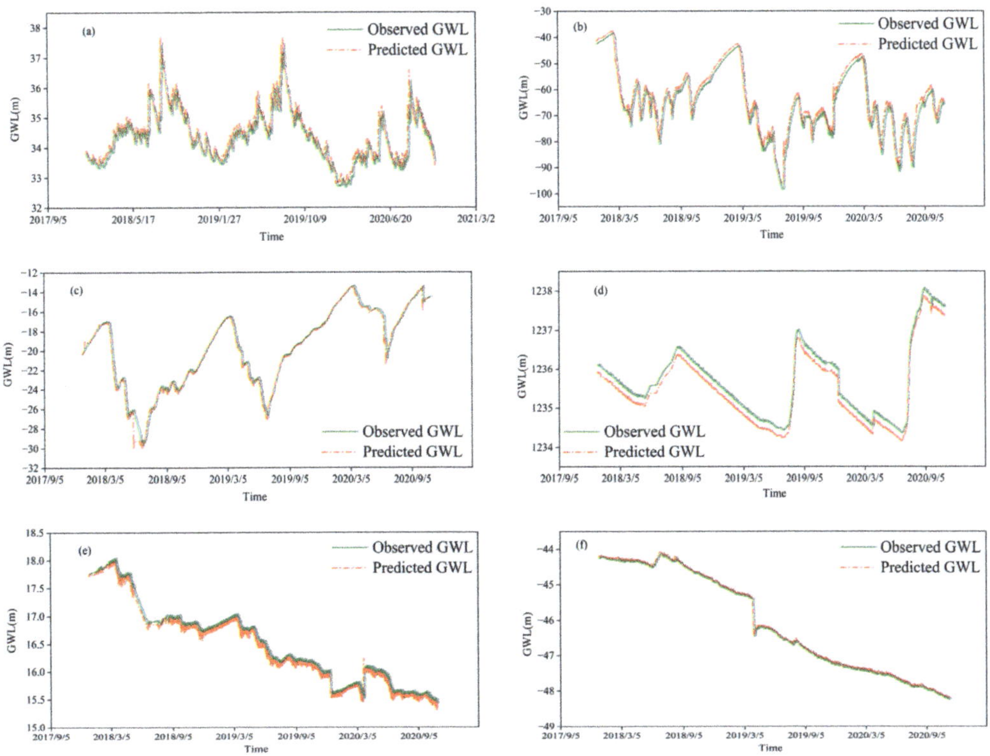

Figure 7. Results of hourly GWL simulation (m) in six stations using the SVM model during 2018–2020 in the training and testing periods: (**a**) Huimazhai station, (**b**) Hongmiao station, (**c**) Xiliangdian station, (**d**) Yanmeidong station, (**e**) Wangduxiancheng station, and (**f**) XincunIIIzu station.

Table 2 shows the simulation results. Generally, the simulation results were better for the dynamically increasing and decreasing stations. For example, the RMSE, R^2, and NSE values of the Yanmeidong station in the testing period were 0.193 m, 0.998, and 0.984, respectively, indicating that SVM was well suited to these two types of monitoring stations.

Considering the dynamically fluctuating stations, the RMSE, R^2, and NSE values of the Huimazhai station in the training period were 0.253 m, 0.953, and 0.921, respectively, while the accuracy in the testing period was markedly lower, with NSE being 24.9% lower in the testing period than in the training period, indicating that SVM was not effective at capturing the nonlinear relationship of dynamically fluctuating stations. Overall, SVM was not suitable for predicting GWL in the Hebei Plain.

Table 2. Results of different performance indicators of the SVM model during the training and testing periods at each site.

Station	Training			Testing		
	RMSE	R^2	NSE	RMSE	R^2	NSE
Huimazhai	0.253	0.953	0.921	0.396	0.757	0.691
Hongmiao	2.299	0.98	0.967	3.823	0.867	0.804
Xiliangdian	0.298	0.995	0.994	0.511	0.915	0.908
Yanmeidong	0.204	0.998	0.909	0.193	0.998	0.984
Wangduxiancheng	0.076	0.992	0.985	0.071	0.929	0.808
XincunIIIzu	0.052	0.999	0.998	0.045	0.990	0.940

3.2. GWL Prediction Using the LSTM Model

During LSTM modeling, the model filtered GWL features through the output gate, saved the useful features, and discarded the useless features to obtain current moment contextual information, which greatly enriched the information in the vector. This also implied that the current moment contextual information Ht was only one part of the global information Ct. Further, we used the LSTM model to simulate the six selected stations. The GWL in the training and testing periods is shown in Figure 8. The GWL prediction results were suboptimal for the dynamically fluctuating stations, including another overestimation of the peak value in the testing period shown in Figure 8a. However, compared with the SVM model, the results of the dynamically increasing station, shown in Figure 8d, improved, with the predicted GWL closer to the observed GWL, although the peak value was overestimated.

Table 3 shows the simulation results. The RMSE, R^2, and NSE values of the Yanmeidong station in the testing period were 0.116 m, 0.996, and 0.994, respectively. Notably, the results from the testing period were better than those from the training period for this station, indicating that the observed and predicted values in the verification period were highly consistent, thus proving the effectiveness of the LSTM model for the dynamically increasing stations. The RMSE, R^2, and NSE values of the Huimazhai station in the training period were 0.263 m, 0.868, and 0.868, respectively. Compared to the R^2 value of the SVM model, the R^2 value of the LSTM model was 12.7% higher. Nevertheless, for the dynamically fluctuating stations, both the peaks and troughs of the predicted GWL were more than the observed GWL; therefore, the prediction results were suboptimal. The overall R^2 value of the LSTM model was >0.85, indicating that it was a good model for predicting the GWL in the Hebei Plain.

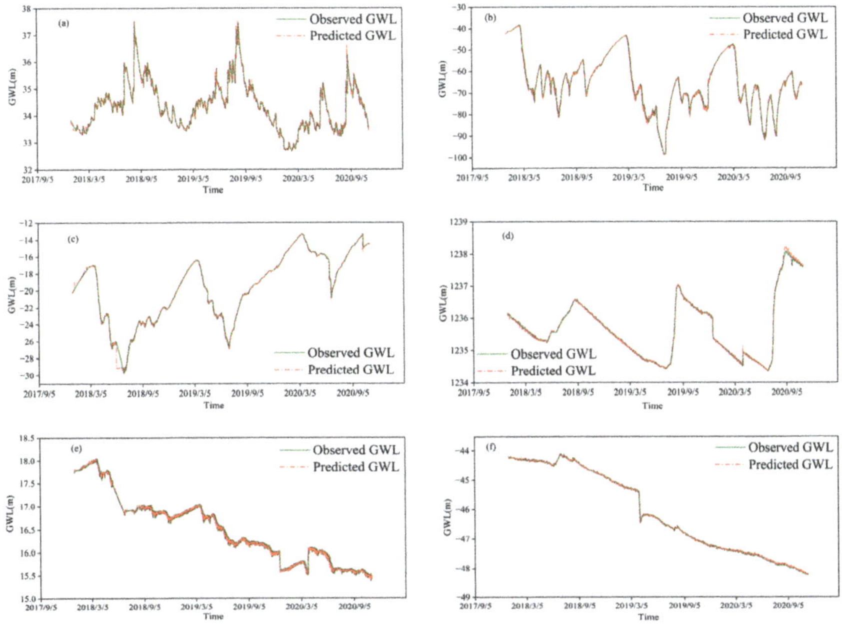

Figure 8. Results of the hourly GWL simulation (m) in six stations using the LSTM model during 2018–2020 in the training and testing periods: (**a**) Huimazhai station, (**b**) Hongmiao station, (**c**) Xiliangdian station, (**d**) Yanmeidong station, (**e**) Wangduxiancheng station, and (**f**) XincunIIIzu station.

Table 3. Results of the different performance indicators of LSTM model during training and testing periods at each monitoring station.

Station	Training			Testing		
	RMSE	R^2	NSE	RMSE	R^2	NSE
Huimazhai	0.192	0.955	0.955	0.263	0.868	0.864
Hongmiao	1.581	0.985	0.984	1.771	0.958	0.958
Xiliangdian	0.244	0.996	0.996	0.338	0.961	0.96
Yanmeidong	0.053	0.994	0.994	0.116	0.996	0.994
Wangduxiancheng	0.049	0.994	0.994	0.036	0.953	0.95
XincunIIIzu	0.037	0.999	0.999	0.028	0.987	0.976

3.3. GWL Prediction Using the MLP Model

The MLP model includes the input, output, and hidden layers, and each node (neuron) in a layer is connected to every node in the following layer. We improved the basic three-layer MLP by adding two linear hidden layers and the ReLU activation function after the first hidden layer. Subsequently, we simulated the selected stations in the Hebei Plain. Notably, this model solved the shortcomings of the SVM and LSTM models for dynamically fluctuating stations (Figure 9a,b). The predicted trend was satisfactory and higher in the testing period than in the training period. However, in the stations with dynamically increasing trends (Figure 8c), further improvement in the prediction results of peak and trough values was not possible.

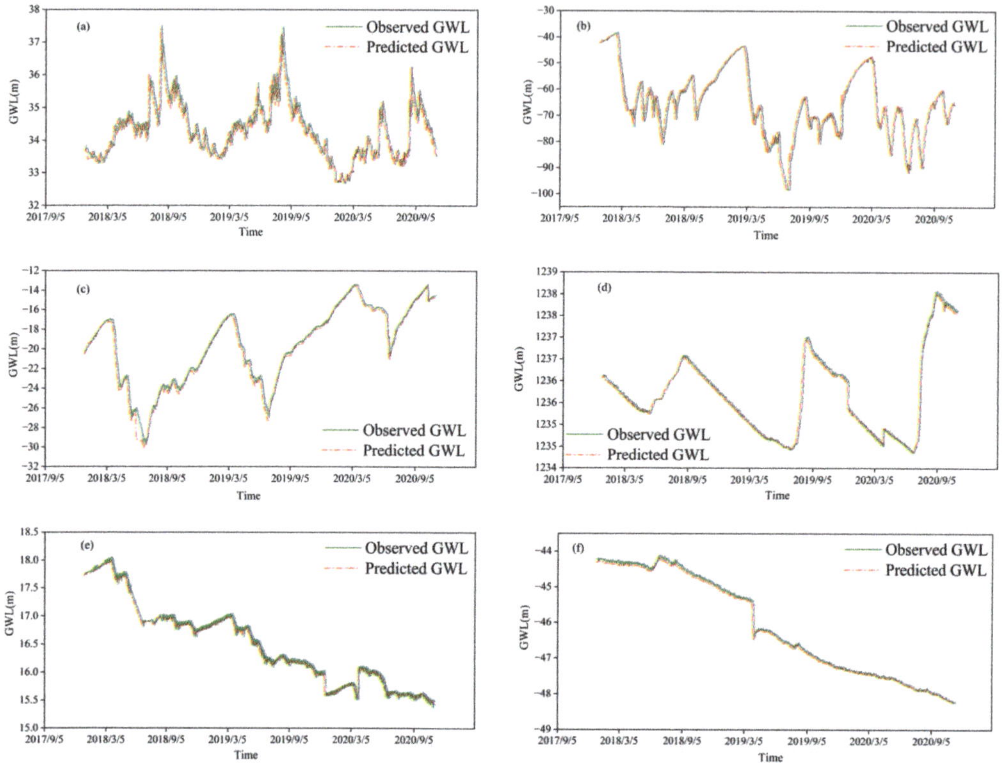

Figure 9. Results of hourly GWL simulation (m) in six stations using the MLP model during 2018–2020 during the training and testing periods: (**a**) Huimazhai station, (**b**) Hongmiao station, (**c**) Xiliangdian station, (**d**) Yanmeidong station, (**e**) Wangduxiancheng station, and (**f**) XincunIIIzu station.

Table 4 shows the simulation results. The RMSE, R^2, and NSE values of the Yanmeidong station in the testing period were 0.08 m, 0.998, and 0.997, respectively, which were the best simulation results. The results for the dynamically fluctuating stations (Huimazhai and Hongmiao) were also satisfactory, especially for the Huimazhai station, with RMSE, R^2, and NSE values of 0.201 m, 0.959, and 0.95 in the training period and 0.128 m, 0.979, and 0.968 in the testing period, respectively. Thus, the results in the testing period were better than those in the training period. Overall, the RMSE values of the various station types were <0.6, and NSE and R^2 values were >0.96, thus making this a highly suitable model for predicting the GWL in the Hebei Plain.

Table 4. Results of different performance indicators of the MLP model during the training and testing periods at each station.

Station	Training			Testing		
	RMSE	R^2	NSE	RMSE	R^2	NSE
Huimazhai	0.201	0.959	0.95	0.128	0.979	0.968
Hongmiao	1.419	0.988	0.987	0.514	0.997	0.996
Xiliangdian	0.347	0.999	0.991	0.295	0.987	0.969
Yanmeidong	0.033	0.998	0.998	0.08	0.998	0.997
Wangduxiancheng	0.041	0.997	0.996	0.028	0.969	0.97
XincunIIIzu	0.051	0.999	0.998	0.014	0.995	0.994

3.4. GWL Prediction Using the GRU Model

Compared with LSTM, GRU has a simpler gated structure and fewer parameters and faster convergence. The Ct in GRU already contained Ht, and there was a trade-off between current unit information and previous global information while generating moment contextual information. Therefore, replacing I_t with 1-Zt can expose all information globally. The GRU model for each of the selected stations in the training and testing periods is shown in Figure 10. The model retained the testing results of the MLP model for dynamically fluctuating stations, but the testing period results for the dynamically increasing station (c) were better than the results of the other models.

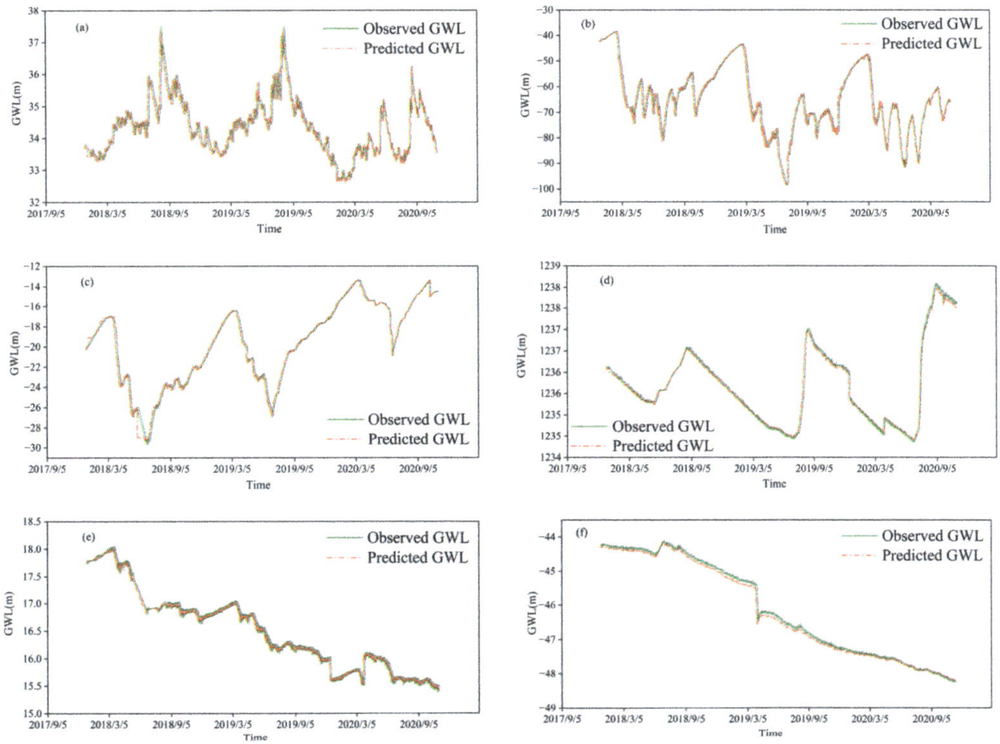

Figure 10. Results of hourly GWL simulation (m) in six stations using the GRU model during 2018–2020 training and testing periods: (**a**) Huimazhai station, (**b**) Hongmiao station, (**c**) Xiliangdian station, (**d**) Yanmeidong station, (**e**) Wangduxiancheng station, and (**f**) XincunIIIzu station.

Table 5 shows the corresponding simulation results. The best training period results were observed in the XincunIIIzu station, with RMSE, R^2, and NSE values of 0.081 m, 0.999, and 0.996, respectively, while the best testing period results were observed in the Yanmeidong station, with RMSE, R^2, and NSE values of 0.098 m, 0.998, and 0.996, respectively. The GRU model results were notably better for dynamically fluctuating stations than those of the first three models. Furthermore, the model maintained the training and testing results for the other station types, thus making it the best model for predicting the GWL in the Hebei Plain.

Table 5. Results of different performance indicators of the GRU model during the training and testing periods at each station.

Station	Training			Testing		
	RMSE	R^2	NSE	RMSE	R^2	NSE
Huimazhai	0.182	0.959	0.959	0.08	0.988	0.987
Hongmiao	1.449	0.987	0.987	0.518	0.996	0.996
Xiliangdian	0.229	0.996	0.996	0.123	0.995	0.995
Yanmeidong	0.04	0.998	0.996	0.098	0.998	0.996
Wangduxiancheng	0.041	0.996	0.996	0.033	0.961	0.96
XincunIIIzu	0.081	0.999	0.996	0.027	0.995	0.978

3.5. Model Comparison

The SVM model had the lowest simulation accuracy for the selected stations, which may be due to the four kernel functions of the SVM. The RBF kernel function selected in this study could be modified further. When σ is too small, RBF can overfit, and when σ is too large, the relationship between X_i and X_j will have less overall influence on the model, causing inaccurate predictions. The LSTM model was the third-best model. It included a forget gate, which makes the partial derivative of the current memory unit to the previous memory unit a constant, thereby solving the disappearing gradient problem of RNN. Nevertheless, several input parameters exist, which increase the likelihood of overfitting. The second-best model was MLP. Its highly nonlinear global effect resulted in good accuracy in the training and testing periods, with an NSE value of 0.997 for the Yanmeidong station, but the model had the slowest learning speed, which was suboptimal in terms of time consumption for moment-to-moment predictions. Moreover, the GRU model was the most accurate for the dynamically fluctuating and dynamically increasing stations. As it had a simple gated structure and fewer input parameters, and as it reduced the risk of overfitting, it had a shorter training time (Table 6). Thus, it was the most suitable model for predicting the GWL in the Hebei Plain.

Table 6. Training time comparison of four models with 500 epochs.

Model	SVM	LSTM	GRU	MLP
Time (min)	1081	1660	1251	2694

Figure 11 presents the scatter plots for each station during the study period. The GRU model had the best correlation between the observed and predicted GWL of the four models, with predicted values near the regression function for almost every station. Although the MLP model was accurate for most stations, it did not capture the extreme values of the dynamically increasing sites. Moreover, the SVM and LSTM models only reflected the GWL trend for dynamically fluctuating stations, and they did not accurately predict the specific GWL values.

Finally, we evaluated the performance of the models for each of the stations using Taylor diagrams (Figure 12). The results showed that the GRU model had the best accuracy for both dynamically fluctuating and dynamically increasing stations, followed by the MLP model, which had the best accuracy for dynamically decreasing stations. Comparatively, the SVM model had the poorest performance, as it was the furthest from the "Ref" point for most stations. It was followed by the LSTM model, which performed reasonably.

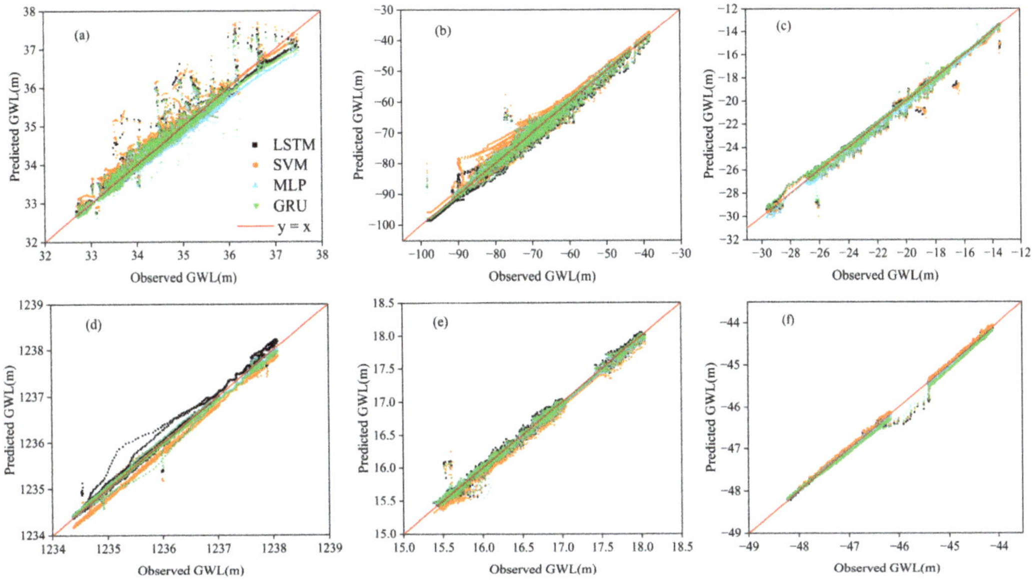

Figure 11. Scatter diagrams of GWL simulations for each site by SVM, LSTM, MLP and GRU models: (**a**) Huimazhai station, (**b**) Hongmiao station, (**c**) Xiliangdian station, (**d**)Yanmeidong station, (**e**) Wangduxiancheng station, and (**f**) XincunIIIzu station.

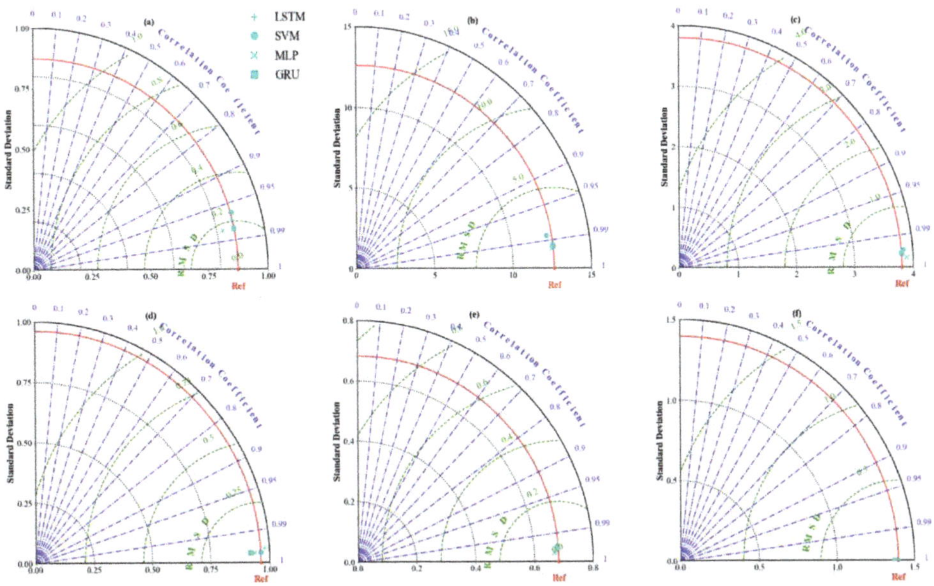

Figure 12. Taylor diagrams of GWL simulations for each site by SVM, LSTM, MLP and GRU models: (**a**) Huimazhai station, (**b**) Hongmiao station, (**c**) Xiliangdian station, (**d**) Yanmeidong station, (**e**) Wangduxiancheng station, and (**f**) XincunIIIzu station.

4. Conclusions

GWL is a crucial indicator for evaluating the health of groundwater resources in the Hebei Plain. This study attempted to model the GWL in the Hebei Plain and predict dynamic changes in the GWL using SVM, LSTM, MLP, and GRU models, and it qualitatively and quantitatively analyzed the training and testing datasets in the modeling process. The main conclusions were as follows:

(1) By comparing the RMSE, R^2, and NSE indicators, we discovered that the GRU model performed the best for dynamically fluctuating and dynamically increasing stations, while the MLP model performed the best for dynamically decreasing stations. The update gate in the GRU model acquired previous moment state information in the current state, which assisted in capturing long-term dependencies in the time series and solved the problem of overfitting to some extent. Moreover, the GRU model not only showed good performance in predicting trends, but it was also better than the other models regarding the training time and capturing extreme values, thus making it the most suitable model for predicting the GWL in the Hebei Plain.

(2) Apart from the different principles of each model, the differences in the simulation results can be attributed to factors such as data segmentation during the modeling process, the length of subsequences, and the uncertainty of model parameters. Moreover, the influence of the different activation functions on the GWL in the different models should also be considered. Furthermore, the training frequency of each model in this study was the same, and adaptive improvements should be made for each model in subsequent studies.

Author Contributions: Conceptualization, Z.W. and C.W.; methodology, Z.W.; software, C.L.; validation, Q.S., W.L. and X.H.; investigation, Z.W., C.W. and T.Q.; resources, Q.S.; data curation, C.W.; writing—original draft preparation, Q.S. and Z.W.; writing—review and editing, C.W.; visualization, Q.S.; supervision, W.L.; project administration, L.Y.; funding acquisition, C.L. All authors have read and agreed to the published version of the manuscript.

Funding: This research was funded by the National Key Research and Development Program of China (2021YFC3000205), Heilongjiang Provincial Applied Technology Research and Development Program (GA19C005), Key R & D Program of Heilongjiang Province (JD22B001), and Independent Research Project of the State Key Laboratory of Simulation and Regulation of Water Cycle in River Basin (SKL2022ZD02).

Data Availability Statement: The data will be available on request.

Acknowledgments: The authors would like to thank the editor and two anonymous reviewers for taking the time to provide their helpful feedback and suggestions.

Conflicts of Interest: The authors declare no conflict of interest.

References

1. Wei, Y.; Sun, B. Optimizing Water Use Structures in Resource-Based Water-Deficient Regions Using Water Resources Input–Output Analysis: A Case Study in Hebei Province, China. *Sustainability* **2021**, *13*, 3939. [CrossRef]
2. Currell, M.J.; Han, D.; Chen, Z.; Cartwright, I. Sustainability of groundwater usage in northern China: Dependence on palaeowaters and effects on water quality, quantity and ecosystem health. *Hydrol. Process.* **2012**, *26*, 4050–4066. [CrossRef]
3. Niu, J.; Zhu, X.G.; Parry, M.A.J.; Kang, S.; Du, T.; Tong, L.; Ding, R. Environmental burdens of groundwater extraction for irrigation over an inland river basin in Northwest China. *J. Clean. Prod.* **2019**, *222*, 182–192. [CrossRef]
4. Gupta, B.B.; Nema, A.K.; Mittal, A.K.; Maurya, N.S. Modeling of Groundwater Systems: Problems and Pitfalls. Available online: https://www.researchgate.net/profile/Atul-Mittal-3/publication/261758986_Modeling_of_Groundwater_Systems_Problems_and_Pitfalls/links/00b495356b45d3464c000000/Modeling-of-Groundwater-Systems-Problems-and-Pitfalls.pdf (accessed on 2 December 2022).
5. Ahmadi, S.H.; Sedghamiz, A. Geostatistical analysis of spatial and temporal variations of groundwater level. *Environ. Monit. Assess.* **2007**, *129*, 277–294. [CrossRef] [PubMed]
6. Chen, H.; Zhang, W.; Nie, N.; Guo, Y. Long-term groundwater storage variations estimated in the Songhua River Basin by using GRACE products, land surface models, and in-situ observations. *Sci. Total Environ.* **2019**, *649*, 372–387. [CrossRef] [PubMed]
7. Sahoo, S.; Russo, T.A.; Elliott, J.; Foster, I. Machine learning algorithms for modeling groundwater level changes in agricultural regions of the US. *Water Resour. Res.* **2017**, *53*, 3878–3895. [CrossRef]

8. Yadav, B.; Ch, S.; Mathur, S.; Adamowski, J. Assessing the suitability of extreme learning machines (ELM) for groundwater level prediction. *J. Water Land Dev.* **2017**, *32*, 103–112. [CrossRef]
9. Xiong, J.; Guo, S.; Kinouchi, T. Leveraging machine learning methods to quantify 50 years of dwindling groundwater in India. *Sci. Total Environ.* **2022**, *835*, 155474. Available online: https://www.sciencedirect.com/science/article/abs/pii/S0048969722025700 (accessed on 3 December 2022). [CrossRef]
10. Pratoomchai, W.; Kazama, S.; Hanasaki, N.; Ekkawatpanit, C.; Komori, D. A projection of groundwater resources in the Upper Chao Phraya River basin in Thailand. *Hydrol. Res. Lett.* **2014**, *8*, 20–26.
11. Thomas, B.F.; Famiglietti, J.S.; Landerer, F.W.; Wiese, D.N.; Molotch, N.P.; Argus, D.F. GRACE groundwater drought index: Evaluation of California Central Valley groundwater drought. *Remote Sens. Environ.* **2017**, *198*, 384–392. [CrossRef]
12. Vapnik, V.N. *The Nature of Statistical Learning Theory*; Springer: New York, NY, USA, 1995; p. 314.
13. Scholkoff, B.; Smola, A.J. *Learning with Kernels: Support Vector Machines, Regularization, Optimization, and Beyond*; MIT Press: Cambridge, MA, USA, 2002; p. 648.
14. Asefa, T.; Kemblowski, M.W.; Urroz, G.; McKee, M.; Khalil, A. Support vectors–based groundwater head observation networks design. *Water Resour. Res.* **2004**, *40*, 11. Available online: https://agupubs.onlinelibrary.wiley.com/doi/abs/10.1029/2004WR003304 (accessed on 3 December 2022). [CrossRef]
15. Yoon, H.; Jun, S.C.; Hyun, Y.; Bae, G.-O.; Lee, K.-K. A comparative study of artificial neural networks and support vector machines for predicting groundwater levels in a coastal aquifer. *J. Hydrol.* **2011**, *396*, 128–138. [CrossRef]
16. Tapak, L.; Rahmani, A.R.; Moghimbeigi, A. Prediction the groundwater level of Hamadan-Bahar plain, west of Iran using support vector machines. *J. Res. Health Sci.* **2013**, *14*, 82–87.
17. Sudheer, C.; Maheswaran, R.; Panigrahi, B.K.; Mathur, S. A hybrid SVM-PSO model for forecasting monthly streamflow. *Neural Comput. Appl.* **2014**, *24*, 1381–1389. [CrossRef]
18. Wang, W.; Xu, D.; Chau, K.; Chen, S. Improved annual rainfall-runoff forecasting using PSO–SVM model based on EEMD. *J. Hydroinformatics* **2013**, *15*, 1377–1390. [CrossRef]
19. Hochreiter, S.; Schmidhuber, J. Long short-term memory. *Neural Comput.* **1997**, *9*, 1735–1780. [CrossRef]
20. Vu, M.T.; Jardani, A.; Massei, N.; Fournier, M. Reconstruction of missing groundwater level data by using Long Short-Term Memory (LSTM) deep neural network. *J. Hydrol.* **2021**, *597*, 125776. [CrossRef]
21. Wunsch, A.; Liesch, T.; Broda, S. Groundwater level forecasting with artificial neural networks: A comparison of long short-term memory (LSTM), convolutional neural networks (CNNs), and non-linear autoregressive networks with exogenous input (NARX). *Hydrol. Earth Syst. Sci.* **2021**, *25*, 1671–1687. [CrossRef]
22. Haykin, S. *Neural Networks and Learning Machines, 3/E*; Pearson Education India: New Delhi, India, 2009. Available online: https://www.pearson.com/en-us/subject-catalog/p/neural-networks-and-learning-machines/P200000003278 (accessed on 4 December 2022).
23. Foddis, M.L.; Montisci, A.; Trabelsi, F.; Uras, G. An MLP-ANN-based approach for assessing nitrate contamination. *Water Supply* **2019**, *19*, 1911–1917. [CrossRef]
24. Ghorbani, M.A.; Deo, R.C.; Karimi, V.; Yaseen, Z.M.; Terzi, O. Implementation of a hybrid MLP-FFA model for water level prediction of Lake Egirdir, Turkey. *Stoch. Environ. Res. Risk Assess.* **2018**, *32*, 1683–1697. [CrossRef]
25. Singh, A.; Imtiyaz, M.; Isaac, R.K.; Denis, D.M. Comparison of soil and water assessment tool (SWAT) and multilayer perceptron (MLP) artificial neural network for predicting sediment yield in the Nagwa agricultural watershed in Jharkhand, India. *Agric. Water Manag.* **2012**, *104*, 113–120. [CrossRef]
26. Jia, X.; Cao, Y.; O'Connor, D.; Zhu, J.; Tsang, D.C.; Zou, B.; Hou, D. Mapping soil pollution by using drone image recognition and machine learning at an arsenic-contaminated agricultural field. *Environ. Pollut.* **2021**, *270*, 116281. [CrossRef] [PubMed]
27. Ijlil, S.; Essahlaoui, A.; Mohajane, M.; Essahlaoui, N.; Mili, E.M.; Van Rompaey, A. Machine Learning Algorithms for Modeling and Mapping of Groundwater Pollution Risk: A Study to Reach Water Security and Sustainable Development (Sdg) Goals in a Mediterranean Aquifer System. *Remote Sens.* **2022**, *14*, 2379. [CrossRef]
28. Cho, K.; Van Merriënboer, B.; Gulcehre, C.; Bahdanau, D.; Bougares, F.; Schwenk, H.; Bengio, Y. Learning phrase representations using RNN encoder-decoder for statistical machine translation. *arXiv* **2014**, arXiv:1406.1078. Available online: https://arxiv.org/abs/1406.1078 (accessed on 4 December 2022).
29. Jeong, J.; Park, E. Comparative applications of data-driven models representing water table fluctuations. *J. Hydrol.* **2019**, *572*, 261–273. [CrossRef]
30. Zhang, D.; Lindholm, G.; Ratnaweera, H. Use long short-term memory to enhance Internet of Things for combined sewer overflow monitoring. *J. Hydrol.* **2018**, *556*, 409–418. [CrossRef]
31. Chen, Y.; Liu, G.; Huang, X.; Chen, K.; Hou, J.; Zhou, J. Development of a surrogate method of groundwater modeling using gated recurrent unit to improve the efficiency of parameter auto-calibration and global sensitivity analysis. *J. Hydrol.* **2021**, *598*, 125726. [CrossRef]

water

MDPI

Article

Ecological Water Demand Estimations for Desert Terminal Lake Survival under Inland River Water Diversion Regulation

Jinqiang Lu [1,2,3], Lingqi Li [1,2,*], Enhui Jiang [1,2], Rong Gan [3], Chang Liu [1,2] and Ya Deng [4]

1 Yellow River Institute of Hydraulic Research, Yellow River Conservancy Commission, Zhengzhou 450003, China
2 Henan Key Laboratory of Ecological Environment Protection and Restoration of Yellow River Basin, Zhengzhou 450003, China
3 School of Water Conservancy Engineering, Zhengzhou University, Zhengzhou 450001, China
4 State Key Laboratory of Hydrology-Water Resources and Hydraulic Engineering, Nanjing Hydraulic Research Institute, Nanjing 210029, China
* Correspondence: lqli@whu.edu.cn

Abstract: Desert terminal lakes are important signals to discern ecological degradation crises, particularly in arid areas where an artificial project of ecological water diversion has designated a quota of river water to prevent lake body shrinkage and protect the ecosystem. Knowledge of the minimum ecological water demand (EWD) is thus necessary to ensure the basic health of lake ecosystems. This study analyzed the spatiotemporal evolution of water boundaries using Landsat satellites data via remote sensing technology from 2002 to 2017 in East Juyan Lake, an inland desert terminal lake of the Heihe River in northwest China. The minimum lake water demand was determined using two estimation methods: the lake-evaporation-oriented EWD method and the minimum water level method. In the latter method, both lake topography (using water-level area curves) and biological survival demands (using bighead carps as indicators) were considered to derive the minimum lake EWD. Water diversion to the lake over the past 15 years has increased the lake's area, but there are still marked intra-annual seasonal variations. The annual minimum lake water demand was suggested to be 54×10^6 m^3/year by comparing the different methods; however, it was not satisfied, and the lake survival was endangered when the occurrence frequency of the annual runoff in the Zhengyixia hydrological station exceeded 65%. This study offered promising directions for inland lake water resource management.

Keywords: lake area; water diversion; ecological water demand (EWD); minimum water level; inland river

Citation: Lu, J.; Li, L.; Jiang, E.; Gan, R.; Liu, C.; Deng, Y. Ecological Water Demand Estimations for Desert Terminal Lake Survival under Inland River Water Diversion Regulation. *Water* **2023**, *15*, 66. https://doi.org/10.3390/w15010066

Academic Editors: Athanasios Loukas and Fernando António Leal Pacheco

Received: 1 October 2022
Revised: 21 December 2022
Accepted: 23 December 2022
Published: 25 December 2022

1. Introduction

Lakes are important territorial freshwater resources, with many functions of regulating river runoff, providing water sources for agriculture, industry, and domestic development, and improving the regional ecological environment [1,2]. Due to changing climate and rapid development of human society, lakes in inland hyper-arid areas face a variety of problems, such as area shrinkage and ecological degradation [3,4]. To address the aquatic ecosystem crisis brought about by a shrinking lake, it is necessary to determine the minimum EWD of lakes [5,6]. Below this threshold, lake ecosystems will be severely damaged.

Lake EWD refers to the water demand to conserve the basic structure and important functions of the lake wetland ecosystem [7,8]. In recent decades, EWD investigation has mainly concerned on river ecosystem, researching the hydrology, hydraulics, ecology, and holistic methods [9,10]. To study the EWD of lakes, the primary methods are the water balance method, the minimum water level method, the function setting method, and the water quality target method [11]. Cao et al. [12] determined the optimal EWD and water level of Baiyangdian Lake according to its ecosystem service value. By studying four natural lakes in the Mediterranean Sea, Petriki et al. [13] found that the hydrological and ecological methods should be combined to evaluate the ecological water regime of lakes. Maihemuti et al. [14] established a lake water balance model and explored the complex nonlinear connections in the lake water level, volume, and boundaries.

East Juyan Lake (101°12′ E–101°19′ E, 42°10′ N–42°20′ N) is an inland terminal lake [15] of the Heihe River. Due to the continuous increase in industrial and agricultural water use in the upstream and midstream regions, East Juyan Lake had dried up six times, it completely dried up in 1992, and the oasis shrank [16,17]. In order to avoid the continuous deterioration of the downstream ecological environment, the national government implemented an ecological water diversion regulation on the mainstream Heihe River in 2000 [18]. In July 2002, East Juyan Lake received water from the upper Heihe River for the first time. East Juyan Lake is in the shape of "a shallow dish", yielding problems, such as a large water surface area, shallow water depth, and high evaporation [19]. Therefore, it is important to determine a minimum EWD that can ensure the continued survival of East Juyan Lake [20].

The primary goals of this study were as follows: (1) to analyze the spatiotemporal tendencies in the water boundaries of East Juyan Lake from 2002 to 2017 using Landsat data, which can help identify the impacts of ecological water diversion regulation on the lake conservation; (2) to estimate the minimum EWD using both hydrological and ecological methods from the three perspectives of lake evaporation, lake topography, and biological impacts, which have been rarely explored before in this typical but significant desert terminal lake; and (3) despite long-term benefits of artificial water diversion, to determine the minimum condition of annual water diversion to maintain the lake alive based on the minimum EWD estimations. This study is expected to provide references for lake management and inland river regulation of ecological water diversion projects in arid areas.

2. Study Area and Materials

2.1. Study Area

East Juyan Lake is situated 44 km north of the city of Ejina Banner and is the terminal lake of the Heihe River, China (Figure 1a). This study area has an extremely dry climate and fragile ecological environment, heavily depending on the water supply from the Heihe River that springs from the Qilian Mountains. At present, there are about 20,000 permanent residents in Ejina Banner, who depend on the Heihe River and local groundwater for more than 97% of their water [21].

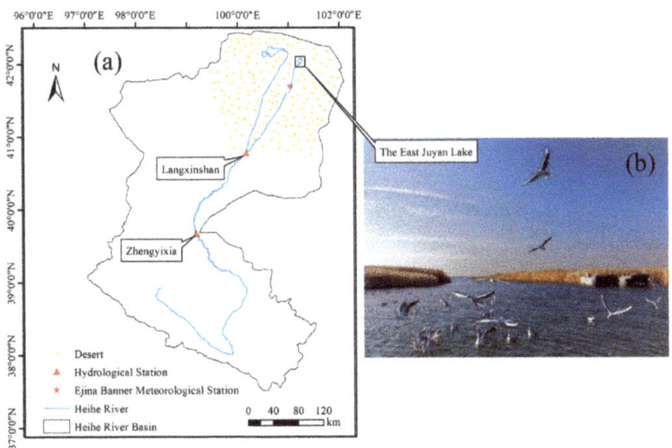

Figure 1. Study area: (**a**) the location of East Juyan Lake and (**b**) a photograph taken in East Juyan Lake.

Ejina Banner has a continental arid climate, with little precipitation and high evaporation [22] and is a typical hyperarid area. According to the calculations by averaging the observed meteorological data of Ejina Banner over years 2002–2017, the annual average precipitation was only 35.0 mm, the annual average temperature was 8.6 °C, and the annual average reference evaporation was approximately 3500 mm, which was one hundred times greater than the annual average precipitation. A project was constructed in the Heihe River basin in 2002, which delivers runoff water to the downstream area between the Zhengyixia and Langxinshan hydrological stations. The downstream Heihe River includes the two tributaries: the West River and the East River at the Langxinshan site, and the East River flowed into East Juyan Lake in the desert regions. Currently, a number of bird habitats are found in the center of East Juyan Lake, marking the gradual improvement of the ecological environment and the continuous enrichment of biodiversity (Figure 1b).

2.2. Data Source

This study was conducted primarily using Landsat remote sensing data and local meteorological observations. The Landsat series of satellite remote sensing images can accurately describe the seasonal fluctuations in lake water surfaces [23]. The Landsat series of satellites includes a total of eight satellites, and the sensors on the Landsat satellites include a multispectral scanner (MSS), a thematic mapper (TM), and so on [24]. Landsat 5, 7, and 8 were used to obtain remote sensing image data of East Juyan Lake from 2002 to 2017. Simple geometric correction and registration preprocessing were performed on the remote sensing images. The polygonal area of the lake was generated by the visual interpretation method. The lake area was calculated using GIS technology to project 128 images by selecting monthly scenes in January and from April to November. The glacial period is from December to March each year. The quality of the selected images was well controlled with no cloud cover over the study area. The item attributes and band information of the selected satellites are shown in Table 1. Landsat data can be freely obtained from online data sharing from NASA (http://glovis.usgs.gov/) (accessed on 20 August 2021).

Table 1. Properties of each item of the Landsat satellite.

System	Launch Time	Sensor	Cycle	Resolution	Number of Bands
5	1984	MSS/TM	16 days	80/30	7
7	1999	ETM+	16 days	Panchromatic 15/Multispectral 30	8
8	2013	OLI	16 days	Panchromatic 15/Multispectral 30	9

Observation data of various meteorological factors, including precipitation and evaporation, were collected from the Ejina Banner meteorological station from 2002 to 2017. The Ejina Banner meteorological station is located southwest of the oasis in Ejina Banner at 41.95° N, 101.07° E. The weather station describes the climate conditions of East Juyan Lake to some extent.

3. Methods

3.1. Lake Evaporation Estimations

East Juyan Lake is a closed lake, and the loss of lake water by evaporation corresponds to the highest share of water consumption [25]. The sum of the lake leakage, groundwater flows, and precipitation recharge in the study area was less than 1% of the evaporation [8] and was thus ignored. The lake water demand mainly offsets water consumption from evaporation.

The lake evaporation is mainly controlled by the lake area, atmospheric pressure, temperature, wind speed, and saturated vapor pressure lock [26]. According to previous studies, Li et al. [27] used the Penman–Monteith method to calculate the evaporation of East

Juyan Lake; Liu et al. [28] calculated the lake evaporation in the lower Heihe River using E601 pan evaporation. In this study, the empirical formula in Equations (1)–(4) adopted by [20,29] was used to calculate the lake evaporation, which is suitable for data-lacking areas, and the results were compared with those estimated from [27,28].

$$E = E_a \cdot A \quad (1)$$

Here, A is the lake area (km^2) obtained by the interpretation of Landsat satellite imagery and arranged in descending order to acquire area values under the 50%, 75%, and 95% percentiles for the period of 2002–2017. E_a is the water surface evaporation per unit area (mm/d), which is further estimated by Equation (2).

$$E_a = \left[0.1 + 0.24\left(1 - U^2\right)^{0.5}\right](e_0 - e_{150})V^{\left(\frac{0.85V}{v+2}\right)} \quad (2)$$

Here, U is the relative humidity, V is the local wind speed (m/s), e_0 is the saturation vapor pressure (hPa), and e_{150} represents the saturated vapor pressure at 1.5 m above the lake surface (hPa). e_0 can be expressed as $e_{0\text{water}}$ at the water surface and determined by Equation (3).

$$e_{0water} = 6.11 \times 10^{\frac{7.45t}{235+t}} \quad (3)$$

Here, t is the lake surface temperature (°C). When the lake freezes, e_0 can be expressed as $e_{0\text{ice}}$ and determined by Equation (4).

$$e_{0\text{ice}} = 6.11 \times 10^{\frac{9.5t}{265+t}} \quad (4)$$

3.2. *Minimum Water Level Method*

The minimum water level method [30] estimates the water demand of lakes by determining the minimum water level and area. The ecological water level is the minimum water level required to sustain the health and basic functions of the lake ecosystem without serious degradation. Therefore, the lake area A (km^2) corresponding to this minimum water level was taken in the extreme case of 95% percentiles. Considering the basic characteristics of the lake, the formula for calculating the minimum lake volume W (10^6 m^3/year) needed to meet ecological processes is:

$$W = (H_{min} - H) \cdot A \quad (5)$$

where H_{min} is the minimum ecological water level elevation (m), and H is the lake bottom elevation (m) with consideration of the inapparent impacts of sediment deposition.

In this study, the two commonly used estimation methods of lake minimum ecological water level were used: the lake morphology analysis method and the biological minimum space requirement method.

3.2.1. Lake Morphology Analysis

The lake morphology method [31] considers that the lake water and topography subsystems are the most basic parts of the lake ecosystem. To maintain a healthy lake ecosystem, the lake water level and morphometry, which are closely associated with changes in lake areas, must not be severely degraded. Lake water level is an indicator reflecting the fluctuation in the lake hydrological and topographic systems, and the lake area is an indicator reflecting the lake functions. Because of the nonlinearity between the water level and the area of lake surface, when the water level varies, the reduction in the lake area could be different for each unit of reduction in the lake water level.

The first-order derivative of the function that water level changes with lake area can indicate the fluctuation degree of water level under the impacts of varying lake area, i.e., the slope of the function curve of Equation (6). When the lake area decreases by one unit, both the lake volume and the corresponding water level will decrease. When the curve slope reaches its maximum, the corresponding water level is considered to reach its greatest change with lake area. This sensitive water level to lake area is usually a relatively low value that could cause negative influences on local biological habitat, species diversity, and desert eco-tourism resources [25] and is thus artificially set to be the target minimum water level in this method. The second-order derivative can represent the concavity and convexity of the fitting functions for water-level area curves $f(\cdot)$. When the second-order derivative in the domain is 0, the corresponding value is the extreme or inflection point in the domain of the first-order derivative [14,32]. The formula is as follows:

$$H_0 = f(A_0) \tag{6}$$

$$\frac{d^2H_0}{dA_0{}^2} = 0 \tag{7}$$

where A_0 is the lake area (km^2), and H_0 is the lake water level elevation (m).

3.2.2. Biological Space Minimum Requirement Method

The lake water level is used as an indicator of the living space of lake organisms. The ecological water level depends on the living space demand of various biological species in the lake. The biological minimum space requirement method [32] uses an aquatic organism as the indicator from the perspective of the minimum biological demand for living space to determine the minimum ecological water level of the lake. Among many species, fish are a commonly used indicator to reflect the situation of an aquatic ecosystem given the advantage of their sensitivity to low water levels and their representability of other types of organisms [31,32]. Therefore, this method describes the biological space minimum requirement by taking the minimum ecological water level of fish as the baseline. The minimum water depth required by the fish plus the elevation of the lake bottom elevation is the minimum ecological water level [33], which is expressed as follows:

$$H_{min} = H_1 + H \tag{8}$$

where H_1 is the water depth (m) threshold for fish to survive in the study area. According to the comprehensive considerations of both natural water level data [18] and previous environmental flow records monitored by the local ecological conservation center [21,25], H_1 was chosen from the water depth interval of 1.50 m~2.00 m which meet the survival needs of fish and most other aquatic organisms in the East Juyan Lake.

4. Results and Discussion

4.1. Temporal Tendencies in Lake Areas

The inter-annual water surface area of the East Juyan Lake showed an obvious uptrend (Figure 2a). The lake almost dried up briefly from June to August in 2003. Since 2004, East Juyan Lake has not dried again. In 2002 and 2003, the water surface area of East Juyan Lake was primarily affected by the amount of water diverted to the lake and did not show marked periodic changes. After 2004, the lake area began to change regularly. The area of the lake changed from 17.33 km^2 in 2002 to 64.80 km^2 in 2017, with an average annual increase of 2.97%. The maximum lake area appeared in 2017 at 70.11 km^2, and the minimum value appeared in 2004 only 10.8 km^2, while the multiyear average area was 45.72 km^2. The three different tendencies in lake areas appeared over time. The marked upward trend was from 2002 to 2005; then, the trend was smaller from 2005 to 2016, and there was a major uptick in 2017. Between 2011 and 2015, the average lake area was maintained at approximately 54 km^2, and the average annual water inflow into the lake during this period (68×10^6 m^3/year) increased markedly compared with the previous period from 2002 to 2010 (52×10^6 m^3/year). However, the inter-annual variation in the wetland area after 2010 was small, and the overall growth rate was only 0.64 km^2/year. In general, the current annual water inflow can only maintain the existing area to avoid marked shrinkage.

In terms of the monthly changes in lake areas (Figure 2b), The ice period lasts from November to March, and the lake area remains stable. After April, the area of the lake begins to decrease, reaches its smallest value in the year from July to August and reaches its highest value between October and November. After summer, the weather gradually becomes warmer, the evaporation on the water surface increases, and the upstream water does not flow into the desert terminal lake. At this time, the area of the lake gradually decreases. After autumn, the upper reaches releases water, and the lake area begins to increase.

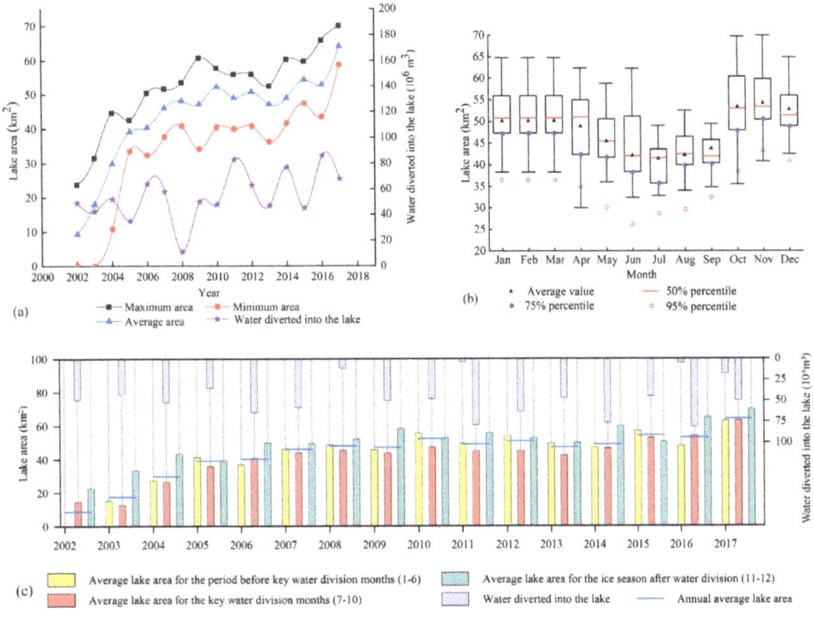

Figure 2. (**a**) Annual variations of the lake area in the East Juyan Lake and water diverted into the lake from the Heihe River, (**b**) monthly variations of East Juyan Lake area (Box-Whisker plot), and (**c**) periodical changes in the lake area and water diverted into the lake, with a whole calendar year divided into the three time periods: the period before key water division months (1–6), key water division months (7–10), and ice season after water division (11–12).

The response of the lake areas to the water diverted into the lake was quite affected by the scale of the continuous implementation of the ecological water diversion. The lake area expanded rapidly with runoff into the lake, which was sensitive to the inflow water volume. In some years, the amount of water diverted to the lake was inversely proportional to the lake area, which indicated that there was a lag time between water diversion and lake area change. Figure 2c helps illustrate the time-lag effects by presenting the periodical variations of the lake area and water diverted into the lake, with a whole calendar year divided into the three time periods: the period before key water division months (1–6), key water division months (7–10), and ice season after water division (11–12). For example, the inflow in 2004 increased to 52 $\times$ 10^6 m^3/year, which represented the expansion of the annual average lake area in 2005. The possible reason might be that most water entering the lake from the Heihe River is during July to October when the critical water division plan starts and meanwhile compensates for large lake evaporation loss. The lowest lake area during July to September in a year gradually increased and reached the highest value of lake area until the annual key water diversion period was completed in October (Figure 2b). Then, in November, the inflow begins to freeze, and lake evaporation decreases as the weather becomes cold [21] with little reduction in lake area. Therefore, despite long-term benefits of artificial water diversion increases, an increase in the lake area was not obvious after a whole calendar year, and risks of lake survival always existed.

4.2. Spatial Variations in Lake Boundaries

From the perspective of the overall shape of the lake (Figure 3), East Juyan Lake did not dry up throughout the successive years since 2004, and its southern boundary expanded markedly. In 2009, the boundary expansion on the east and west sides of the lake was larger. In 2013, the lake boundaries were generally unchanged. In 2017, the shape of the lake remained basically fixed, and the southern boundary expanded marginally. In the early stage of water inflow because the lake was dry before, the difference between the largest and smallest areas of the year was large, and there were marked spatial shape differences. With the continuous implementation of ecological water diversion projects, the intra-annual largest area and the smallest area gradually increased, and the small temporary lake was found in the south, in the recent years, during the high-water levels from 2014 to 2017.

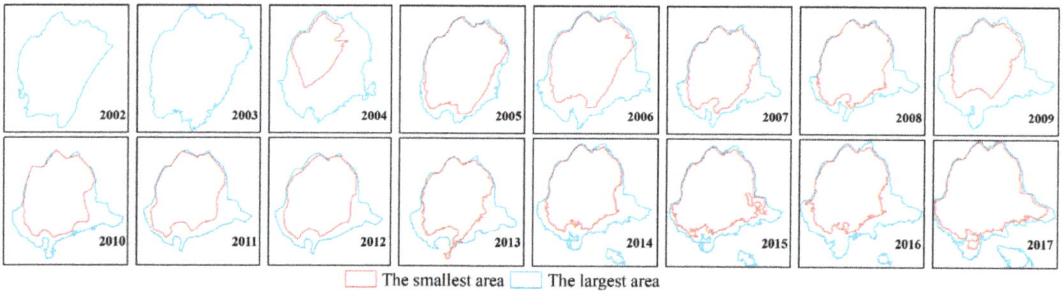

Figure 3. The intra-annual largest area and the smallest area of East Juyan Lake.

4.3. EWD Estimation Results

4.3.1. EWD by Lake Evaporation

According to the water surface area calculated under the 50%, 75%, and 95% percentiles for East Juyan Lake, the corresponding lake evaporation water consumption and EWD are shown in Table 2.

Table 2. The EWD by lake evaporation in East Juyan Lake under different percentiles.

Percentile (%)	Lake Area (km^2)	EWD by Lake Evaporation (10^6 m^3)		
		(1)	(2)	(3)
		Based on Empirical Formula	Based on E601 Pan Evaporation	Based on Penman–Monteith Formula
50	47.88	77	75	78
75	44.40	71	69	72
95	34.60	54	54	56

The lake evaporation results estimated based on the empirical formula were between and similar to those converted from the E601 pan evaporation and the Penman–Monteith formula. In the case of 50%, maintaining a lake surface area of 47.88 km^2 required an EWD of 77 $\times$ 10^6 m^3/year. In the case of 75%, maintaining a lake surface area of 44.40 km^2 required an EWD of 71 $\times$ 10^6 m^3/year. In the case of 95%, to maintain the lake surface area of 34.60 km^2, the EWD was 54 $\times$ 10^6 m^3/year.

4.3.2. Minimum Water Level Method Considering the Lake Morphology

The nonlinear connection in the water level and the surface area of East Juyan Lake is shown in Figure 4, from which the minimum ecological water level of the lake can be calculated.

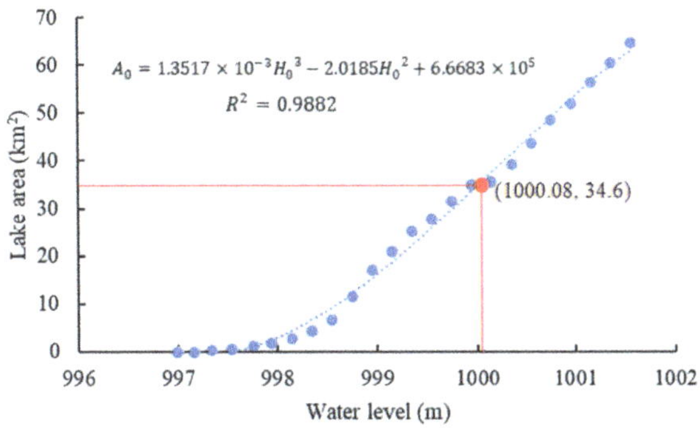

Figure 4. The fitting results for scatters of the water level and surface area of East Juyan Lake.

The lake water level (H_0) and the surface area (A_0) was fitted as follows:

$$A_0 = 1.3517 \times 10^{-3}H_0{}^3 - 2.0185H_0{}^2 + 6.6683 \times 10^5 \quad (9)$$

By solving Equation (9), the minimum ecological water level $H_{\min}$ elevation of East Juyan Lake was 1000.08 m, and the lake bottom elevation H was 997.00 m. Based on the minimum water level method, when the multiyear average area of East Juyan Lake was 34.60 km^2, the EWD was calculated to be 106.57 $\times$ 10^6 m^3/year in combination with Equation (5).

4.3.3. Minimum Water Level Method Considering the Biological Space Minimum Demand

Bighead carps (*Aristichthys nobilis*) were listed as a protected fish species in East Juyan Lake and were thus selected as the representative fish indicator in this study. In line with the findings in the previous reference [33], who compared the water change cycle method and ecological evolution method, the lowest ecological water level for East Juyan Lake was chosen to be 1.75 m, which was also consistent with the actual survey of the local wetland conservation center [21,25]. Using the minimum water level method based on the biological space minimum demand, the EWD was calculated to be 61 $\times$ 10^6 m^3/year.

4.4. Comparison of Different Methods

The minimum EWD values for the three different methods are summarized from the highest to the lowest as follows: 106.57 $\times$ 10^6 m^3/year from the lake morphology method, 61 $\times$ 10^6 m^3/year from the biological minimum space requirement, and 54 $\times$ 10^6 m^3/year from the lake evaporation.

The basic principle of the empirical formula is to establish the empirical fitting-curves between the evaporation and the meteorological elements observed on the ground in the absence of measured data. The lake morphology method determined the ecological water level by studying the fitting of the water level and area that belonged to a semiempirical method. This method showed the advantage in less requirement of detailed ecological data and containing more specific lake information. The disadvantage of this method was that it did not reflect the seasonal changes, with a lack of ecosystem mechanisms. This method is suitable for small lakes with low and relatively stable water levels and no marked environmental pollution. The biological minimum space requirement method belongs to the category of the habitat method and has advantages in the ecological mechanism of lake-related organisms, but the difficulty is in obtaining the required biological data.

Compared with the other two methods, lake evaporation has a more complete theoretical system and has been widely verified to give a good estimation agreement with observation measurements [34,35]. Therefore, using the lake evaporation method, the annual minimum EWD of East Juyan Lake was suggested to be 54×10^6 m^3/year. Some comparative results were previously reported. For example, Zhang et al. [36] adopted the water mass balance model to simulate the impact of ecological water transfer on East Juyan Lake and found that the annual water inflow should be, on average, maintained at 61×10^6 m^3/year to maintain an annual average lake area, which was lower than the estimated EWD of 77×10^6 m^3/year at the 50% scenario (Table 2). Li et al. [37] indicated that the minimum annual evaporative water consumption in East Juyan Lake was 58×10^6 m^3/year, which was close to the results of 54×10^6 m^3/year at the 95% scenario (Table 2) in this study.

The dynamic lake changes are primarily reflected by variations in lake surface areas or water levels. Compared with manual monitoring methods, remote sensing technology has overcome the difficulties in field investigation and inconsistencies in data. In recent years, Landsat has often been used in lake area extraction [24,38]. Therefore, lake area data obtained from Landsat imagery in this study could be reliable.

4.5. Lake Survival Risk Based on Minimum Lake Water Demand

The annual mean runoff values of the Zhengyixia and Langxinshan hydrological stations were 1.17×10^9 m^3/year and 0.65×10^9 m^3/year (Figure 5), respectively. In the period of 2002–2010, only 33% of the annual runoff from the two hydrographic stations exceeded the multiyear average annual runoff, while during the period 2011–2017, 71% of the annual runoff volume exceeded the multiyear average value. Therefore, the Heihe River was recently in its water-abundant period and enabled the delivery of sufficient water for the ecological restoration in East Juyan Lake. The annual average precipitation, another water source in East Juyan Lake, was only 34.15 mm and was generally stable. Thus, the water supply from local precipitation could be negligible.

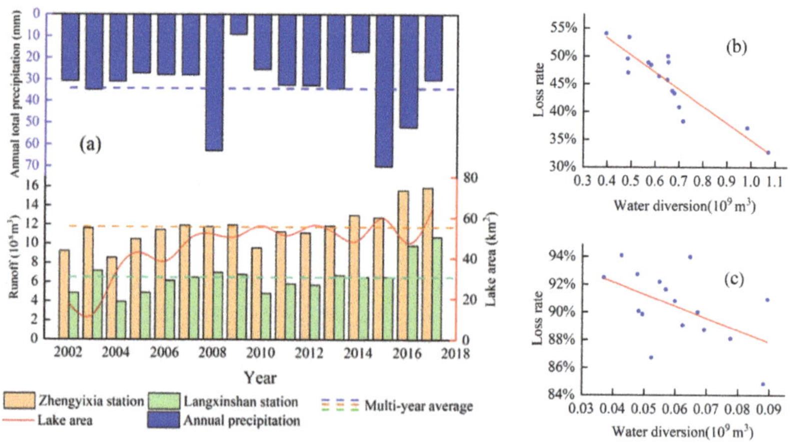

Figure 5. (**a**) Changes in runoff, precipitation, and area trends. The relationship between the quantity and the loss rate of water diversion for (**b**) Zhengyixia and Langxinshan hydrological stations and (**c**) Langxinshan hydrological stations and East Juyan Lake.

The average loss rate, which included water delivery losses and water withdrawal in the study area, decreased from the highest at 54% to 32% from 2002 to 2017 with the gradual increase in the water delivery volume from the Zhengyixia to the Langxinshan sections. In the section from Langxinshan to East Juyan Lake, the loss rate increased to an average of 90%, and the minimum loss rate was 85%, largely due to the intense regional evaporation. Although the water delivery continued to increase, the loss rate was still high. The currently wider lake water boundaries were demonstrated to be attributed to the more abundant incoming water under inland river water diversion regulation. Thus, with the change in the hydrological cycle, once the probability of dry or continuous dry weather increases, the favourable inflow conditions to keep East Juyan Lake from drying up no longer exist, and the risk of drying up East Juyan Lake likely remains high.

To ensure the minimum EWD of 54×10^6 m^3/year for the survival of East Juyan Lake, the incoming water from the Langxinshan hydrological station would be at least 460×10^6 m^3/year, which was estimated by adding both the maximum water delivery losses and maximum water withdrawal in the river-flowing regions between the Langxinshan station and East Juyan Lake for the period 2002–2017. Then, the annual runoff of Zhengyixia station (as the first controlled hydrological station in the downstream Heihe River) should discharge at least 930×10^6 m^3/year to satisfy this requirement of 460×10^6 m^3/year in the Langxinshan station, according to the well-fitting relationship between the annual runoff of Zhengyixia and Langxinshan stations from 1980 to 2017 ($R^2 > 0.96$) (Figure 6a). In other words, when the runoff frequency in the Zhengyixia station exceeded 65%, the East Juyan Lake might encounter again the danger of drying up, as shown by the Pearson-III type probability distribution curve (Figure 6b), which has been proved to be suitable in hydrological frequency analysis for most watershed in China [39].

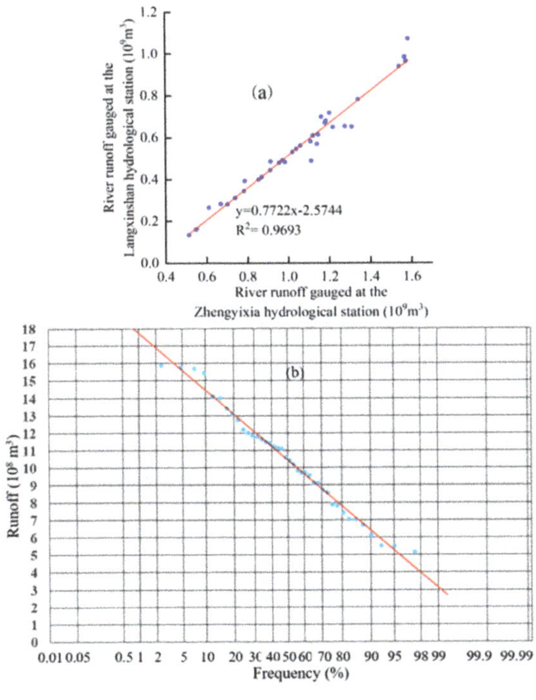

Figure 6. (**a**) The relationship between Zhengyixia and Langxinshan hydrological station. (**b**) The runoff frequency curve of Zhengyixia hydrological station.

5. Conclusions

This study used Landsat series data as a primary information source, combined local meteorological data, and comprehensively used geographic information technologies to analyze the spatiotemporal tendencies in water boundaries of East Juyan Lake over decades. Using hydrological and ecological methods, the minimum EWD of East Juyan Lake was estimated. Since 2002, under river water diversion regulation for East Juyan Lake, the lake area has been increasing. In 2017, the lake area had reached 70 km^2 and exhibited marked seasonal changes. The lake area was the largest during the period of October to November and the smallest from July to August. The shape of the lake continued to change over time and was still expanding. Based on the results of the lake evaporation methods used, the minimum EWD of East Juyan Lake was suggested to be 54×10^6 m^3/year. When the frequency of runoff discharge from the Zhengyixia hydrological station exceeds 65% based on a Pearson-III type probability distribution fitting, the discharged water from the Zhengyixia station could not satisfy the minimum water demand of 460×10^6 m^3/year in the downstream Langxinshan station, and finally, the East Juyan Lake might dry up again.

In addition, how to coordinate the ecological needs around East Juyan Lake and maintain the health of its ecosystem are key problems that must be addressed. Future research should couple the interactions between the ecological and hydrological factors and social factors to explore the impact of different inflow schemes on ecological restoration.

Author Contributions: Methodology and data curation, J.L.; writing—original draft preparation, J.L. and L.L.; writing—review and editing, L.L. and R.G.; funding acquisition, L.L. and E.J.; Resources and software, C.L. and Y.D. All authors have read and agreed to the published version of the manuscript.

Funding: This study was supported jointly by the National Key Research and Development Program of China (2021YFC3200401), the National Natural Science Foundation of China (42041004), the Young Elite Scientist Sponsorship Program of China Association for Science and Technology (YESS20200273), and the Basic Scientific Research Special Fund of Central Nonprofit Research Institutes (HKY-JBYW-2020-04).

Institutional Review Board Statement: Not applicable.

Informed Consent Statement: Not applicable.

Data Availability Statement: The used data in this study are not publicly available due to privacy/ethical restrictions.

Conflicts of Interest: The authors declare no conflict of interest.

References

1. Ye, X.C.; Xu, C.Y.; Zhang, Q.; Yao, J.; Li, X.H. Quantifying the human induced water level decline of China's largest freshwater lake from the changing underlying surface in the lake region. *Water Resour. Manag.* **2018**, *32*, 1467–1482. [CrossRef]
2. Pakzad, S.; Keshtkar, A.R.; Keshtkar, H.; Atashi, H.; Afazli, A. Impact of lake surface changes on climate fluctuation within a lake-affected region. *Environ. Earth Sci.* **2021**, *80*, 160. [CrossRef]
3. Zhou, H.; Chen, Y.; Perry, L.; Li, W. Implications of climate change for water management of an arid inland lake in Northwest China. *Lake Reserv. Manag.* **2015**, *31*, 202–213. [CrossRef]
4. Sajedipour, S.; Zarei, H.; Oryan, S. Estimation of environmental water requirements via an ecological approach, a case study of Bakhtegan Lake. *Iran. Ecol. Eng.* **2017**, *100*, 246–255. [CrossRef]
5. Liu, J.L.; Yang, Z.F. Ecological and environmental water demand of the lakes in the Haihe–Luanhe Basin of North China. *J. Environ. Sci.* **2002**, *14*, 234.
6. Liu, D.; Wang, X.; Zhang, Y.L.; Yan, S.J.; Cui, B.S.; Yang, Z.F. A landscape connectivity approach for determining minimum ecological lake level, implications for lake restoration. *Water* **2019**, *11*, 2237. [CrossRef]
7. Fu, Y.C.; Leng, J.W.; Zhao, J.Y.; Na, Y.; Zou, Y.P.; Yu, B.J.; Fu, G.S.; Wu, W.Q. Quantitative calculation and optimized applications of ecological flow based on nature–based solutions. *J. Hydrol.* **2021**, *598*, 126216. [CrossRef]
8. Zhang, Z.D.; Zheng, Y.; Han, F.; Xiong, R.; Feng, L. Recovery of an endorheic lake after a decade of conservation efforts, Mediating the water conflict between agriculture and ecosystems. *Agric. Water Manag.* **2021**, *256*, 107107. [CrossRef]
9. Jia, H.F.; Ma, H.T.; Wei, M.J. Calculation of the minimum ecological water requirement of an urban river system and its deployment, a case study in Beijing central region. *Ecol. Model.* **2011**, *222*, 3271–3276. [CrossRef]
10. Cui, Y.; Zhang, Q.; Chen, X.H.; Jiang, T. Research progress on theories and methods of ecological water demand. *J. Lake Sci.* **2010**, *22*, 465–480.
11. Cui, B.S.; Zhao, X.; Yang, Z.F. Calculation of the minimum ecological water demand of lakes based on the principle of ecohydrology. *Acta Ecol. Sin.* **2005**, *7*, 1788–1795.
12. Cao, T.G.; Yi, Y.J.; Liu, H.X.; Yang, Z.F. Integrated ecosystem services–based calculation of ecological water demand for a macrophyte–dominated shallow lake. *Glob. Ecol. Conserv.* **2020**, *21*, 2351–9894. [CrossRef]
13. Petriki, O.; Zervas, D.; Doulgeris, C.; Bobori, D. Assessing the ecological water level, the case of four Mediterranean Lakes. *Water* **2020**, *12*, 2977. [CrossRef]

14. Maihemuti, B.; Aishan, T.; Simayi, Z.; Alifujiang, Y.; Yang, S.T. Temporal scaling of water level fluctuations in shallow lakes and its impacts on the lake eco–environments. *Sustainability* **2020**, *12*, 3541. [CrossRef]
15. Khan, S.U.; Khan, I.; Zhao, M.; Chien, H.P.; Fahad, S. Spatial heterogeneity of ecosystem services, a distance decay approach to quantify willingness to pay for improvements in Heihe River Basin ecosystems. *Environ. Sci. Pollut. Res.* **2019**, *26*, 25247–25261. [CrossRef] [PubMed]
16. Xu, F.; Li, H.; Bao, H.J. Performance comparisons of land institution and land regulation systems on water area decrease. *Habitat. Int.* **2018**, *77*, 12–20. [CrossRef]
17. Qin, S.; Gao, G.Y.; Lv, Y.H.; Wang, S.; Jiang, X.H.; Fu, B.J. River flow is critical for vegetation dynamics, Lessons from multi–scale analysis in a hyper–arid endorheic basin. *Sci. Total Environ.* **2017**, *603–604*, 290–298.
18. Xiao, S.C.; Peng, X.M.; Tian, Q.Y. Climatic and human drivers of recent lake–level change in East Juyan Lake, China. *Reg. Environ. Chang.* **2016**, *16*, 1063–1073. [CrossRef]
19. Yu, T.F.; Si, J.H.; Qi, F.; Xi, H.Y.; Chu, Y.W.; Kai, L. Simulation of pan evaporation and application to estimate the evaporation of Juyan Lake.; Northwest China under a hyper–arid climate. *Water.* **2017**, *9*, 952. [CrossRef]
20. Hu, S.; Ma, R.; Sun, Z.Y.; Ge, M.Y.; Zeng, L.L.; Huang, F.; Bu, J.W.; Wang, Z. Determination of the optimal ecological water conveyance volume for vegetation restoration in an arid inland river basin.; Northwestern China. *Sci. Total Environ.* **2021**, *788*, 147775. [CrossRef] [PubMed]
21. Li, L.; Jiang, E.; Yin, H.; Wu, K.; Dong, G. Ultrashort-term responses of riparian vegetation restoration to adjacent cycles of ecological water conveyance scheduling in a hyperarid endorheic river basin. *J. Environ. Manag.* **2022**, *320*, 115803. [CrossRef]
22. Xiao, S.C.; Xiao, H.L. Radial growth of Tamarix ramosissima responds to changes in the water regime in an extremely arid region of northwestern China. *Environ. Geol.* **2007**, *54*, 543–551. [CrossRef]
23. Zhang, C.H.; Lv, A.F.; Zhu, W.B.; Yao, G.B.; Qi, S.S. Using multisource satellite data to investigate lake area.; water level.; and water storage changes of terminal lakes in ungauged regions. *Remote Sens.* **2021**, *13*, 3221. [CrossRef]
24. Zhang, Z.X.; Chang, J.; Xu, C.Y.; Zhou, Y.; Wu, Y.H.; Chen, X.; Jiang, S.S.; Duan, Z. The response of lake area and vegetation cover variations to climate change over the Qinghai–Tibetan Plateau during the past 30 years. *Sci. Total Environ.* **2018**, *635*, 443–451. [CrossRef]
25. Yang, Q.; Xiao, H.L.; Zhao, L.J.; Yang, Y.G.; Li, C.Z.; Zhao, L.A.; Yin, L. 2011 Hydrological and isotopic characterization of river water.; groundwater.; and groundwater recharge in the Heihe River basin.; northwestern China. *Hydrol. Process.* **2011**, *25*, 1271–1283. [CrossRef]
26. Zhang, H.; Zhang, L.; Zhao, C.Y. Ecological water requirement estimation of the rump lake in an extreme arid region of East Juyanhai. *Acta Ecol. Sin.* **2014**, *34*, 2102–2108.
27. Li, Z.; Li, Z.; Xu, Z.; Zhou, X. Temporal variations of reference evapotranspiration in Heihe River Basin of China. *Hydrol. Res.* **2013**, *44*, 904–916. [CrossRef]
28. Liu, X.; Yu, J.; Wang, P.; Zhang, Y.C.; Du, C.Y. Lake Evaporation in a Hyper-Arid Environment, Northwest of China—Measurement and Estimation. *Water* **2016**, *8*, 527. [CrossRef]
29. Ye, Z.X.; Chen, S.F.; Zhang, Q.F.; Liu, Y.C.; Zhou, H.H. Ecological water demand of Taitema Lake in the lower reaches of the Tarim River and the Cherchen River. *Remote Sens.* **2022**, *14*, 832. [CrossRef]
30. Doulgeris, C.; Koukouli, P.; Georgiou, P.; Dalampakis, P.; Karpouzos, D. Assessment of minimum water level in lakes and reservoirs based on their morphological and hydrological features. *Hydrology* **2020**, *7*, 83. [CrossRef]
31. Xu, Z.X.; Wang, H.; Dong, Z.C.; Tang, K. Research on the minimum ecological water demand in the Nansi Lake area. *J. Hydraul. Eng.* **2006**, *7*, 784–788.
32. Xu, Z.X.; Chen, M.J.; Dong, Z.C. Calculation method for the minimum ecological water level of lakes. *Acta Ecol. Sin.* **2004**, *10*, 2324–2328.
33. Tao, J.; Li, X.; Zuo, Q. Comparison of calculation methods for ecological water demand of lakes and application of examples. *South North Water Transf. Water Conserv. Sci. Technol.* **2022**, *20*, 365–374.
34. Shang, S.P.; Shang, S.H. Simplified Lake surface area method for the minimum ecological water level of lakes and wetlands. *Water* **2018**, *10*, 1056. [CrossRef]
35. Healey, N.C.; Rover, J.A. Analyzing the effects of land cover change on the water balance for case study watersheds in different forested ecosystems in the USA. *Land* **2022**, *11*, 316. [CrossRef]
36. Zhang, M.M.; Wang, S.; Gao, G.Y.; Fu, B.J.; Ye, Z.X.; Shen, Q. Exploring responses of lake area to river regulation and implications for lake restoration in arid region. *Ecol. Eng.* **2019**, *128*, 18–26. [CrossRef]
37. Li, B.; Zhang, Y.C.; Wang, P.; Du, C.Y.; Yu, J.J. Estimating dynamics of terminal lakes in the second largest endorheic river basin of northwestern China from 2000 to 2017 with Landsat imagery. *Remote Sens.* **2019**, *11*, 1164. [CrossRef]
38. Cao, H.Y.; Han, L.; Liu, Z.H.; Li, L.Z. Monitoring and driving force analysis of spatial and temporal change of water area of Hongjiannao lake from 1973 to 2019. *Ecol. Inform.* **2021**, *61*, 101230. [CrossRef]
39. Yue, S.; Pilon, P. Probability distribution type of Canadian annual minimum streamflow. *Hydrol. Sci. J.* **2005**, *50*, 438. [CrossRef]

water

MDPI

Article

A Dam Construction Event Recorded by High-Resolution Sedimentary Grain Size in an Outflow-Controlled Lake (Hulun Lake, China)

Hongbin Gao [1,2], Rui Zhang [1,2], Gang Wang [1,2], Yanru Fan [1,2,*], Xinfeng Zhu [1,2], Junfeng Wu [1,2] and Li Wu [3]

1 College of Municipal and Environmental Engineering, Henan University of Urban Construction, Pingdingshan 467000, China
2 Henan Province Key Laboratory of Water Pollution Control and Rehabilitation Technology, Pingdingshan 467000, China
3 State Key Lab of Hydrology-Water Resources and Hydraulic Engineering, Nanjing Hydraulic Research Institute, Nanjing 210098, China
* Correspondence: gaohongbin0922@126.com; Tel.: +86-185-3755-7690

Abstract: The distribution of sediment grain size can record past environmental conditions and human activity. In this study, radioisotope of ^{210}Pb and ^{137}Cs and a grain size of a 41 cm core in Hulun Lake were applied to reconstruct the high-resolution sedimentation history. The profiles of the grain size of the lake sediments show that silt (4–63 μm) was the largest contribution with an average content of 84.05%, and the second largest contribution was sand (>63 μm) with an average content of 15.68%. The median grain size and the mean grain size in the whole sediment core was 22.39 μm and 36.85 μm, respectively. Correlations of the sedimentological variables with instrumental measurements were also analyzed. The peak–trough value of the mean grain size of the sediments in Hulun Lake can reflect the magnitude of rainfall intensity and river discharge. The clay and silt contents at a depth of approximately 32–38 cm was different from other depths throughout the core, which showed continuous maxima with an average content of 0.35% and 94.08%. These changes in grain size correspond to the period of dam construction in 1963–1970. Therefore, the sediment grain size of Hulun Lake effectively recorded the dam-building activity.

Keywords: sedimentary grain size; dam construction; precipitation; paleoenvironmental reconstruction; Hulun Lake

Citation: Gao, H.; Zhang, R.; Wang, G.; Fan, Y.; Zhu, X.; Wu, J.; Wu, L. A Dam Construction Event Recorded by High-Resolution Sedimentary Grain Size in an Outflow-Controlled Lake (Hulun Lake, China). *Water* **2022**, *14*, 3878. https://doi.org/10.3390/w14233878

Academic Editors: Qiting Zuo, Fuqiang Wang, Jiaqi Zhai, Xiuyu Zhang, Dunxian She, Lei Zou, Rong Gan and Zengliang Luo

Received: 6 November 2022
Accepted: 25 November 2022
Published: 28 November 2022

1. Introduction

To better understand past climatic and environmental changes beyond the available short instrumental record, a high-resolution geological archive of lake sediment can provide important clues [1–4]. In addition, detailed reconstructions of the history of environmental changes may improve our understanding of potential future environmental changes.

Lake sediment grain size has been used widely for paleoenvironmental reconstructions, such as of regional climate change and lake level fluctuations [5], dust storms [6], and human activities [7], since it provides important information about the depositional environmental, and migration processes. However, the interpretation of past environmental conditions from sediment core grain size is difficult because of the complexity of sediment processes [8]. Numerous studies have shown that the characteristics of sediment grain size have different interpretations on different time scales [9]. For long-term scales and low resolution (10^2 a or 10^3 a) studies, sediment grain size indicates the history of changes in lake levels, which can be interpreted as coarser grains being easily transported and deposited because of a fall in the lake level and the shrinkage of lake area, while finer grains are difficult to deposit because of the strong hydraulic disturbance at low lake levels during dry periods. The opposite can be illustrated during wet periods. However, for short time scales and high resolution (a or 10 a) studies, large coarse sediment grain sizes indicate wet years with high regional precipitation, because the increased regional precipitation causes increases in the erosion intensity of the watershed. In contrast, a decrease in coarse grain size and an increase in fine grain size indicates drought years with low rainfall [7,10].

In addition, the distribution of sediment grain size in the sediment core can also reflect information on recent human activity, such as artificial dam construction, water conservancy projects, and land use changes in the surrounding lake watershed [11,12]. Previous studies have shown that artificial dam construction has a significant impact on the deposition and transport of grains in water bodies [11,13]. The rivers entering the lake usually carry grains, including coarse and fine grains. Some of the relatively fine grains and a smaller amount of the large grains could be carried out with the outflow, and the remaining coarse and fine grains are deposited at the bottom of the lake. However, only a small amount of the fine grains could be carried out with the outflow, and most of the coarse grains and the remaining fine grains were deposited at the bottom of the lake when the dam was built due to the decreasing sediment transport capacity, which resulted in a relatively large proportion of the fine grains being deposited at the bottom at this time. Therefore, the palaeoenvironmental implications of grain size cannot be routinely applied to palaeolimnological studies, and credible conclusions can be obtained only after comprehensively analyzing the historical meteorological observations, process-related information, and the extent of all factors in sedimentary records.

Hulun Lake, situated in a semiarid area in the northeastern part of Inner Mongolia, China, is an important water body maintaining the grassland ecological environment of the lake basin. Currently, the lake faces a serious ecological crisis, namely, eutrophication, a shrinking area, and a gradual impact on the ecological function in the lake ecosystem [14]. However, few studies on the environmental changes of Hulun Lake and its influencing factors have been performed due to insufficient historical instrumental data, and difficulties in conducting field experiments because of the cold weather and inconvenient traffic.

A continuous sediment record could preserve in Hulun Lake because there was no historic dredging. Therefore, the reconstruction and interpretation of the proxies archived in the lake sediment provide a feasible approach for obtaining a better understanding of the past climatic and environmental changes in the lake. Previous studies for the reconstruction of sediment proxy records in Hulun Lake have mainly focused on topics with a long time scale and low resolution. Hu [15] found that muddy sediments deposited during a high-water level period (corresponding to a humid climate) have comparatively high κ (magnetic susceptibility) values in Hulun Lake sediment. In contrast, the sandy sediments deposited during the low-water level period (corresponding to an arid climate) have low κ values. Wen [16] described changes in the vegetation and climate of the East Asian monsoon margin during the Holocene, applied pollen-assemblage data from a sediment core of Hulun Lake, and revealed that the changes in the monsoon precipitation on the millennial to centennial scales are related to ocean–atmosphere interactions in the tropical Pacific. Xiao [17] used a lognormal distribution function fitting method of sediment grain size distributions, and the results indicated that the lake levels fluctuated in response to the intensity of monsoonal precipitation. Higher percentages of nearshore components accompanied by more sand-fraction proportions and coarser median grain sizes reflect lower lake stands.

Little research has been conducted on lake sediment with high resolution to determine the past deposition processes and environmental changes in Hulun Lake. In addition, previous findings on the interpretation of proxies of lake sediment from Hulun Lake were applied on a long time scale, and with low resolution [15–17]; it is necessary to advance the determination of whether the results can be used for the reconstruction of lake environments on short time scales.

The objective of this study was to interpret the proxies of sediment grain size in Hulun Lake. In addition, we combined the use of radioisotope analyses of ^{210}Pb and ^{137}Cs with grain size analyses to reconstruct the high-resolution sedimentation history and interpret the lake deposition process and paleoenvironment. Furthermore, the correlation of the sedimentological variables with instrumental measurements (local precipitation and lake inflow discharge) were analyzed. The results of the proxy interpretation provide insights into the past and potential future environmental changes in and around the lake.

2. Materials and Methods

2.1. Study Area

Hulun Lake (48°31′–49°20′ N, 116°58′–117°48′ E), located in the northeast part of Inner Mongolia, China (Figure 1), is the largest lake in northern China, with a maximum surface area of 2339 km^2 and a maximum water depth of 8 m [18]. The lake adjoins the Greater Khingan Mountains and the Mongolian Plateau, which is a grassland-type lake wetland ecosystem with biodiversity and ecological functions found in cold and arid regions across the world. There are two rivers that control the main input sources of Hulun Lake, in which the Kherlen River is from Mongolia and the Urshen River is derived from Beier Lake. Another artificial river named the Xinkai River lies in the northern part of the lake and is an intermittent river that flows out when the lake elevation exceeds 543.4 m.a.s.l. According to records, this artificial river was built with the approval of the government due to the rising lake water level threatening the production of coal mines downstream. A dam was constructed to block the water outlet from the artificial river started build on 15 June 1965 to drained outward through the river on 8 September 1971 [18]. The lake water could not be discharged because the dam blocked the outlet of the river during the construction stage of the project, so that the lake became a closed lake. In addition, the water level was high during this period.

Since the water level has declined sharply in the last 10 years, Hulun Lake has become a closed lake without an outlet. The study area is located in a semiarid area, and the climate around the Hulun Lake basin is controlled by westerly winds and the East Asian monsoon. Its mean annual precipitation is 247–319 mm, most of which occurs in summer. The mean annual air temperature is 0.3 °C, and the mean annual evaporation reaches 1400–1900 mm, which is 5–6 times the mean annual precipitation [19].

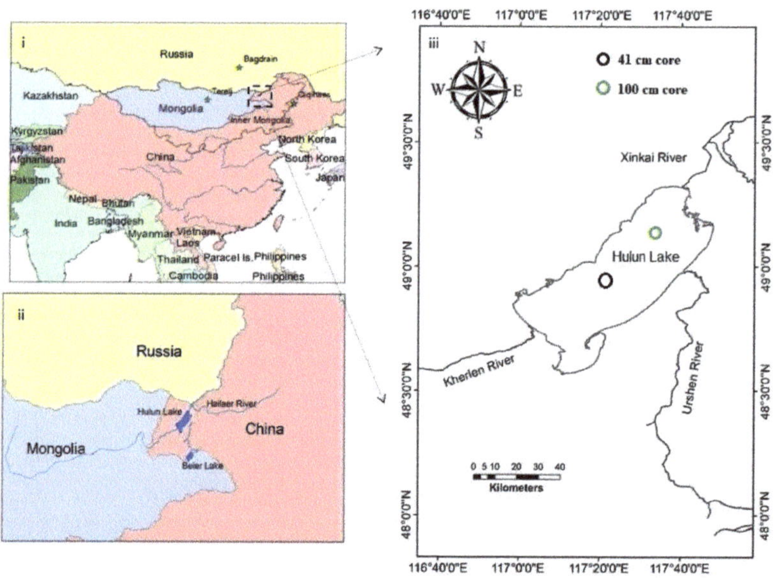

Figure 1. Location of study area: (**i**) study area on a country-scale map; (**ii**) whole Hulun Lake basin; (**iii**) core sampling site (black circle indicates the location of the 41 cm core of this study, while the green circle indicates the location of the 100 cm core sampled by Liang (2017) from the Inner Mongolia water environment group [20]).

2.2. Sampling

A 41 cm-long sediment core was obtained at the deepest site, with a 5.6 m water depth (Figure 2) in Hulun Lake, China, in July 2015 using a Glew Corer [21]. The core samples were sliced immediately in 1 cm intervals on board. Then, 41 subsamples were stored in sealed bags in an ice cooler and transferred to a refrigerator (<4 °C) after being transported to the laboratory.

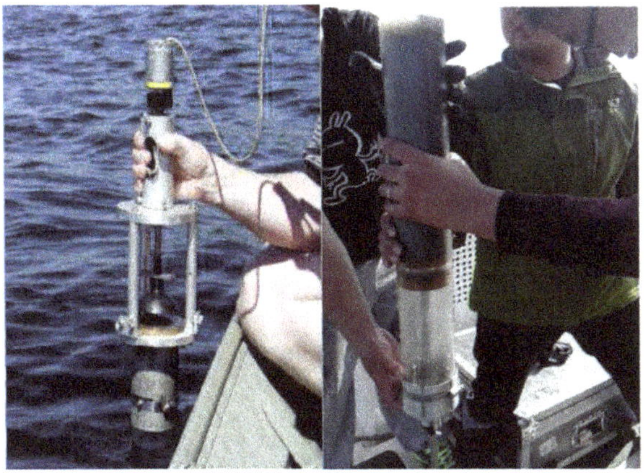

Figure 2. Sampling of sediment core in Hulun Lake.

To compare the results, another sediment core of Hulun Lake was employed as a referenced sample core (100 cm), which was sampled northeast of the core sampling site in this study in March 2015 by Liang (2017) from the environment group of Inner Mongolia Agricultural University Water [20] (Figure 1).

2.3. Experiments and Methods

2.3.1. Sediment Core Chronology

To determine the age of the sediment core, ^{137}Cs and ^{210}Pb chronologies were conducted by gamma spectrometry at the Nanjing Institute of Geography and Limnology Chinese Academy of Sciences. The profile of ^{210}Pb dating was calculated by the constant initial concentration model (CIC), which was described previously [22,23]. Based on the constant initial concentration model (CIC), the $^{210}Pb_{ex}$ activity C (Bq/g) in a layer at depth z' (cm) can be expressed as:

$$C = C_0 \cdot \exp\left(-\frac{\lambda}{S} Z'\right) \quad (1)$$

where C_0 (Bq/g) is the ^{210}Pbex activity at the top of the sediment core, λ is the ^{210}Pb radioactive decay constant, $0.031a^{-1}$, and S is the sedimentation rate (cm $year^{-1}$).

Correspondingly, the sedimentary age (t) at a certain depth of the sediment can be calculated as:

$$t = \lambda^{-1} ln(C_0/C) = Z'/S \quad (2)$$

The radionuclide ^{210}Pb (half-life of 22.26 years) is a natural radioactive isotope derived from ^{238}U decay. The ^{238}U in the Earth's crust decays to ^{226}Ra and further to ^{222}Rn, which then escapes to the atmosphere and decays to ^{210}Pb. This fallout fraction derived from the atmosphere is termed "unsupported" or "excess" ^{210}Pb ($^{210}Pb_{ex}$). In addition, ^{210}Pb also forms within the sediment itself as a product of the decay of ^{238}U; this fraction in sediment is termed "supported" ^{210}Pb. $^{210}Pb_{ex}$ concentrations can be estimated by subtracting the ^{226}Ra-supported ^{210}Pb concentrations from the total ^{210}Pb concentrations [24,25].

The artificial nuclide ^{137}Cs (half-life of 30.17 years) is a fission product that was introduced into the environment as a result of atmospheric nuclear weapons tests conducted initially in the early 1950s. The highest ^{137}Cs activity may represent the period of maximum radionuclide fallout in the Northern Hemisphere, which is associated with the peak of atomic weapons testing in 1963 [26,27]. In summary, the onset (1950) and peak (1963) of the ^{137}Cs concentration in the sediment depth profile can be used as validation of ^{210}Pb dating results.

2.3.2. Grain Size of the Sediment Core

For pretreatment of the 41 sediment samples for grain size analysis, carbonates were removed by 10% HCl, and organic matter was removed by 30% H_2O_2. Following the HCl and H_2O_2 treatments, the samples were rinsed at least three times with deionized water. Furthermore, the samples were dispersed by adding 10 mL of 0.05 M sodium hexametaphosphate as the dispersing agent for treatment in an ultrasonic vibrator for 15 min. Finally, grain size analysis of the bulk sediment was carried out with a Mastersizer 2000 laser particle analyzer detecting a 0.02–2000 μm size range. Each sample was analyzed twice, and the relative error was no more than 2%. The grain size parameters were calculated following Folk and Ward (1957). The parameters of grain size (median size and mean size) and sediment component (%) of sand (>63 μm), silt (4–63 μm), and clay (<4 μm) were chosen for the analysis.

3. Results

3.1. Sediment Core Chronology

The experimental data for the concentrations of $^{210}Pb_{ex}$ in the Hulun Lake sediment core is plotted as a logarithmic profile versus depth in Figure 3. Using a least-squares weighted fit, a straight trend line can be achieved with a coefficient of determination (R^2) equal to 0.80 (Figure 3). Then, a mean sedimentation rate of approximately 0.72 cm year^{-1} of the core was calculated by Equation (1). Defining the surface sediment as the age of 2015, a chronology frame of the whole core was established, which responded to a 57 years age series from 1958 to 2015 (Figure 4). The experimental data for the concentrations of ^{137}Cs in the sediment at each depth interval is presented in Figure 4. The ^{137}Cs activity that reached a "peak" at 37 cm is associated with the 1963 fallout in Hulun Lake sediment, this mark fits very well with the age series based on ^{210}Pb data, which is at a depth of 37 cm, corresponding to 1963 (Figure 4).

3.2. Characteristics of the Profile Distribution of Sediment Grain Size in Hulun Lake

The results of the sediment grain component (clay, silt, and sand) of Hulun Lake are shown in Figure 5. The profiles of the grain size of the lake sediments show that silt was the component of grain with the largest contribution in the entire sediment core, with an average content of 84.05%, changing from 72.79% at 24 cm to 95.11% at 35 cm. The component with the second largest contribution was sand, with an average content of 15.68%, in which the maximum content of 27.00% was at 24 cm and the minimum content of 4.51% was at 35 cm of the sediment core. The clay content was the lowest of the components in the entire sediment core, with an average of 0.27%.

The contents of different grain sizes in lake sediment reflects different depositional environments. A high sand content may indicate a strong potentiality of sediment transport, while a high clay or silt content indicates a stable depositional environment and weak sediment transport capacity [28]. The sediment core of this study was collected from the center of the lake, where the potentiality of sediment transport is considered poor. The contents of clay and silt in the sediment were much higher than the contents of sand in this study, indicating that the sedimentary environment of the lake was relatively stable in the center of the lake due to the larger water depth and weak sediment transport capacity.

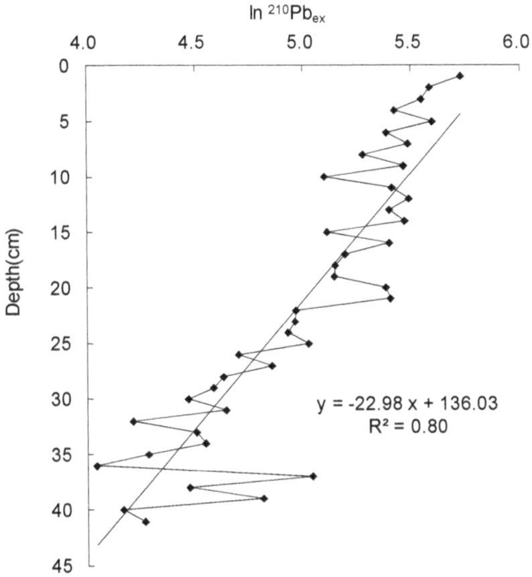

Figure 3. Log plots of ^{210}Pbex with its depth and its correlation in the sediment core.

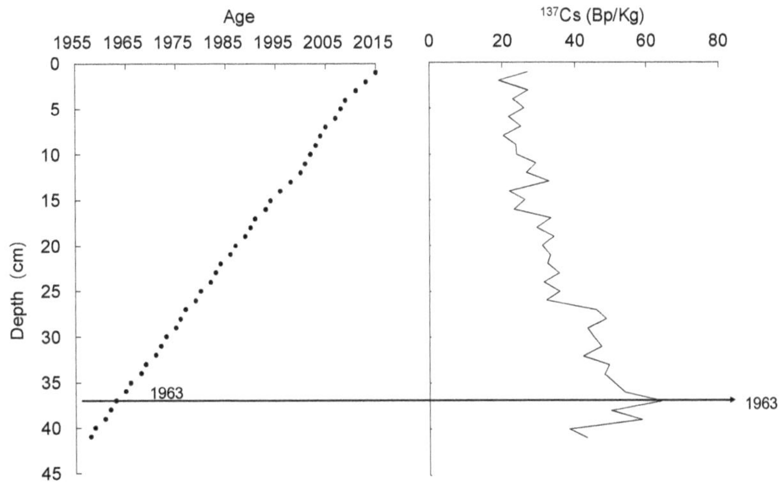

Figure 4. The age profiles determined by ^{210}Pb and ^{137}Cs.

In particular, it can be seen from the grain component profiles that the contents of the grain at a depth of approximately 32–38 cm was different from other depths throughout the core. The clay and silt contents showed continuous maxima with an average content of 0.35% and 94.08%, while the sand contents showed continuous minima with an average content of 5.57% (gray dotted box in Figure 5). These results may indicate that changes occurred in the sediment transport capacity or deposition patterns during this period.

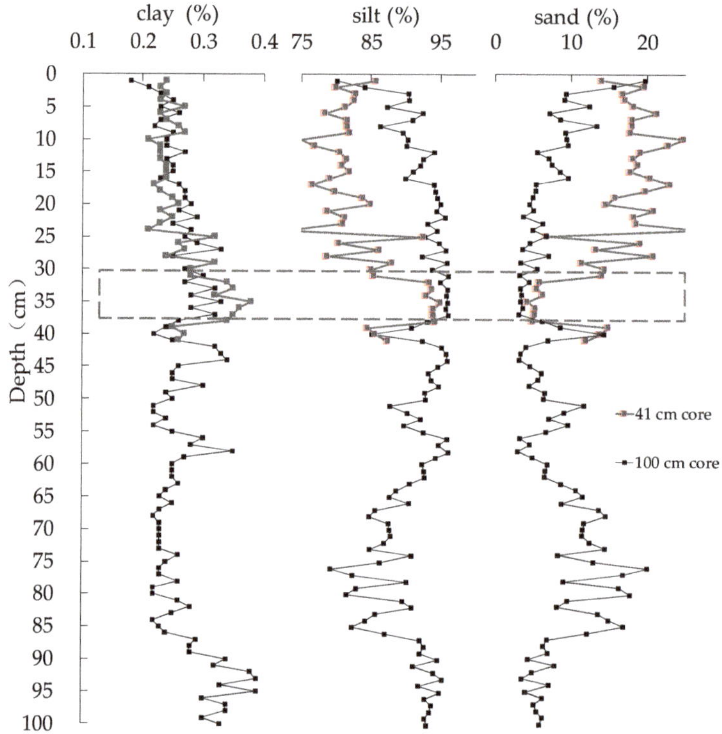

Figure 5. Profiles of the sediment grain components (clay, silt, sand; %) in the 41 core and 100 cm core of Hulun Lake.

To verify the reliability of the grain size proxy analysis results in this study, a referenced 100 cm core was employed for comparison. The grain size classification results of the 100 cm core are shown in Figure 5. In the entire sediment core profile, the average contents of clay, silt, and sand were 0.27%, 91.46%, and 8.27%, respectively. The grain size distribution of the 100 cm core varied widely, however, the characteristics of the grain size distribution at sediment depths of 31–37 cm was different from other depths throughout the core, in which the clay and silt contents showed continuous large values with an average content of 0.30% and 95.99%, while the sand contents showed continuous small values with an average contents of 3.70% (gray dotted box in Figure 5). This is consistent with the distribution characteristics in the range of 32–38 cm of the 41 cm core in this study, and shows that the sediment transport capacity or deposition process changed at this stage.

The profiles of the median and mean grain size of the 41 cm core are shown in Figure 6, which shows that the average value of the median grain size in the whole sediment core was 22.39 μm, with a maximum value of 30.91 μm at a depth of 24 cm and a minimum value of 14.81 μm at a depth of 35 cm. The average value of the mean grain size in the entire sediment core was 36.85 μm, with a maximum value of 51.71 μm at a depth of 24 cm and a minimum value of 22.64 μm at a depth of 35 cm. Compared to the median grain size, the mean grain size is larger than the median grain size, indicating that fine grains are the main component in the sediment of Hulun Lake. Similarly, the distribution of grain size at depths of 32–38 cm showed continuous minima with an average median size of 16.13 μm and an average mean size of 24.30 μm.

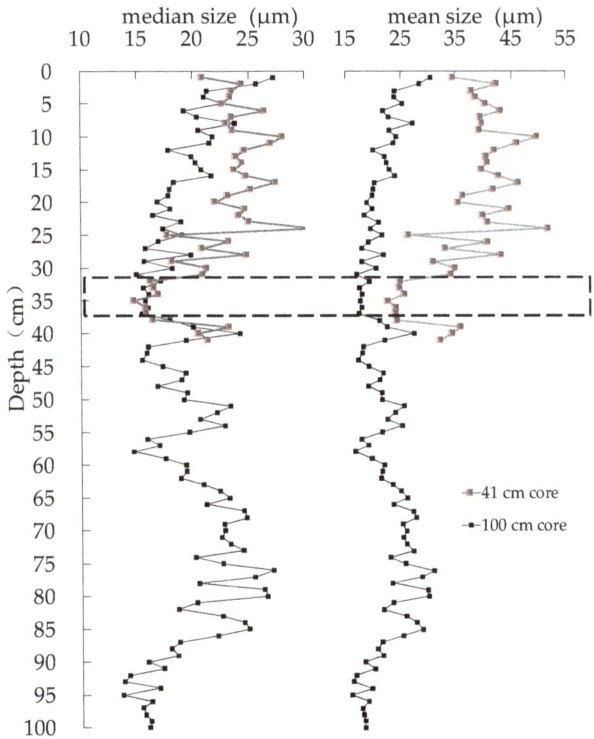

Figure 6. Profiles of the sediment grain parameters (median size and mean size) in the 41 core and 100 cm sediment core of Hulun Lake.

Compared to the grain size parameters of the 100 cm core (Figure 6), the mean and median grain sizes are smaller than those of the 41 cm core. This could be related to the sampling area of the core, where the sediment transport capacity and distance from the estuary are different. More studies are needed to further demonstrate the result. In addition, comparing the 100 cm core, it can be found that the characteristics of the mean and median grain size distributions at 31–37 cm are different, showing continuous minimum values throughout the core profiles. This is consistent with the distribution characteristics of the 41 cm core in this study.

4. Discussion

4.1. The Relationship between Precipitation and Sedimentary Grain Size

The grain size composition of lake sediments in arid and semiarid regions is mainly affected by the action of water and wind. As previously discussed, the distribution of the grain size of lake sediments at a short scale and high resolution is related to the variation in rainfall in the basin, which mainly depends on the intensity and quantity of the water source that feeds the lake. The impact of wind and waves may be relatively small due to the large lake surface and the increasing water depth. To explore the relationship between the sediment grain size of Hulun Lake and rainfall in the basin, the rainfall data from meteorological stations in the Hulun Lake basin during the past 60 years were compared to the changes in sediment grain size.

According to the chronology determined by the ^{210}Pb and ^{137}Cs of the 41 cm core, the mean grain size of the sediments and the hydrological elements (precipitation and river discharge) of the lake at the corresponding times are plotted in Figure 7. It should be noted here that the age of the sediment core was determined based on the calculated average deposition rate of 0.72 cm/a. However, the amount of debris entering the lake in different time periods and the depositional environment were different. Therefore, a discrepancy of one to two years between the hydrological element data and sediment chronology may have occurred, which was reported in another study [7].

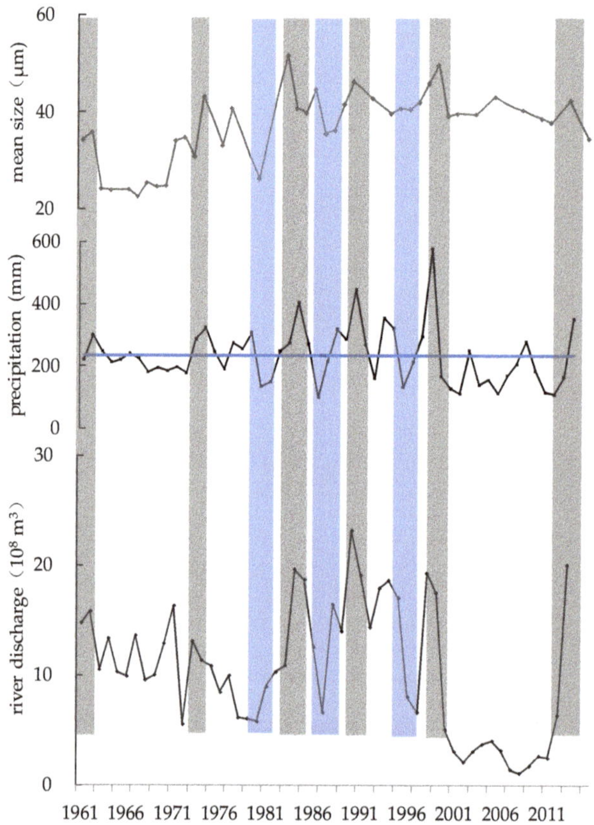

Figure 7. Comparisons the mean grain size in the lake sediment with the precipitation and river discharge records (gray boxes indicate the peaks of precipitation correspond to peaks of the mean grain size, blue boxes indicate the troughs in the precipitation correspond to the troughs of the mean grain size, blue horizontal line is the annual average precipitation).

The consistent tendency between precipitation and mean grain size varies in peaks and troughs, as shown in Figure 7. The peaks of precipitation in 1962, 1974, 1984, 1990, 1998, and 2013 correspond to peaks of the mean grain size of the sediment core in 1962, 1974, 1983, 1990, 1999, and 2013, respectively (gray boxes). Similarly, the troughs in the precipitation in 1980, 1986, and 1995 roughly correspond to the troughs of the mean grain size in 1980, 1987, and 1994, respectively (blue boxes). Although the peaks and troughs of the mean grain size in the sediments do not correspond fully to the peaks and troughs of precipitation, the errors are within the range of one to two years. As discussed above, the radionuclides were calculated using the average deposition rate, and a discrepancy of one to two years of the chronology is acceptable. Therefore, the peak–trough value of the mean grain size of the sediments in Hulun Lake can reflect the magnitude of rainfall intensity, indicating that the sediment grain size is affected by regional rainfall changes. In wet periods, the grains entering the lake water body from the ground may increase due to the increased erosion intensity on the ground with heavy rainfall. In addition, the coarse grains can also reach the center of the lake, resulting in strong sediment transport capacity. In contrast, the grains entering the lake water body from the ground may decrease due to low rainfall, and coarse grains may reach the center of the lake with difficulty because of insufficient sediment transport capacity. Therefore, the grain size of the sediment can reflect the amount of past rainfall in the lake basin, and then the past climatic conditions can be interpreted using the lake sediment.

4.2. The Relationship between River Discharge and Sedimentary Grain Size

Similarly, comparing the river discharge over the past 50 years to the changes in mean grain size indicates that the changes in the peak and trough values are similar to the corresponding relationship between precipitation and mean grain size. As shown in Figure 7, the peak values of river discharge in 1962, 1973, 1984, 1990, 1998, and 2013 correspond to the peak values of the mean grain size of sediment cores in 1962, 1974, 1984, 1990, 1998, and 2013, respectively (gray box). Similarly, the trough of river discharge in 1980, 1987, and 1996 corresponds to the valleys of the mean grain size of the sediment cores in 1980, 1987, and 1994, respectively (blue box). This is mainly because the river discharge in the Hulun Lake basin is related to rainfall, which determines the amount of river discharge. Therefore, the change in sediment grain size in Hulun Lake can also indicate the amount of river discharge and allows to interpretation of past climatic conditions.

4.3. A Dam Construction Event Recorded by Sedimentary Grain Size

The stratigraphic distribution of the grain size characteristics of Hulun Lake sediment shows that intervals between 32 and 38 cm are different from other depths, continuously presenting the minimum value in the whole core. As shown in Figure 7, the precipitation and river discharge in this period are nearly equivalent to the annual average precipitation and average river discharge, but the changes in the mean grain size of the sediment do not correspond to the precipitation and river discharge here. Therefore, the different changes in grain size distribution in this period may be controlled by factors other than the change in hydrological conditions. In addition, the distribution characteristics of the grain size parameters of the 100 cm core northwest of the lake also continued to show the minimum values in the whole core at the same depth, indicating that the changes in these two cores at 32–38 cm were affected by the same factor.

The changes in grain size were found at a depth of approximately 32–38 cm, which corresponds to the period of approximately 1963–1970. During this period, the dam was built to block the water outlet from the lake. Construction of the dam may have significantly affected the sedimentation processes in Hulun Lake. In order to demonstrate the difference in grain size distribution between the lake with and without a dam, the sediment grain size frequency curves at 32–38 cm and other depths were selected for comparison, as shown in Figure 8. The distribution of the 32–38 cm sedimentary section during the period when the dam was constructed was different, and its peak appeared earlier than at other sedimentary depths, indicating that the part occupied by fine grains increased. This result is consistent with the previous findings proposed by Nahm [7], highlighting that large grains were difficult to transfer due to a lack of hydraulic gradient generated by the flow when the outlet was blocked by artificial dam construction. In addition, smaller grains could not be flow out of the lake because there was no outflow, resulting in an increase in the grain size of the clay content in the sediment while the sand content decreased. Therefore, the distribution of the sediment grain size of Hulun Lake clearly shows that the lake was affected by dam-building activities, and the lake sediment effectively recorded the information of this human activity. It is rather remarkable that not only did the sedimentary environment of the lake change, but also a large amount of sediment was deposited on the bottom of the lake due to the building dam.

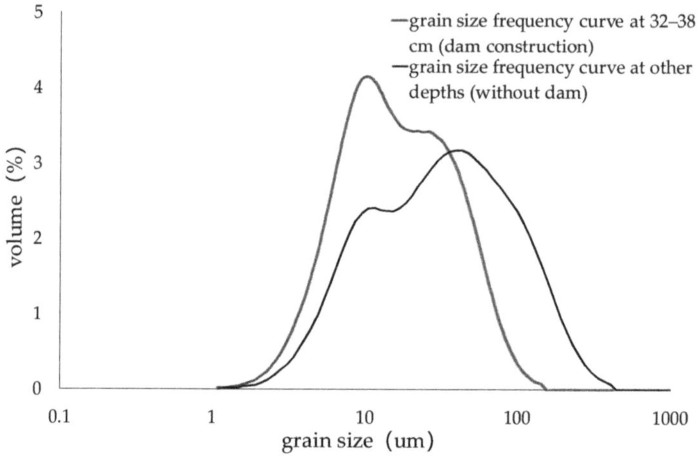

Figure 8. The sketch map of sediment grain size frequency curves at 32–38 cm of the 41 cm core corresponding to the period of dam construction and other depths corresponding to the period without dam.

5. Conclusions

In the present study, we combined the use of radioisotope analyses of ^{210}Pb and ^{137}Cs with grain size analyses to interpret the lake deposition process and historical environment. Furthermore, the correlation of the sedimentological variables with instrumental measurements were analyzed. The results of the proxy interpretation provide insights into the past and potential future environmental changes in and around the lake. The grain size of the sediment can reflect the amount of past rainfall in the lake basin and river discharge into the lake, and then the past climatic conditions can be interpreted using the lake sediment. The changes in grain size at a depth of approximately 32–38 cm correspond to the period of dam construction in 1963–1970. Therefore, the sediment grain size of Hulun Lake effectively recorded the dam-building activity. The results of this study show that the sediments in the lake are well suited for high-resolution paleoenvironmental investigations. The credible conclusions of the palaeoenvironmental implications of grain size in lakes can be obtained by comprehensively analyzing the sedimentation process-related information.

Author Contributions: H.G. and Y.F. designed and performed research. H.G., R.Z. and G.W. wrote the paper. X.Z., J.W. and L.W. assisted experiment and provided comments. All authors have read and agreed to the published version of the manuscript.

Funding: This research was funded by Natural Science Foundation of Henan province (No. 222300420106), National College Students' Innovation and Entrepreneurship Training Program (No. 202211765003), Open-ended Fund of State Key Lab. Of Hydrology-Water Resources and Hydraulic Engineering, Hohai University (No. 2017490611), Henan Provincial Department of Education Key Project (No. 19A210008, 19B570001, 21B610002), Science and Technology Project of Henan Province (No. 212102310277), Project of Young Core Instructor of Henan University of Urban Construction (No. YCJQNGGJS202103).

Data Availability Statement: The data that support the findings of this study are available from the authors upon reasonable request.

Acknowledgments: Comments from the anonymous reviewers are appreciated.

Conflicts of Interest: The authors declare no conflict of interest.

References

1. Espinoza-Encinas, I.R.; Israde-Alcántara, I.; Domínguez-Vázquez, G.; Zárate-del Valle, P.F.; Huerta-Magaña, C.G. A 15,000-yr paleo-environmental record from Lake Chapala, west-central Mexico. *J. Paleolimnol.* **2022**, *68*, 377–393. [CrossRef]
2. Roberts, L.R.; Holmes, J.A.; Leng, M.J.; Sloane, H.; Horne, D.J. Effects of cleaning methods upon preservation of stable isotopes and trace elements in shells of *Cyprideis torosa* (Crustacea, Ostracoda): Implications for palaeoenvironmental reconstruction. *Quat. Sci. Rev.* **2018**, *189*, 197–209. [CrossRef]
3. Wang, J.; Wu, J.; Zeng, H. Sediment record of abrupt environmental changes in Lake Chenpu, upper reaches of Yellow River Basin, north China. *Environ. Earth Sci.* **2014**, *73*, 6355–6363. [CrossRef]
4. Raja, P.; Achyuthan, H.; Geethanjali, K.; Kumar, P.; Chopra, S. Late Pleistocene Paleoflood Deposits Identified by Grain Size Signatures, Parsons Valley Lake, Nilgiris, Tamil Nadu. *J. Geol. Soc. India* **2018**, *91*, 547–553. [CrossRef]
5. Liu, J.; Wang, Y.; Li, T.; Tian, F.; Yang, J. Comparison of grain-size distributions between nearshore sections and a deep-water sediment core from Dali Lake, North China, and inferred Holocene lake-level changes. *J. Paleolimnol.* **2016**, *56*, 123–135. [CrossRef]
6. Qiang, M.; Liu, Y.; Jin, Y.; Song, L.; Huang, X.; Chen, F. Holocene record of eolian activity from Genggahai Lake, northeastern Qinghai-Tibetan Plateau, China. *Geophys. Res. Lett.* **2014**, *41*, 589–595. [CrossRef]
7. Nahm, W.-H.; Lee, G.H.; Yang, D.-Y.; Kim, J.-Y.; Kashiwaya, K.; Yamamoto, M.; Sakaguchi, A. A 60-year record of rainfall from the sediments of Jinheung Pond, Jeongeup, Korea. *J. Paleolimnol.* **2009**, *43*, 489–498. [CrossRef]
8. Colman, S.M.; Baucom, P.C.; Bratton, J.F.; Cronin, T.M.; McGeehin, J.P.; Willard, D.; Zimmerman, A.R.; Vogt, P.R. Radiocarbon Dating, Chronologic Framework, and Changes in Accumulation Rates of Holocene Estuarine Sediments from Chesapeake Bay. *Quat. Res.* **2002**, *57*, 58–70. [CrossRef]
9. Chen, J.A.; Wan, G.; Zhang, D.D.; Zhang, F.; Huang, R. Environmental records of lake sediments at different time scales: Sediment grain size as an example. *Sci. China Ser. D Earth Sci.* **2004**, *47*, 954–960. [CrossRef]
10. Brown, S.; Bierman, P.; Lini, A.; Thompson Davis, P.; Southon, J. Reconstructing lake and drainage basin history using terrestrial sediment layers: Analysis of cores from a postglacial lake in New England, USA. *J. Paleolimnol.* **2002**, *28*, 219–236. [CrossRef]
11. Gupta, H.; Kao, S.-J.; Dai, M. The role of mega dams in reducing sediment fluxes: A case study of large Asian rivers. *J. Hydrol.* **2012**, *464–465*, 447–458. [CrossRef]
12. Yang, Y.-P.; Zhang, M.-J.; Li, Y.-T.; Fan, Y.-Y. Trend and dynamic cause of sediment particle size on the adjacent continental shelf of the Yangtze Estuary. *China Ocean Eng.* **2016**, *30*, 992–1003. [CrossRef]
13. Watson, E.B.; Pasternack, G.B.; Gray, A.B.; Goñi, M.; Woolfolk, A.M. Particle size characterization of historic sediment deposition from a closed estuarine lagoon, Central California. *Estuar. Coast. Shelf Sci.* **2013**, *126*, 23–33. [CrossRef]
14. Chen, X.; Chuai, X.; Yang, L.; Zhao, H. Climatic warming and overgrazing induced the high concentration of organic matter in Lake Hulun, a large shallow eutrophic steppe lake in northern China. *Sci. Total Environ.* **2012**, *431*, 332–338. [CrossRef] [PubMed]
15. Hu, S.; Wang, S.; Appel, E.; Ji, L. Environmental mechanism of magnetic susceptibility changes of lacustrine sediments from Lake Hulun, China. *Sci. China Ser. D Earth Sci.* **2000**, *43*, 534–540. [CrossRef]
16. Wen, R.; Xiao, J.; Chang, Z.; Zhai, D.; Xu, Q.; Li, Y.; Itoh, S.; Lomtatidze, Z. Holocene climate changes in the mid-high-latitude-monsoon margin reflected by the pollen record from Hulun Lake, northeastern Inner Mongolia. *Quat. Res.* **2010**, *73*, 293–303. [CrossRef]
17. Xiao, J.; Chang, Z.; Wen, R.; Zhai, D.; Itoh, S.; Lomtatidze, Z. Holocene weak monsoon intervals indicated by low lake levels at Hulun Lake in the monsoonal margin region of northeastern Inner Mongolia, China. *Holocene* **2009**, *19*, 899–908. [CrossRef]
18. Xu, Z.J.; Jiang, F.Y.; Zhao, H.W.; Zhang, Z.B.; Sun, L. *Annals of Hulun Lake*; Jilin Literature and History Publishing House: Jilin, China, 1989. (In Chinese)
19. Wen, R.; Xiao, J.; Chang, Z.; Zhai, D.; Xu, Q.; Li, Y.; Itoh, S. Holocene precipitation and temperature variations in the East Asian monsoonal margin from pollen data from Hulun Lake in northeastern Inner Mongolia, China. *Boreas* **2010**, *39*, 262–272. [CrossRef]
20. Liang, L. Environment and Climate Evolution Recorded in Sediment of Hulun Lake since Middle-Late Holocene. Ph.D. Thesis, Environment Group, Inner Mongolia Agricultural University, Hohhot, China, 2017. (In Chinese)
21. Glew, J.R.; Smol, J.P.; Last, W.M. Sediment core collection and extrusion. In *Tracking Environmental Change Using Lake Sediments*; Last, W.M., Smol, J.P., Eds.; Springer: Dordrecht, The Netherlands, 2002; pp. 73–105.
22. Krishnaswamy, S.; Lal, D.; Martin, J.M.; Meybeck, M. Geochronology of lake sediments. *Earth Planet. Sci. Lett.* **1971**, *11*, 407–414. [CrossRef]
23. Robbins, J.A.; Edgington, D.N. Determination of recent sedimentation rates in Lake Michigan using Pb-210 and Cs-137. *Geochim. Cosmochim. Acta* **1975**, *39*, 285–304. [CrossRef]
24. Appleby, P.G.; Nolan, P.J.; Giford, D.W.; Godfrey, M.J.; Oldfeld, F.; Anderson, N.J.; Battarbee, R.W. ^{210}Pb dating by low background gamma counting. *Hydrobiologia* **1986**, *143*, 21–27. [CrossRef]
25. Appleby, P.G.; Oldfield, F. Applications of lead-210 to sedimentation studies. In *Uranium-Series Disequilibrium: Application to Earth, Marine, Environmental Sciences*; Ivanovich, M., Harmon, R.S., Eds.; Clarendon Press: Oxford, UK, 1992; pp. 731–778.
26. Ritchie, J.C.; McHenry, J.R. Application of Radioactive Fallout Cesium-137 for Measuring Soil Erosion and Sediment Accumulation Rates and Patterns: A Review. *J. Environ. Qual.* **1990**, *19*, 215–233. [CrossRef]
27. Andersen, T.J.; Mikkelsen, O.A.; Møller, A.L.; Pejrup, M. Deposition and mixing depths on some European intertidal mudflats based on ^{210}Pb and ^{137}Cs activities. *Cont. Shelf Res.* **2000**, *20*, 1569–1591. [CrossRef]
28. Yin, Y.; Fang, N.Q.; Wang, Q.; Nie, H.G.; Qing, Z.L. Magnetic susceptibility of lacustrine sediments and its environmental significance: Evidence from Napahai lake, northwestern Yunnan, China. *Sci. Geogr. Sin.* **2002**, *22*, 413–419. (In Chinese)

water

MDPI

Article

Characterization of Dissolved Organic Matter of Sediments in Urban Streams Using EEMs–PARAFAC and Absorption Spectroscopy: A Case Study in Wuhan, China

Hui Zhang [1,2], Zufan Liu [1,2], Jing Xu [1,2], Jun Yang [3], Xiang Zhang [1,2,*] and Shiyong Tao [1,2,*]

1 State Key Laboratory of Water Resources and Hydropower Engineering Science, Wuhan University, Wuhan 430072, China
2 Hubei Key Laboratory of Water System Science for Sponge City Construction, Wuhan University, Wuhan 430072, China
3 Aquatic Eco-Health Group, Fujian Key Laboratory of Watershed Ecology, Key Laboratory of Urban Environment and Health, Institute of Urban Environment, Chinese Academy of Sciences, Xiamen 361021, China
* Correspondence: zhangxiang@whu.edu.cn (X.Z.); taoshiyong@whu.edu.cn (S.T.)

Abstract: Urbanization has notably changed the characteristics and functions of watershed ecosystems worldwide, influencing the characteristics of chromophoric dissolved organic matter (CDOM) and dissolved organic matter (DOM) of sediments in urban streams. In this study, the biogeochemical characteristics of 42 water samples and the optical absorption and excitation–emission matrix spectra (EEMs) of 14 sediment samples collected from 14 urban streams in Wuhan were systematically examined. In addition, five water samples and one sediment sample were collected in Mulan Lake as a reference for non-urban areas. The a_{254} values of sediments in urban streams ranged widely (25.7–197.6 m^{-1}), and the mean (116.32 ± 60.5 m^{-1}) was significantly higher than the reference (51.52 m^{-1}), indicating clear individual differences and a higher concentration of CDOM. Two humus-like components and one tryptophan-like component were effectively identified by parallel factor analysis (PARAFAC). The fluorescence index (FI)/biological index (BIX) of DOM of sediments in urban streams was mostly within 1.4–1.7/0.8–1.0, indicating a compound of both allochthonous and autochthonous sources. Compared with the reference, lower FI and BIX and higher humification index (HIX) revealed a higher allochthonous input and humification degree of DOM of sediments in urban streams. Spearman's correlation analysis and redundancy analysis demonstrated that heavy metals and other water quality parameters had a considerable impact on CDOM concentrations and DOM components. This study could support the use of DOM as an effective tool to monitor the water environment and provide insights into future water pollution management strategies.

Keywords: dissolved organic matter; urban streams; EEMs–PARAFAC; water quality; correlation analysis

Citation: Zhang, H.; Liu, Z.; Xu, J.; Yang, J.; Zhang, X.; Tao, S. Characterization of Dissolved Organic Matter of Sediments in Urban Streams Using EEMs–PARAFAC and Absorption Spectroscopy: A Case Study in Wuhan, China. *Water* **2022**, *14*, 3181. https://doi.org/10.3390/w14193181

Academic Editors: Qiting Zuo, Fuqiang Wang, Jiaqi Zhai, Xiuyu Zhang, Dunxian She, Lei Zou, Rong Gan, Zengliang Luo and Roohollah Noori

Received: 24 August 2022
Accepted: 6 October 2022
Published: 10 October 2022

1. Introduction

Dissolved organic matter (DOM) is a heterogeneous compound of organic matters, mainly composed of proteins, humus and other aromatic and aliphatic organic compounds [1,2]. DOM is ubiquitous in surface water, pore water, sediments and other natural environments [3,4]. DOM is also considered as a tracer in biogeochemical cycling, which plays an important role in the course of material circulation and energy exchange [5]. DOM characteristics are associated with their source and environment conditions. In general, there are two main sources of DOM, including allochthonous inputs and autochthonous production [6]. Allochthonous DOM represents the terrigenous input materials, and autochthonous DOM predominantly originates from the microbial decomposition of organic matter and macrophytes and algae production [7,8]. DOM has been discussed in studies of biogeochemical processes in estuaries, reservoirs and lakes, due to its importance in element cycling as well as ecosystem function [9–11]. Chromophoric dissolved organic matter (CDOM) refers to the fraction of DOM based on the absorption of ultraviolet and photosynthetically active radiation [12]. In general, light absorption by CDOM can promote not only the primary production of aquatic ecosystems, but also photochemically induced transformations in natural waters [13].

During recent years, rapid industrialization and increased urbanization have brought about a variety of urban water problems, including an impact on aquatic ecosystems and the deterioration of water quality in urban streams [14]. The increase of impervious surface in urban catchments changes the hydrological process of urban streams, and enhanced water disturbance promotes the interactions between sediment and overlying water, especially under heavy rain conditions [15]. Many pollutants accumulated in sediments are then released into the overlying waters, among which the speciation, transformation and characterization of DOM have become a hot topic in water environmental protection studies [16]. Urbanization also has an effect on the components and characteristics of DOM in the water environment [17]. Previous studies have explored the interactions between DOM and water pollutants, such as TN and TP, and revealed the relationship between DOM in natural waters and water quality indexes [18]. DOM content in water has been reported to be regulated by sediment release, which provides a perspective on the response of water quality to sediment characteristics [19]. Previous studies have suggested that DOM from surface sediments and overlying waters exchanges at the water–sediment interface, increasing the proportion of autochthonous DOM [2]. Therefore, systematic analysis of DOM in sediments is helpful to provide scientific basis and management strategy for the control of water environmental pollution.

Wuhan is a typical city suffering from urban water problems, such as eutrophication, decreased self-purification capacity and water ecological degradation, which is a threat to the balance of aquatic ecosystems [20]. Point (mainly domestic sewage) and non-point pollution sources caused by runoff washing the roads in urban areas lead to spatial distribution differences of water pollution in Wuhan. Previous studies in Wuhan have mostly investigated the common water quality parameters of urban lakes [21]. Our team has also studied geochemical and isotopic characteristics of urban streams during the late spring of 2019 and revealed the fragmentation of urban hydrological connectivity [22]. Some studies have explored the water environment carrying capacity of urban lakes [23], and others have examined the distribution of organic compounds in the Yangtze River [24]. However, few studies have focused specifically on DOM of sediments in urban streams, especially using fluorescence spectroscopy, which can reflect water quality and the characteristics and variations of fluorescent components simultaneously. Analyzing DOM is a rapid and effective way to monitor the current situation and source of water pollution, especially when non-point pollution is hard to detect. The main purposes of this research were to (1) characterize properties of sediments in urban streams, including the CDOM absorption properties, DOM components and their abundance, etc.; (2) reveal the effect of urbanization on CDOM properties and DOM composition; (3) explore the correlations between water quality, CDOM and DOM in Wuhan.

2. Materials and Methods

2.1. Site Description and Sample Collection

Wuhan (113°41′–115°05′ E and 29°58′–31°22′ N), the capital of Hubei Province, is located in central China and the middle reaches of the Yangtze River (Figure 1). Precipitation is abundant and unevenly distributed within the year, with an annual average rainfall of 1140–1265 mm [25]. As a new first-tier city in China, the resident population of Wuhan is increasing year by year, reaching 12.45 million in 2020 [26]. The urban core consists of seven districts and accounts for approximately 57.89% of Wuhan's population. As an important national transportation junction, Wuhan is obviously a modern metropolis with rapid economic development and intense human activities. Wuhan has an inland area of 2217.6 km^2, accompanied by two major rivers (the Yangtze and Han rivers) and plenty of lakes, reservoirs and urban streams (shallow waterbodies and canals) that are interconnected [25]. Due to the development of commerce, tourism and industrial activities, water environment problems, including water pollution, eutrophication and the reduction of biodiversity, have gradually become serious in the urban streams in Wuhan.

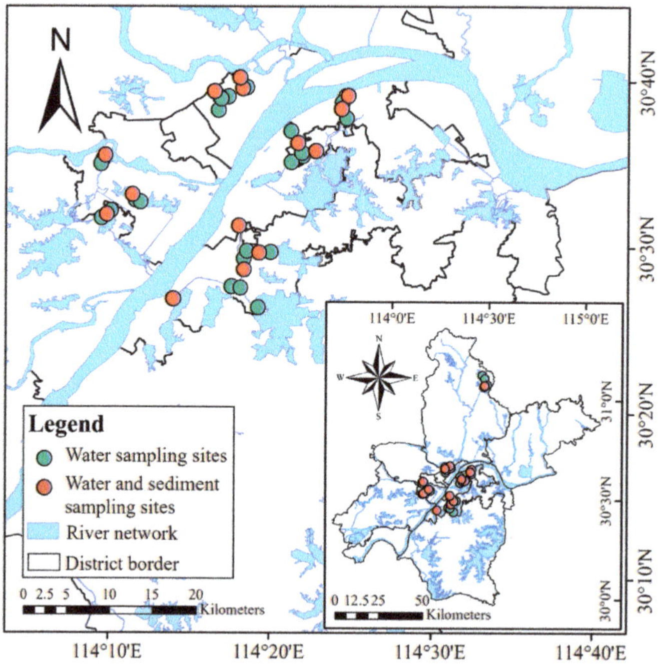

Figure 1. Location map of the study area, and the water and sediment sampling sites.

An urban stream and sediment sampling program was conducted during a rainless period in mid-August, 2020. For each urban stream, three water samples were sampled along the stream flow. Owing to the weak fluidity and relative stability of sediments, only one sediment sample was taken by the grab sampler from each urban stream. In total, 42 water samples and 14 sediment samples were collected in urban streams of the central regions, including five districts. In addition, samples from Mulan Lake, far from the central districts, were also collected as the reference for non-urban areas, totaling five water samples and one sediment sample. In general, the urban streams in this study were shallow in depth, ranging from 0.2 m to 2.2 m, averaging 1.15 m. The samples were quickly collected and then transported to the laboratory in dark and cold conditions.

2.2. Sample Pretreatment and Biogeochemical Parameter Measurement

Water temperature (Temp), pH, dissolved oxygen (DO), turbidity (Turb), salinity (Sal), chlorophyll a (Chl-a), oxidation-reduction potential (ORP), and electrical conductivity (EC) were measured in situ using a Hydrolab DS5 multi-probe water quality analyzer (Hach company, Loveland, CO, USA). Concentrations of heavy metals were determined using an inductively coupled plasma mass spectrometer (ICP-MS) (Nexion 350, PerkinElmer, Waltham, MA, USA) using the sample measurement method described in detail in our previous work [22].

Sediment samples were screened through 100 mesh after natural air drying and grinding. The contents of carbon and nitrogen (C and N) in the sediments were then determined by an elemental analyzer (Elementar Vario Macro cube, Hanau, Germany). Sediments DOM were extracted by preparing 200 g/L suspended solution, stirred continuously for 24 h at a velocity of 70 r/min and a constant temperature of 25 °C without light. After centrifuging at a speed of 7500 r/min for 5 min and standing for 30 min, the obtained solution was then filtered with 0.22 μm polyethersulfone filter membrane [19]. CDOM and DOM were measured using the filtered supernatant, and the analysis methods are described in detail in Sections 2.3 and 2.4.

2.3. CDOM Absorption Analysis

A Shimadzu UV-3600 UV-Vis spectrophotometer was used to determine the UV-Vis spectra of filtered supernatant. The CDOM absorption coefficient $a(\lambda)$ was expressed as Equation (1) [13,27]:

$$a(\lambda) = 2.303D(\lambda)/L \quad (1)$$

where $D(\lambda)$ is the absorbance after deducting the corresponding absorbance of Milli-Q water, and L is the cuvette length (m). The absorption coefficient of CDOM at 254 nm (a_{254}) was used to quantify the content of CDOM, with higher a_{254} indicating higher concentration of CDOM.

The spectral slope parameter S was computed by Equation (2) [27]:

$$a(\lambda) = a(\lambda_0)e^{(S(\lambda_0-\lambda))} \quad (2)$$

where λ_0 is the reference wavelength (440 nm). $S_{275\text{–}295}$ and $S_{350\text{–}400}$ refer to the spectral slopes of 275–295 nm and 350–400 nm, respectively. The humification signal of CDOM is inversely proportional to S. The lower the $S_{275\text{–}295}$ value, the stronger the terrestrial humic acid signal [28].

The calculation formula of the spectral slope ratio (S_R) is shown in Equation (3) and is sensitive to the characteristics of CDOM [27]. A lower S_R value indicates higher DOM molecular weight and greater aromaticity [27,29].

$$S_R = \frac{S_{275-295}}{S_{350-400}} \quad (3)$$

2.4. DOM Fluorescence Measurement

The excitation–emission matrix spectra (EEMs) of filtered supernatant were measured using a FS5 fluorescence spectrometer (Edinburgh Instruments, Scotland) [30]. The EEMs of Milli-Q water were measured to remove the interferences of water Raman scattering peaks. Moreover, the fluorescence intensity of 3D fluorescence spectra was calibrated as a Raman unit (R.U.) by the ratio of fluorescence integral intensity of Milli-Q water at the wavelength of 350 nm (E_x) and 371–428 nm (E_m).

Three fluorescent indexes, including the fluorescence index (FI), biological index (BIX) and humification index (HIX), were computed to further distinguish the source of DOM and quantify humification degree. The calculation formulae are shown in Equations (4)–(6) [31–33]:

$$\mathrm{FI} = \frac{F_{\mathrm{E_x}=370nm,\mathrm{E_m}=470nm}}{F_{\mathrm{E_x}=370nm,\mathrm{E_m}=520nm}} \quad (4)$$

$$\mathrm{BIX} = \frac{F_{\mathrm{E_x}=310nm,\mathrm{E_m}=380nm}}{F_{\mathrm{E_x}=310nm,\mathrm{E_m}=430nm}} \quad (5)$$

$$\mathrm{HIX} = \frac{A_{\mathrm{E_x}=254nm,\mathrm{E_m}=435-480nm}}{A_{\mathrm{E_x}=254nm,\mathrm{E_m}=300-345nm}} \quad (6)$$

where F is the fluorescence intensity and A is the integrated intensity. FI was usually applied to analyze the source of DOM, with higher FI indicating more autochthonous DOM from microbial decomposition [34]. BIX was generally used to estimate the relative contribution of autochthonous DOM, and HIX was used to characterize the humification degree [35,36].

2.5. Data Analysis

Three fluorescent components of EEMs of sediments were identified by parallel factor analysis (PARAFAC) using the "DOMFluor" toolbox on the MATLAB platform, after Raman and Rayleigh scattering of spectral data were removed. For the accuracy and reliability of the results obtained, the optimal fluorescence fraction was extracted through half-split verification and residual examination. Redundancy analysis (RDA) and Spearman's correlation analysis were performed on the standardized PARAFAC components, calculated fluorescent indexes and biogeochemical parameters, using Canoco5 and R 4.2.0, respectively.

3. Results and Discussion

3.1. Biogeochemical Characteristics

The biogeochemical parameters of water and sediment samples in urban streams and the reference are presented in Table 1.

Table 1. Biogeochemical parameters of water and sediment samples in urban streams and the reference.

Sampling Types	Parameters	Urban Streams		Reference	
		Min–Max	Mean ± SD	Min–Max	Mean ± SD
Waters	pH	6.53–9.52	7.31 ± 0.65	7.84–8.15	7.99 ± 0.10
	Temp (°C)	27.33–35.40	30.10 ± 2.10	31.09–0.31	31.09 ± 0.31
	DO (mg/L)	0.00–15.44	4.51 ± 4.00	7.24–7.73	7.49 ± 0.16
	Turb (NTU)	6.40–122.20	27.82 ± 20.41	6.30–7.00	6.58 ± 0.28
	Sal	0.08–0.74	0.19 ± 0.13	0.20–0.20	0.02 ± 0.00
	Chl-a (μg/L)	ND–41.25	6.11 ± 10.72	ND	ND
	ORP (mV)	59.0–409.0	336.4 ± 64.1	333.0–420.0	395.0 ± 31.9
	EC (μS/cm)	181.3–1394.0	383.9 ± 243.5	70.2–73.3	71.08 ± 1.13
Sediments	C (%)	0.50–6.26	3.29 ± 1.57	6.4–6.4	6.4 ± 0.00
	N (%)	0.15–0.47	0.35 ± 0.10	0.72–0.72	0.72 ± 0.00
	C/N	2.00–21.60	9.35 ± 4.41	8.23–8.23	8.86 ± 0.00

Note: ND refers to not detected.

The pH of urban streams ranged from weakly acidic to weakly alkaline. Clearly, physicochemical parameters of urban streams in Wuhan varied greatly, especially DO, Turb and EC. The mean concentrations of Chl-a (6.11 ± 10.72 μg/L) in urban streams were relatively low compared with the content of Chl-a of urban lakes previously reported in Wuhan, with a mean of 114.56 μg/L [37]. Some studies have also shown that Chl-a concentrations in urban rivers were significantly higher, compared to peri-urban rivers [38], which is in agreement with our research. Increased impervious surfaces lead to urban rainfall runoff being discharged into urban streams rather than seeping directly downward [39]. Enriching nutrients such as nitrogen and phosphorus in urban streams increase the Chl-a content and contribute to algae bloom. Less than half of urban streams (46.7%) had DO concentrations higher than 3 mg/L (Grade III surface water quality standards). The concentration of DO in the non-urban area was higher, with an average close to 7.5 mg/L (Grade I). As confirmed by the previous study, urban streams that received point source pollution, mainly through sewage effluent, had a lower value of DO [40]. The impact of domestic sewage, industrial wastewater and runoff pollution on urban streams may be serious, as water volume of these streams is not large [39]. The mean ORP in urban streams was 336.4 ± 64.1 mV, slightly less than the reference water (395.0 ± 31.9 mV), revealing a lower oxidability. Compared with the EC of the reference water (71.08 ± 1.13 μS/cm), urban streams were clearly less purified, with a mean EC of 383.9 ± 243.5 μS/cm. The C/N of sediments was within a wide range, demonstrating that sediments collected were considerably different in physical properties.

Heavy metal in natural surface water is generally at low values. However, with the development of agriculture and industries in urban areas, a mass of wastewater derived from anthropogenic activities has carried lots of heavy metals into surface water, resulting in severe heavy metal pollution in some urban streams [41]. Therefore, many studies pay attention to the types and contents of heavy metals when investigating the urban water environment. In this research, the content of heavy metals has also been examined. The results demonstrated that concentrations of heavy metals in urban streams were mostly higher than those in the reference, except for Cr (Figure 2). Overall, urban streams in Wuhan contain metals at relatively low levels, with the concentrations of Ni, Cu, Cr, Pb lower than 10 μg/L.

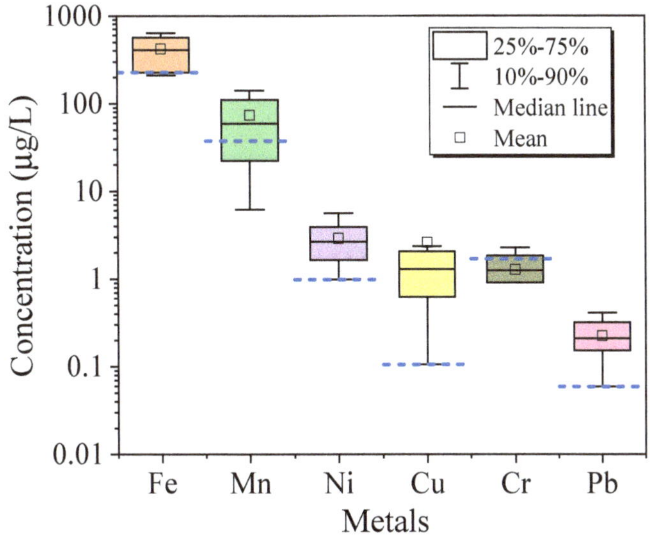

Figure 2. Variations in the concentrations of heavy metals (μg/L) in the urban streams. The blue dashed line refers to the reference level.

3.2. CDOM Absorption Characteristics

As demonstrated in Figure 3, absorbance decreases exponentially with increased wavelength, except for the absorption peaks in 200–230 nm and 260–280 nm. There was no absorption peak in the CDOM absorption spectra of sediments in most urban streams, consistent with that of natural waterbodies [27]. The CDOM absorption coefficients of sediments in most urban streams were apparently higher than the reference, with high and flat curves for urban streams. It is well known that inorganic ions have significant UV absorption at wavelengths lower than 230 nm [42]. CDOM absorption spectra of sediments in some urban streams showed a similar strong adsorption peak. Low intensity absorption peaks in the 260–280 nm wavelength range may be induced by an unconjugated group with lone pair electrons, such as carboxyl [42].

Among 15 sediment samples, a_{254} ranged from 25.7 m^{-1} to 197.6 m^{-1}, with an average of 116.32 ± 60.5 m^{-1} for urban streams and 51.52 m^{-1} for the reference, demonstrating that CDOM contents of sediments in most urban streams were significantly higher than that in the reference. This may be attributed to the large magnitude of conjugated structure and high degree of humification of CDOM in sediments from urban streams. The variation of CDOM of sediments in urban streams normally corresponds to those of urban waters [18]. In a previous investigation, values of a_{254} of natural waters from rivers were reported to be higher than those of lakes and coastal marine environments, with means of 50.5 m^{-1}, 13.2 m^{-1} and 6.4 m^{-1}, although all at a relatively low level [43]. Moreover, a_{254} of sediments in urban streams exhibited considerable variability, even in the same district.

A previous study has reported that higher amounts of CDOM were observed in urban waters, due to high population density, water pollution and an increase of surrounding artificial surfaces [5]. These were also thought to be the factors leading to the variation and higher concentration of CDOM of sediments in the study area.

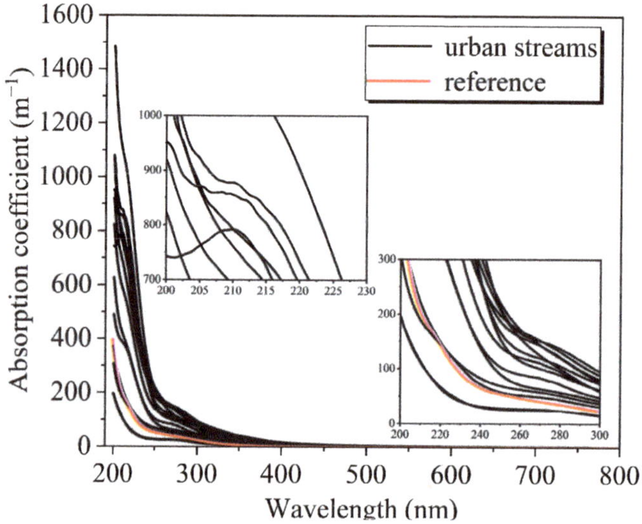

Figure 3. UV-Vis absorption spectra of CDOM in sediments from urban streams and the reference, with special curves in the upper left and lower right corner.

The $S_{275\text{–}295}$ varied greatly, from 14.6 μm^{-1} to 24.9 μm^{-1}, and the mean was 17.88 ± 2.66 μm^{-1} for urban streams and 18.8 μm^{-1} for the reference. $S_{275\text{–}295}$ of sediments in urban streams ranged widely, indicating differences in their content, molecular weight and sources of CDOM. Since $S_{275\text{–}295}$ was also an indicator of DOM sources, sediments of urban streams were inferred to contain different amounts of allochthonous DOM from terrigenous substances (mainly humus-like matter with a large molecular weight) [44].

The mean value for S_R (0.93 ± 0.13) of sediments in urban streams was higher than in the reference (0.88), indicating a lower molecular weight of CDOM. S_R, representing the structural changes of DOM, is inversely correlated with the molecular weight of DOM. A small S_R indicates that DOM is mostly newly generated (photocatalytic oxidation or microbial activities) or mainly imported from external sources. On the contrary, a larger S_R demonstrates that DOM is mainly endogenous (self-decomposition of microorganisms in water) or that photobleaching is strong [28]. The fact that S_R of sediments in urban streams ranged extensively apparently indicates enormously different DOM sources, mainly autochthonous and allochthonous, respectively.

3.3. Variations and Characteristics of DOM

Based on the PARAFAC model, three-dimensional fluorescence spectra analysis and split half validation were carried out on the sediment samples, and three DOM components were eventually identified (Table 2 and Figure 4).

Table 2. Descriptions of the three fluorescence components identified by PARAFAC analysis.

Component	E_x (nm)	E_m (nm)	Type
C1	260, 365	475	Humic-like acid
C2	325	402	Humic-like (fulvic) acid
C3	275	334	Protein-like (tryptophan) component

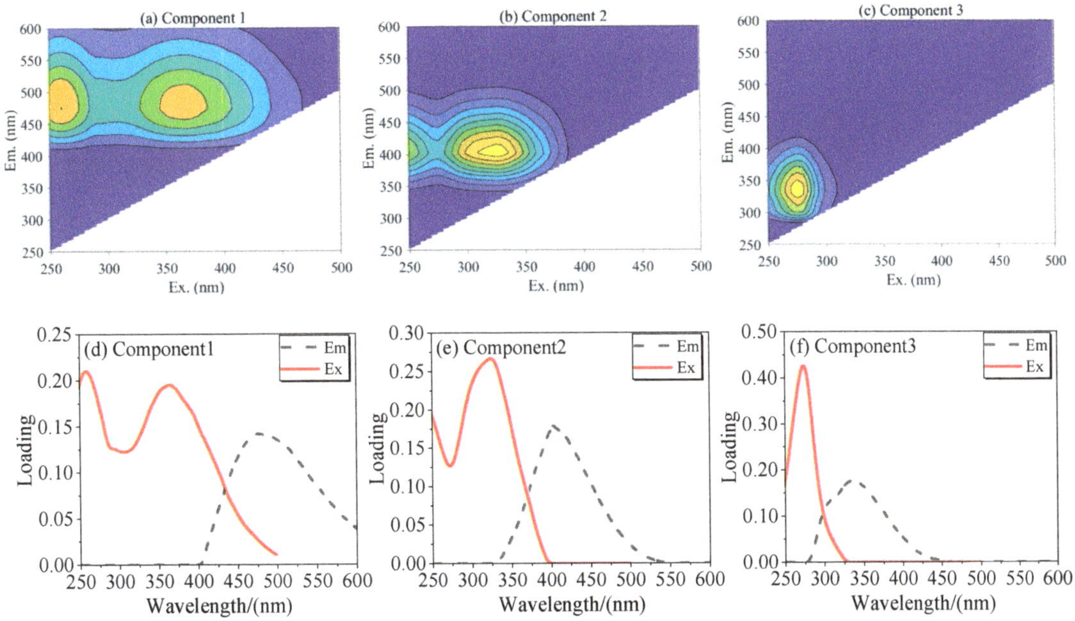

Figure 4. Identification of the three components extracted from the PARAFAC model. (**a**–**c**): contour plots of C1, C2 and C3. (**d**–**f**): excitation and emission loadings of C1, C2 and C3.

Component1 (C1) demonstrated two fluorescence peaks at an E_x/E_m wavelength of 260 nm (365 nm)/475 nm, consistent with humic-like peaks, as previously reported [6,45]. Component2 (C2), with a primary fluorescence peak at an E_x/E_m wavelength of 325 nm/402 nm, was similar to fulvic-like acid in the visible region, a humus-like substance with a relatively small molecular weight. The fulvic acid was considered to be derived from the degradation and transformation of macromolecular humus-like substances [46]. Component3 (C3) showed a primary fluorescence peak with E_x/E_m wavelength of 275 nm/334 nm and was considered as a tryptophan-like matter, related to carboxyl functional groups and aromatic protein structures generated by microbial degradation [4,47]. C2/C1 is the ratio of humic-like acid and fulvic-like acid fluorescence peaks, which is often used to characterize the composition of humus and reflect the degree of humus aggregation to a certain extent. C2/C1 of sediments in urban streams were in a wide range and mostly lower than the reference, with means of (1.53 $\pm$ 0.36) and 1.88, respectively, accounting for the variation of the humus aggregation degree and worse quality of sediments in urban streams.

The maximum fluorescence intensity (F_{max}) of components extracted, based on the PARAFAC model, can be used to characterize their abundance (Figure 5a). The total fluorescence intensity (F_T) and component fluorescence intensity of sediments in urban streams were significantly higher than those in the reference. It is worth noting that F_T of sediments in urban streams was nearly four times that of the reference, with means of 2.51 $\pm$ 0.97 R.U. and 0.65 R.U., respectively, indicating a higher DOM concentration of sediments in urban streams. A large amount of nutrients and organic matter such as humus-like substances from runoff input were found to be carried into urban streams by point source or non-point source pollution, resulting in higher allochthonous organic matter in waterbodies [48]. Therefore, it is reasonable to believe that the external inputs influenced by the process of urbanization increase the DOM content of sediments in urban streams. In general, the concentration of C3 of sediments in most urban sampling sites was the highest and ranged widely, with a distinct difference in the spatial distribution. Although slightly lower than C3, the content of C2 was rather uniformly distributed. Moreover, the concentration of C1 was clearly the lowest, demonstrating that the humus of DOM of sediments in most urban streams was mainly composed of micromolecules.

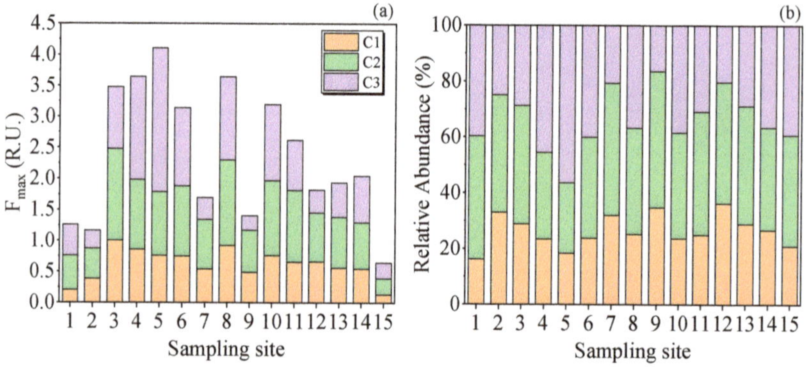

Figure 5. Distribution of the (**a**) abundance and (**b**) relative abundance of three PARAFAC components of the sediment samples in Wuhan; 1–14 refer to the urban streams, 15 refers to the reference.

The relative abundance of three fluorescence components is displayed in Figure 5b. The average relative abundance of C2 of sediments in urban streams was highest, C3 was the next highest and C1 was the lowest. Humus-like substances (C1 + C2) were significantly higher than protein-like matter (C3), with means of 66.81% and 33.19%, respectively. Humus-like components are commonly considered to be mainly related to external inputs, including runoff input and farmland water regression [49]. Protein-like components are commonly supposed to be derived from domestic sewage and also probably originate from microbial degradation of algae or aquatic plant residues [50]. The relative abundance of humus-like and protein-like components were possibly associated with land use, population density and the economic system, as there were obvious individual differences in each sampling sediment. Research in Jilin Province has also found that urban waters have higher relative abundance of humic-like acid (74%), accompanied by lower relative abundance of protein-like substances (26%) [5]. In this study, more than 71% of sediments in urban streams were characterized by lower relative abundance of C3 and higher relative abundance of C1 + C2 compared to the reference, indicating that DOM of sediments is influenced by urbanization, with increasing humus content and decreasing tryptophan content. The results above indicate that DOM of sediments in urban streams was affected by both allochthonous and autochthonous origins, with the terrestrial inputs being slightly higher.

FI varied from 1.53 to 1.76, distributed within the range of 1.4–1.9, indicating the simultaneous presence of terrigenous and biogenic DOM (Figure 6). Most BIX of sediments in urban streams was within 0.8–1.0, demonstrating the moderate autochthonous inputs. Nonetheless, BIX of some sediment samples was less than 0.8 or higher than 1.0, indicating distinct spatial distribution differences and multiple sources of DOM of sediments in urban streams. The range of HIX was from 1.04 to 5.30, providing evidence of various degrees of humification, since HIX represents DOM humification degree. Significantly higher HIX was observed compared to the reference, demonstrating that DOM of sediments in urban streams were more humified and aromatic. This discovery was consistent with a previous study, which reported that urban waterbodies had a wide range of fluorescence indexes, due to differences in land use types [51]. The influence of allochthonous and autogenous sources on DOM of surface sediments has also been found in a past case study [52]. The lower FI and lower BIX, along with higher HIX, than the reference indicates that DOM of sediments in urban streams in Wuhan were more affected by the terrestrial input and less generated by aquatic organisms with higher degree of humification.

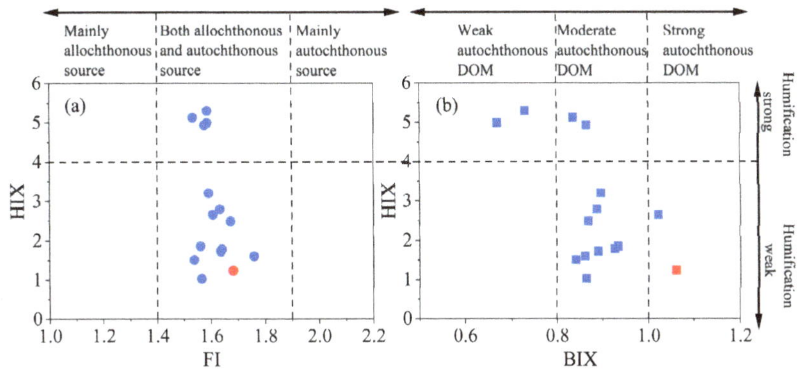

Figure 6. Fluorescence indexes of (**a**) FI and HIX (**b**) FI and BIX of DOM of sediments. The blue points refer to urban sampling sites, and the red ones were the reference.

3.4. Relationship between DOM Optical Parameters and Water Environmental Factors

The degradation of water ecosystems has always been a hot issue in urban rivers and also a threat to the sustainable development of water resources, making it essential to monitor and control water quality [14]. In general, surface water varies rather greatly under different scenarios. As a cumulative environmental medium, sediment could reflect the situation in past periods and the influence on future water quality. The content and components of DOM were influenced by the sources and concentrations of some biogeochemical factors, so the relationship among them has been widely studied in some oceans, rivers and lakes [11,18,53]. The RDA of optical characteristics and biogeochemical factors is shown in Figure 7. C1 was significantly positively correlated with C2, a_{254}, BIX, Chl-a, Sal and EC. C2 and a_{254} share the same correlation with C1. C3 was positively correlated with both BIX and FI. HIX was significantly positively correlated with Temp, C/N, Turb, DO, pH and ORP, and negatively correlated with FI and HIX.

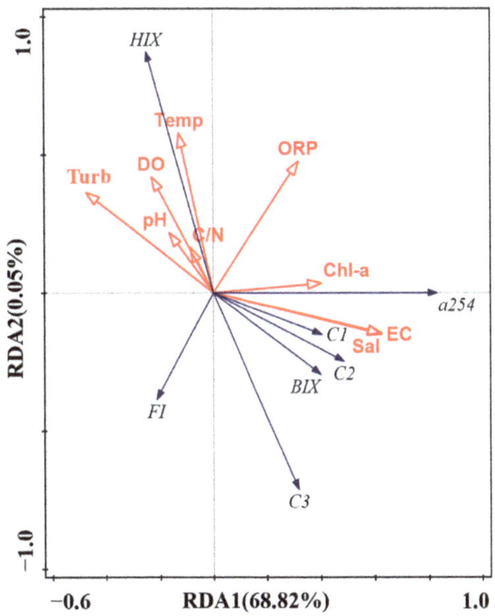

Figure 7. RDA of spectral parameters and biogeochemical factors.

Since C1 and C2 were positively correlated and had similar correlations with other factors, it could be concluded that they might be obtained from common allochthonous sources. The strong positive correlation between C1, C2 and a_{254}, and relatively weak correlation between C3 and a_{254}, reiterated that humic-like acid was the major component of DOM. The non-fluorescent organic matter in phytoplankton can be transformed and degraded into humus-like fluorescent substances by bacteria, contributing to the positive correlation between Chl-a, C1 and C2. Some ions have been shown to be able to complexate or chelate with soil humus, which explains the positive correlation between EC and C1, as well as C2 [54]. Since Sal and EC are interchangeable, they showed the same correlation with C1 and C2. The positive correlation between C3, FI and BIX demonstrates that tryptophan-like substances might be closely associated with the proportion of autochthonous DOM and allochthonous DOM. Protein-like substances with small molecular weight have lower C/N ratios, which fully explains the negative correlation between C3 and C/N. The inverse correlations between HIX and FI, and between HIX and BIX, indicate the higher fluorescence intensity, the more allochthonous DOM and the lower the degree of humification. A previous study has reported that DO was regarded as an oxidant in accelerating the generation of organic matters in waterbodies [55]. Due to the positive correlation between HIX and other physicochemical parameters, we can conclude that the degree of humification of DOM of sediments was partly influenced by overlying water. Further research is still required to reveal the relationships between DOM of sediments and water quality.

The relationship between DOM components of sediments and heavy metals was explored by Spearman's correlation analysis (Figure 8). The results showed positive correlations between most heavy metals, indicating that they might be derived from a similar origin of water pollution. However, Cr was negatively correlated with all other metals, which might be due to the different source of some industrial wastewater around the sampling sites. Interestingly, all three components were positively correlated with most metals, especially Ni, Fe and Pb. A previous investigation has confirmed the toxic metal (Ni^{2+}, Pb^{2+}) binding affinity of DOM, and that the metal complexing capacity was influenced by DOM concentrations, structure and components [56]. It is considered that the existence of DOM in surface waters might reduce the precipitation of heavy metals by inhibiting the formation of carbonate and hydroxide precipitations, and then enhancing their migration ability [57,58]. However, when heavy metals and macro-molecular DOM form insoluble chelates, the migration ability of heavy metals may be decreased instead, which would lead to the precipitation of DOM as well as metals [59]. Some studies have also considered that the existence of metals may promote the aggregation of humic substances by decreasing the intermolecular repulsion [60]. The presence of such processes might be also influenced by salinity, as it was found in RDA analysis that EC and Sal in surface water had a significantly positive correlation with three components of sediments. Therefore, heavy metals in overlying water may enhance the accumulation of DOM in sediments. Heavy metals, especially Fe and Mn, with relatively higher concentration levels, were not significantly correlated with FI and BIX, because heavy metals in urban streams were mainly from allochthonous sources (point and non-point water pollution), while the sources of DOM included allochthonous and autochthonous inputs. In addition, the relatively weak negative correlation between Fe/Mn and HIX indicates that heavy metals have a slight negative influence on the humification process, i.e., the biological transformation process. Overall, the pollution by heavy metals was not serious and interactions between heavy metals and DOM were not obvious in urban streams in Wuhan.

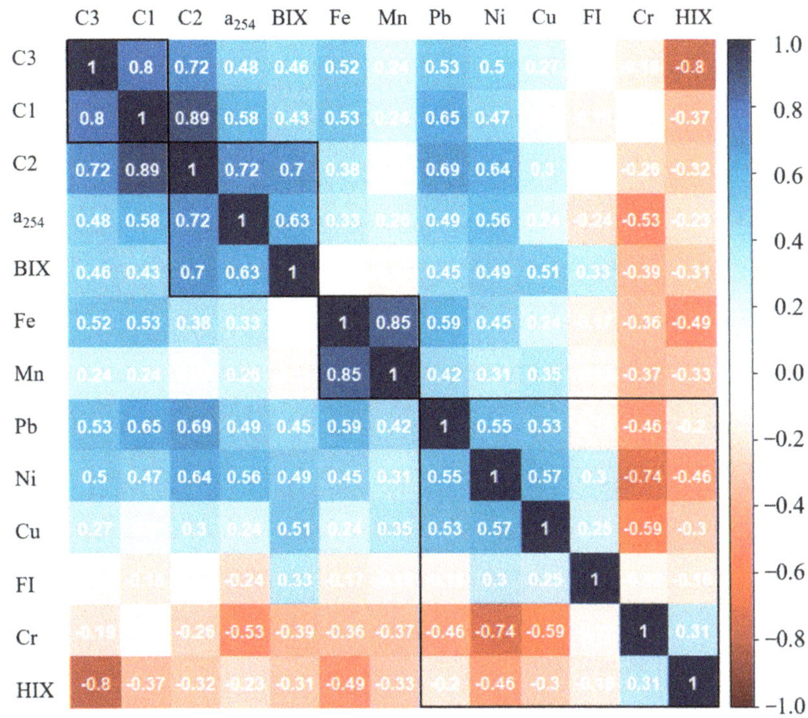

Figure 8. Spearman's correlation analysis results between heavy metals and (C)DOM optical characteristics of sediments. Blue color means positive correlations and the red refers to negative correlations.

4. Conclusions

In this study, the biogeochemical characteristics of waterbodies and (C)DOM of sediments in urban streams in Wuhan have been investigated. The results showed that CDOM and DOM concentrations of sediments in urban streams ranged widely and were at relatively high levels, with a_{254} (116.32 $\pm$ 60.5 m^{-1}) and F_T (2.51 $\pm$ 0.97 R.U.) nearly twice and four times higher than the reference (51.52 m^{-1}, 0.65 R.U.), respectively. The fluorescence components of sediments in urban streams identified by the EEM–PARAFAC model included humic-like acid (C1), fulvic-like acid (C2) and protein-like substance (C3). The calculated results of FI, BIX and HIX demonstrated that DOM of sediments in urban streams had allochthonous inputs and autochthonous production simultaneously and a high degree of humification. The relative abundance of humus-like acid (66.81%) was much higher than that of protein-like matter (33.19%), indicating a higher allochthonous input. Comparison of the urban and reference samples revealed that urbanization played a remarkable role in the increase of CDOM and DOM contents and the variation of DOM components and sources. RDA and Spearman's correlation analysis showed that DOM optical characteristics of sediments in urban streams were associated with most heavy metals and other water quality parameters. This study could support the use of the EEM–PARAFAC method to identify diverse sources of DOM, and it promotes a better understanding of DOM characteristics of sediments in urban streams and their relationship with biogeochemical parameters.

Author Contributions: Conceptualization, X.Z.; methodology, J.X. and Z.L.; software, H.Z.; validation, H.Z. and S.T.; formal analysis, H.Z. and Z.L.; investigation, H.Z., Z.L. and S.T.; resources, J.X. and J.Y.; writing—original draft preparation, H.Z.; writing—review and editing, S.T. and X.Z.; visualization, H.Z. and S.T.; supervision, S.T. and X.Z.; project administration, J.X., J.Y. and X.Z.; funding acquisition, J.X., J.Y. and X.Z. All authors have read and agreed to the published version of the manuscript.

Funding: This research was funded by the Strategic Priority Research Program of the Chinese Academy of Sciences (No. XDA23040302), the National Natural Science Foundation of China (No. 41890823) and the Key Research and Development Project of Hubei Province (No. 2021BCA128).

Institutional Review Board Statement: Not applicable.

Informed Consent Statement: Not applicable.

Data Availability Statement: The data that support the findings of this study are available from the authors upon reasonable request.

Acknowledgments: Comments from the anonymous reviewers are appreciated.

Conflicts of Interest: The authors declare no conflict of interest.

References

1. Zhang, Y.; Zhang, E.; Yin, Y.; Van Dijk, M.A.; Feng, L.; Shi, Z.; Liu, M.; Qin, B. Characteristics and Sources of Chromophoric Dissolved Organic Matter in Lakes of the Yungui Plateau, China, Differing in Trophic State and Altitude. *Limnol. Oceanogr.* **2010**, *55*, 2645–2659. [CrossRef]
2. Schmidt, F.; Koch, B.P.; Goldhammer, T.; Elvert, M.; Witt, M.; Lin, Y.S.; Wendt, J.; Zabel, M.; Heuer, V.B.; Hinrichs, K.U. Unraveling Signatures of Biogeochemical Processes and the Depositional Setting in the Molecular Composition of Pore Water DOM across Different Marine Environments. *Geochim. Cosmochim. Acta* **2017**, *207*, 57–80. [CrossRef]
3. Para, J.; Coble, P.G.; Charrière, B.; Tedetti, M.; Fontana, C.; Sempéré, R. Fluorescence and Absorption Properties of Chromophoric Dissolved Organic Matter (CDOM) in Coastal Surface Waters of the Northwestern Mediterranean Sea, Influence of the Rhône River. *Biogeosciences* **2010**, *7*, 4083–4103. [CrossRef]
4. Coble, P.G. Marine Optical Biogeochemistry: The Chemistry of Ocean Color. *Chem. Rev.* **2007**, *107*, 402–418. [CrossRef] [PubMed]
5. Lyu, L.; Liu, G.; Shang, Y.; Wen, Z.; Hou, J.; Song, K. Characterization of Dissolved Organic Matter (DOM) in an Urbanized Watershed Using Spectroscopic Analysis. *Chemosphere* **2021**, *277*, 130210. [CrossRef] [PubMed]
6. Stedmon, C.A.; Thomas, D.N.; Granskog, M.; Kaartokallio, H.; Papadimitriou, S.; Kuosa, H. Characteristics of Dissolved Organic Matter in Baltic Coastal Sea Ice: Allochthonous or Autochthonous Origins? *Environ. Sci. Technol.* **2007**, *41*, 7273–7279. [CrossRef]
7. Bertilsson, S.; Jones, J.B. Supply of Dissolved Organic Matter to Aquatic Ecosystems: Autochthonous Sources. In *Aquatic Ecosystems: Interactivity of Dissolved Organic Matter*; Academic Press: New York, NY, USA, 2003; pp. 3–24. [CrossRef]
8. Pagano, T.; Bida, M.; Kenny, J.E. Trends in Levels of Allochthonous Dissolved Organic Carbon in Natural Water: A Review of Potential Mechanisms under a Changing Climate. *Water* **2014**, *6*, 2862–2897. [CrossRef]
9. Fellman, J.B.; Petrone, K.C.; Grierson, P.F. Source, Biogeochemical Cycling, and Fluorescence Characteristics of Dissolved Organic Matter in an Agro-Urban Estuary. *Limnol. Oceanogr.* **2011**, *56*, 243–256. [CrossRef]
10. Khan, A.L.; Sokol, E.R.; McKnight, D.M.; Saunders, J.F.; Hohner, A.K.; Summers, R.S. Phytoplankton Drivers of Dissolved Organic Material Production in Colorado Reservoirs and the Formation of Disinfection By-Products. *Front. Environ. Sci.* **2021**, *9*, 1–13. [CrossRef]
11. Kothawala, D.N.; Stedmon, C.A.; Müller, R.A.; Weyhenmeyer, G.A.; Köhler, S.J.; Tranvik, L.J. Controls of Dissolved Organic Matter Quality: Evidence from a Large-Scale Boreal Lake Survey. *Glob. Chang. Biol.* **2014**, *20*, 1101–1114. [CrossRef] [PubMed]
12. Song, K.; Shang, Y.; Wen, Z.; Jacinthe, P.A.; Liu, G.; Lyu, L.; Fang, C. Characterization of CDOM in Saline and Freshwater Lakes across China Using Spectroscopic Analysis. *Water Res.* **2019**, *150*, 403–417. [CrossRef]
13. Singh, S.; D'Sa, E.J.; Swenson, E.M. Chromophoric Dissolved Organic Matter (CDOM) Variability in Barataria Basin Using Excitation-Emission Matrix (EEM) Fluorescence and Parallel Factor Analysis (PARAFAC). *Sci. Total Environ.* **2010**, *408*, 3211–3222. [CrossRef] [PubMed]
14. Chen, Q.; Mei, K.; Dahlgren, R.A.; Wang, T.; Gong, J.; Zhang, M. Impacts of Land Use and Population Density on Seasonal Surface Water Quality Using a Modified Geographically Weighted Regression. *Sci. Total Environ.* **2016**, *572*, 450–466. [CrossRef]
15. Yi, L.; Qiulin, L.; Kun, Y.; Wenqing, X.; Xiaolu, Z.; Chunxue, S.; Yuanting, X.; Yan, Z.; Chao, Z. Thermodynamic Analysis of Air-Ground and Water-Ground Energy Exchange Process in Urban Space at Micro Scale. *Sci. Total Environ.* **2019**, *694*, 133612. [CrossRef]
16. Aguilar-Alarcón, P.; Gonzalez, S.V.; Simonsen, M.A.; Borrero-Santiago, A.R.; Sanchís, J.; Meriac, A.; Kolarevic, J.; Asimakopoulos, A.G.; Mikkelsen, Ø. Characterizing Changes of Dissolved Organic Matter Composition with the Use of Distinct Feeds in Recirculating Aquaculture Systems via High-Resolution Mass Spectrometry. *Sci. Total Environ.* **2020**, *749*, 142326. [CrossRef]

17. Lambert, T.; Perolo, P.; Escoffier, N.; Perga, M.E. Enhanced Bioavailability of Dissolved Organic Matter (DOM) in Human-Disturbed Streams in Alpine Fluvial Networks. *Biogeosciences* **2022**, *19*, 187–200. [CrossRef]
18. Kikuchi, T.; Fujii, M.; Terao, K.; Jiwei, R.; Lee, Y.P.; Yoshimura, C. Correlations between Aromaticity of Dissolved Organic Matter and Trace Metal Concentrations in Natural and Effluent Waters: A Case Study in the Sagami River Basin, Japan. *Sci. Total Environ.* **2017**, *576*, 36–45. [CrossRef] [PubMed]
19. Zhou, Y.; Zhao, C.; He, C.; Li, P.; Wang, Y.; Pang, Y.; Shi, Q.; He, D. Characterization of Dissolved Organic Matter Processing between Surface Sediment Porewater and Overlying Bottom Water in the Yangtze River Estuary. *Water Res.* **2022**, *215*, 118260. [CrossRef] [PubMed]
20. Su, L.; Wang, S.; Ji, R.; Zhuo, G.; Liu, C.; Chen, M.; Li, H.; Zhang, L. New Insight into the Role of FDOM in Heavy Metal Leaching Behavior from MSWI Bottom Ash during Accelerated Weathering Using Fluorescence EEM-PARAFAC. *Waste Manag.* **2022**, *144*, 153–162. [CrossRef] [PubMed]
21. Dou, Y.; Yu, X.; Liu, L.; Ning, Y.; Bi, X.; Liu, J. Effects of Hydrological Connectivity Project on Heavy Metals in Wuhan Urban Lakes on the Time Scale. *Sci. Total Environ.* **2022**, *853*, 158654. [CrossRef]
22. Tao, S.; Zhang, X.; Xu, J.; Pan, G.; Gu, F. Anthropogenic Impacts on Isotopic and Geochemical Characteristics of Urban Streams: A Case Study in Wuhan, China. *Environ. Sci. Pollut. Res.* **2021**, *28*, 39186–39198. [CrossRef] [PubMed]
23. Ding, L.; Chen, K.L.; Cheng, S.G.; Wang, X. Water Ecological Carrying Capacity of Urban Lakes in the Context of Rapid Urbanization: A Case Study of East Lake in Wuhan. *Phys. Chem. Earth* **2015**, *89–90*, 104–113. [CrossRef]
24. Wang, F.; Xia, X.; Sha, Y. Distribution of Phthalic Acid Esters in Wuhan Section of the Yangtze River, China. *J. Hazard. Mater.* **2008**, *154*, 317–324. [CrossRef]
25. Du, N.; Ottens, H.; Sliuzas, R. Spatial Impact of Urban Expansion on Surface Water Bodies-A Case Study of Wuhan, China. *Landsc. Urban Plan.* **2010**, *94*, 175–185. [CrossRef]
26. Zhang, J.; Lai, J.J.; He, Z.G.; Deng, C.F.; Ying, X.L. *Wuhan Statistical Yearbook*; China Statistics Press: Beijing, China, 2021.
27. Helms, J.R.; Aron, S.; Ritchie, J.D.; Minor, E.C. Absorption Spectral Slopes and Slope Ratios as Indicators of Molecular Weight, Source, and Photobleaching of Chromophoric Dissolved Organic Matter. *Limnol. Oceanogr.* **2008**, *53*, 955–969. [CrossRef]
28. Fichot, C.G.; Benner, R. The Spectral Slope Coefficient of Chromophoric Dissolved Organic Matter ($S_{275-295}$) as a Tracer of Terrigenous Dissolved Organic Carbon in River-Influenced Ocean Margins. *Limnol. Oceanogr.* **2012**, *57*, 1453–1466. [CrossRef]
29. Spencer, R.G.M.; Butler, K.D.; Aiken, G.R. Dissolved Organic Carbon and Chromophoric Dissolved Organic Matter Properties of Rivers in the USA. *J. Geophys. Res. Biogeosci.* **2012**, *117*, G03001. [CrossRef]
30. Wang, J.; Xu, J.; Xia, J.; Wu, F.; Zhang, Y. A Kinetic Study of Concurrent Arsenic Adsorption and Phosphorus Release during Sediment Resuspension. *Chem. Geol.* **2018**, *495*, 67–75. [CrossRef]
31. Mcknight, D.M.; Boyer, E.W.; Westerhoff, P.K.; Doran, P.T.; Kulbe, T.; Andersen, D.T. Spectrofluorometric Characterization of Dissolved Organic Matter for Indication of Precursor Organic Material and Aromaticity. *Limnol. Oceanogr.* **2001**, *46*, 38–48. [CrossRef]
32. Korak, J.A.; Dotson, A.D.; Summers, R.S.; Rosario-Ortiz, F.L. Critical Analysis of Commonly Used Fluorescence Metrics to Characterize Dissolved Organic Matter. *Water Res.* **2014**, *49*, 327–338. [CrossRef] [PubMed]
33. Ohno, T. Fluorescence Inner-Filtering Correction for Determining the Humification Index of Dissolved Organic Matter. *Environ. Sci. Technol.* **2002**, *36*, 742–746. [CrossRef] [PubMed]
34. Clark, M.G.; Biagi, K.M.; Carey, S.K. Optical Properties of Dissolved Organic Matter Highlight Peatland-like Properties in a Constructed Wetland. *Sci. Total Environ.* **2022**, *802*, 149770. [CrossRef] [PubMed]
35. Shirina, M.; Jang, I.; Lee, J.; Bin, H.; Jin, H.; Park, J. Synergistic Effects of Urban Tributary Mixing on Dissolved Organic Matter Biodegradation in an Impounded River System. *Sci. Total Environ.* **2019**, *676*, 105–119. [CrossRef]
36. Speetjens, N.J.; Tanski, G.; Martin, V.; Wagner, J.; Richter, A.; Hugelius, G.; Boucher, C.; Lodi, R.; Knoblauch, C.; Koch, B.P.; et al. Dissolved Organic Matter Characterization in Soils and Streams in a Small Coastal Low-Arctic Catchment. *Biogeosciences* **2022**, *19*, 3073–3097. [CrossRef]
37. Yang, X.; Jiang, Y.; Deng, X.; Zheng, Y.; Yue, Z. Temporal and Spatial Variations of Chlorophyll a Concentration and Eutrophication Assessment (1987–2018) of Donghu Lake in Wuhan Using Landsat Images. *Water* **2020**, *12*, 2192. [CrossRef]
38. Tang, J.; Li, X.; Cao, C.; Lin, M.; Qiu, Q.; Xu, Y.; Ren, Y. Compositional Variety of Dissolved Organic Matter and Its Correlation with Water Quality in Peri-Urban and Urban River Watersheds. *Ecol. Indic.* **2019**, *104*, 459–469. [CrossRef]
39. Mladenov, N.; Parsons, D.; Kinoshita, A.M.; Pinongcos, F.; Mueller, M.; Garcia, D.; Lipson, D.A.; Grijalva, L.M.; Zink, T.A. Groundwater-Surface Water Interactions and Flux of Organic Matter and Nutrients in an Urban, Mediterranean Stream. *Sci. Total Environ.* **2022**, *811*, 152379. [CrossRef]
40. Li, X.; Yu, Y.; Fan, H.; Tang, C. Intense Denitrification and Sewage Effluent Result in Enriched ^{15}N in N_2O from Urban Polluted Rivers. *J. Hydrol.* **2022**, *608*, 127631. [CrossRef]
41. Islam, M.S.; Ahmed, M.K.; Raknuzzaman, M.; Habibullah-Al-Mamun, M.; Islam, M.K. Heavy Metal Pollution in Surface Water and Sediment: A Preliminary Assessment of an Urban River in a Developing Country. *Ecol. Indic.* **2015**, *48*, 282–291. [CrossRef]
42. Pretsch, E.; Bühlmann, P.; Badertscher, M. *Structure Determination of Organic Compounds: Tables of Spectral Data*; Springer: Berlin/Heidelberg, Germany, 2009; ISBN 9783540938095.
43. Xu, H.; Guo, L. Molecular Size-Dependent Abundance and Composition of Dissolved Organic Matter in River, Lake and Sea Waters. *Water Res.* **2017**, *117*, 115–126. [CrossRef]

44. De Stefano, L.G.; Valdivia, A.S.; Gianello, D.; Gerea, M.; Reissig, M.; García, P.E.; García, R.D.; Soto Cárdenas, C.; Diéguez, M.C.; Queimaliños, C.P.; et al. Using CDOM Spectral Shape Information to Improve the Estimation of DOC Concentration in Inland Waters: A Case Study of Andean Patagonian Lakes. *Sci. Total Environ.* **2022**, *824*, 153752. [CrossRef] [PubMed]
45. Stedmon, C.A.; Markager, S.; Bro, R. Tracing Dissolved Organic Matter in Aquatic Environments Using a New Approach to Fluorescence Spectroscopy. *Mar. Chem.* **2003**, *82*, 239–254. [CrossRef]
46. Xu, H.; Guo, L. Intriguing Changes in Molecular Size and Composition of Dissolved Organic Matter Induced by Microbial Degradation and Self-Assembly. *Water Res.* **2018**, *135*, 187–194. [CrossRef] [PubMed]
47. Yu, H.; Song, Y.; Tu, X.; Du, E.; Liu, R.; Peng, J. Assessing Removal Efficiency of Dissolved Organic Matter in Wastewater Treatment Using Fluorescence Excitation Emission Matrices with Parallel Factor Analysis and Second Derivative Synchronous Fluorescence. *Bioresour. Technol.* **2013**, *144*, 595–601. [CrossRef]
48. Romero González-Quijano, C.; Herrero Ortega, S.; Casper, P.; Gessner, M.O.; Singer, G.A. Dissolved Organic Matter Signatures in Urban Surface Waters: Spatio-Temporal Patterns and Drivers. *Biogeosciences* **2022**, *19*, 2841–2853. [CrossRef]
49. Skoog, A.; Wedborg, M.; Fogelqvist, E. Decoupling of Total Organic Carbon Concentrations and Humic Substance Fluorescence in a an Extended Temperate Estuary. *Mar. Chem.* **2011**, *124*, 68–77. [CrossRef]
50. Sankar, M.S.; Dash, P.; Singh, S.; Lu, Y.H.; Mercer, A.E.; Chen, S. Effect of Photo-Biodegradation and Biodegradation on the Biogeochemical Cycling of Dissolved Organic Matter across Diverse Surface Water Bodies. *J. Environ. Sci.* **2019**, *77*, 130–147. [CrossRef]
51. Coble, A.A.; Wymore, A.S.; Potter, J.D.; McDowell, W.H. Land Use Overrides Stream Order and Season in Driving Dissolved Organic Matter Dynamics Throughout the Year in a River Network. *Environ. Sci. Technol.* **2022**, *56*, 2009–2020. [CrossRef]
52. Zhang, H.; Cui, K.; Guo, Z.; Li, X.; Chen, J.; Qi, Z.; Xu, S. Spatiotemporal Variations of Spectral Characteristics of Dissolved Organic Matter in River Flowing into a Key Drinking Water Source in China. *Sci. Total Environ.* **2020**, *700*, 134360. [CrossRef]
53. Yamashita, Y.; Hashihama, F.; Saito, H.; Fukuda, H.; Ogawa, H. Factors Controlling the Geographical Distribution of Fluorescent Dissolved Organic Matter in the Surface Waters of the Pacific Ocean. *Limnol. Oceanogr.* **2017**, *62*, 2360–2374. [CrossRef]
54. Boguta, P.; D'Orazio, V.; Senesi, N.; Sokołowska, Z.; Szewczuk-Karpisz, K. Insight into the Interaction Mechanism of Iron Ions with Soil Humic Acids. The Effect of the PH and Chemical Properties of Humic Acids. *J. Environ. Manag.* **2019**, *245*, 367–374. [CrossRef] [PubMed]
55. McNeill, K.; Canonica, S. Triplet State Dissolved Organic Matter in Aquatic Photochemistry: Reaction Mechanisms, Substrate Scope, and Photophysical Properties. *Environ. Sci. Process. Impacts* **2016**, *18*, 1381–1399. [CrossRef] [PubMed]
56. Rikta, S.Y.; Tareq, S.M.; Uddin, M.K. Toxic Metals (Ni^{2+}, Pb^{2+}, Hg^{2+}) Binding Affinity of Dissolved Organic Matter (DOM) Derived from Different Ages Municipal Landfill Leachate. *Appl. Water Sci.* **2018**, *8*, 1–8. [CrossRef]
57. Blesa, M.; Borghi, E.; Maroto, A.; Regazzoni, A. Adsorption of EDTA and Iron-EDTA Complexes on Magnetite and the Mechanism of Dissolution of Magnetite by EDTA. *J. Colloid Interface Sci.* **1984**, *98*, 295–305. [CrossRef]
58. Bedsworth, W.W.; Sedlak, D.L. Sources and Environmental Fate of Strongly Complexed Nickel in Estuarine Waters: The Role of Ethylenediaminetetraacetate. *Environ. Sci. Technol.* **1999**, *33*, 926–931. [CrossRef]
59. Fan, T.; Yao, X.; Ren, H.; Liu, L.; Deng, H.; Shao, K. Regional-Scale Investigation of the Molecular Weight Distribution and Metal-Binding Behavior of Dissolved Organic Matter from a Shallow Macrophytic Lake Using Multispectral Techniques. *J. Hazard. Mater.* **2022**, *439*, 129532. [CrossRef]
60. Bryan, N.D.; Jones, M.N.; Birkett, J.; Livens, F.R. Aggregation of Humic Substances by Metal Ions Measured by Ultracentrifugation. *Anal. Chim. Acta* **2001**, *437*, 291–308. [CrossRef]

water

MDPI

Article

Analysis of the Runoff Component Variation Mechanisms in the Cold Region of Northeastern China under Climate Change

Shuiqing Liu [1], Zuhao Zhou [1,*], Jiajia Liu [1], Jia Li [2], Pengxiang Wang [3], Cuimei Li [4,*], Xinmin Xie [1], Yangwen Jia [1] and Hao Wang [1]

1 State Key Laboratory of Simulation and Regulation of Water Cycle in River Basins, China Institute of Water Resources and Hydropower Research, Beijing 100038, China
2 Planning, Design and Administration Bureau of South-to-North Water Diversion Project, Ministry of Water Resources, Beijing 100038, China
3 China Three Gorges Corporation, Beijing 100038, China
4 The School of Environmental Science and Engineering, Suzhou University of Science and Technology, Suzhou 215009, China
* Correspondence: zhzh@iwhr.com (Z.Z.); cuimeili@163.com (C.L.); Tel.: +86-138-1156-3728 (Z.Z. & C.L.); Fax: +86-010-6878-5625 (Z.Z. & C.L.)

Citation: Liu, S.; Zhou, Z.; Liu, J.; Li, J.; Wang, P.; Li, C.; Xie, X.; Jia, Y.; Wang, H. Analysis of the Runoff Component Variation Mechanisms in the Cold Region of Northeastern China under Climate Change. *Water* **2022**, *14*, 3170. https://doi.org/10.3390/w14193170

Academic Editor: Alexander Shiklomanov

Received: 11 September 2022
Accepted: 5 October 2022
Published: 8 October 2022

Abstract: Climate change alters hydrological processes in cold regions. However, the mechanisms of runoff component variation remain obscure. We implemented a WEP-N model to estimate monthly runoff in the Songhua River Basin (SRB) between 1956 and 2018. All flow simulations were accurate (NSE > 0.75 and RE < 5%). The annual runoff was attenuated in 1998, and the hydrological series (1956–2018) was divided into base and change periods in that year. Relative to the BS (base scenario), annual production flow reduction was −28.2% under climate change and water use. A multifactor attribution analysis showed that climate change and water use contributed 77.0% and 23.0% to annual runoff reduction, respectively. Decreases in annual surface and base flow explained 62.1% and 35.7% of annual production flow reduction, respectively. The base flow increased by 8.5% and 6.5% during the freezing and thawing periods, respectively. Relative to the BS, groundwater recharge increased by 9.2% and 4.1% during the freezing and thawing periods, respectively, under climate change conditions. Climate change was the dominant factor attenuating production flow. The change in production flow occurred mainly during the non-freeze-thaw period. The decrease in total production flow in the SRB was caused mainly by the decrease in the surface flow, where the reduction in base flow accounted for a relatively small proportion. Production flow attenuation aggravated water shortages. The utilization rate of groundwater resources is far below the internationally recognized alarm line. Therefore, attention should be directed towards certain areas of the SRB and other regions with minimal groundwater exploitation.

Keywords: precipitation; temperature; production flow component; Songhua River Basin; runoff variation; WEP-N model

1. Introduction

Global climate change and human activity significantly affect the hydrological cycle in cold regions [1–4]. Trends in runoff and the factors that influence them regionally include runoff trends that are positive in certain cold areas. Runoff may be influenced by glacier melt in response to increasing temperature. This phenomenon occurs in the Upper Khovd River Basin of Central Asia [5,6]. Runoff may also be influenced by increasing precipitation. This process occurs in most of Russia and on the south slope of the Altai Mountains in Northwestern China [5,7]. Runoff may also be influenced by the increases in precipitation, glacial meltwater, and permanently frozen soil meltwater that occur with rising temperatures. These effects are observed in the Yangtze River source region, the Nagqu River Basin in the southern part of the Qinghai-Tibet Plateau, the north and south slopes of the Tianshan Mountains, and the north slope of Qilian Mountain in Northwestern China [4,5,8–10].

Runoff has been decreasing in response to climate change and human activity in the alpine mountains of Northern Eurasia, Central Asia, South Africa, South America, and elsewhere. Runoff has also decreased as a result of high water consumption in areas such as the inland rivers downstream of the Buqtyrma River Basin in Central Asia [11] and the midstream region of the Heihe River Basin in the arid inland river basins of Northwestern China [12]. Runoff into the rivers of cold regions has been decreasing in response to climate change. Some of these rivers were affected mainly by precipitation. They include the rivers of Mount Kilimanjaro in Tanzania, the tropical Andes of South America, and the south slope of the Altai Mountains. In these cases, the proportions of runoff recharge from glacial meltwater are small [5,13–17]. Certain rivers are affected mainly by temperature, such as those in the Northern Rocky Mountains, where flow attenuation may be caused by a reduction in snowpack accumulation at lower altitudes [18,19]. Certain rivers are affected by both temperature and precipitation, such as those in Northwestern China, the northern part of the Qinghai-Tibet Plateau, and the Songhua River Basin (SRB) in Northeastern China [4,5,20,21]. In regions with few glaciers and minimal rainfall, glacial meltwater decreases and evaporation increases with rising temperature. This phenomenon occurs in the headwater region of the Manas River, the north slopes of the Qilian and Kunlun Mountains, the south slope of the Tianshan Mountains in Northwestern China, and the source regions of the Yellow River Basin in the northern part of the Qinghai-Tibet Plateau [4,5,22,23].

Earlier studies [11,12] have investigated variations in runoff in response to climate change and water use in cold regions. Nevertheless, the mechanisms of runoff component variation are poorly understood. The present study used the Songhua River Basin (SRB) in Northeastern China as an example. We analyzed the mechanisms of runoff component variation during the annual freezing, thawing, and non-freeze-thaw periods of the year based on simulations of the soil freeze-thaw and water cycle processes.

2. Materials and Methods

2.1. Study Area

The SRB is located in Northeastern China between 41°42′–51°38′ N and 119°52′–132°31′ E (Figure 1). Its elevation is in the range of ~50–2700 m. The SRB covers 557,000 km^2 and spans Heilongjiang, Jilin, Liaoning, and Inner Mongolia Provinces. The Songhua River is the main tributary of Heilongjiang, and it has two sources to the north and the south. The Nenjiang River to the north originates from Yilehuli Mountain, which is a branch of the Daxing'an Mountains. The second source to the south originates from Tianchi Lake in Changbai Mountain. Precipitation dominates the hydroclimatologic regime of the area, and there is no glaciation. The SRB is characterized by seasonally frozen soil. The maximum freezing depth is >200 cm in the basin. The longest freezing period is from October to July of the following year [24]. The average annual precipitation, temperature, and runoff in the SRB were 533.18 mm, 2.95 °C, and 629.9 billion m^3, respectively, between 1956 and 2018.

2.2. Data Collection

Several parameters could be directly measured for each basin. (a) The digital elevation method (DEM) was implemented at an accuracy of 30 m. (b) Daily precipitation, temperature, humidity, wind speed, and sunshine hours were compiled for 51 national meteorological stations in the Songhua River Basin. These meteorological data were released by the National Meteorological Information Center (2019). (c) Land use data for 1990, 2000, and 2005 were measured at 30-m resolution and provided by the Institute of Geography of the Chinese Academy of Sciences, Xinjiang, China. (d) Soils and their characteristic properties were derived from the Second National Soil Census (NSCO, 1979). (e) Water use data for 2000–2018 were acquired from the SRB Water Resources Bulletin (http://www.slwr.gov.cn/, (10 July 2022)). Water use data for 1956–1999 were extrapolated from the data for population (1956–2018), irrigated area (1956–2018), gross domestic product (GDP) (1956–2018), and water use (1980–2018). (f) Population, irrigated area, and GDP data for the SRB (1956–2018) were obtained from the Statistical Yearbooks of Heilongjiang, Jilin, Liaoning, and Inner Mongolia Provinces (http://tjj.hlj.gov.cn/tjsj, (10 July 2022), http://tjj.jl.gov.cn/tjsj/tjnj/, (10 July 2022), http://tjj.ln.gov.cn/tjsj/sjcx/ndsj, (10 July 2022), http://tj.nmg.gov.cn/tjyw/jpsj/, (10 July 2022)). (g) Groundwater resources data for 1980–2018 were acquired from the SRB Water Resources Bulletin (http://www.slwr.gov.cn/, (10 July 2022)).

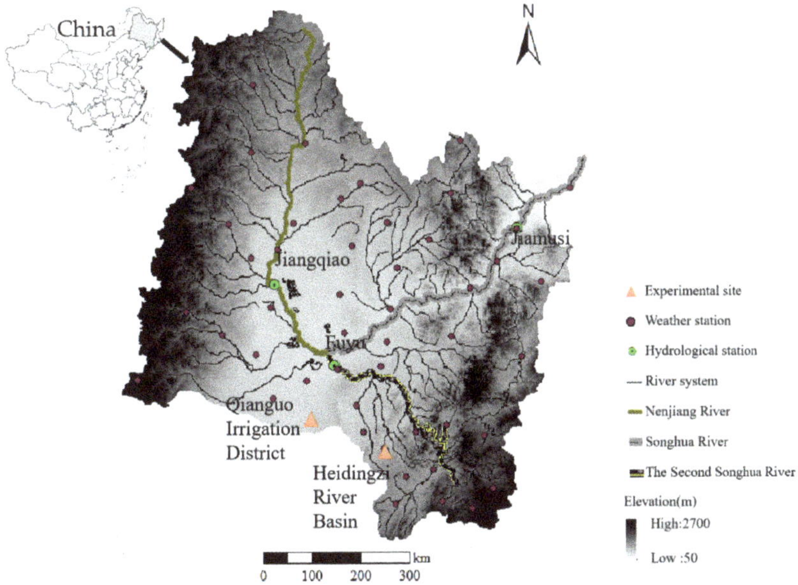

Figure 1. Overview of the SRB.

The model was verified using measured flow data for the Jiangqiao, Fuyu (1956–2000, 2006–2018), and Jiamusi (1956–2018) Stations.

2.3. Research Method

The Mann–Kendall trend and Pettitt mutation analyses [25,26] were used to identifying the trends and detect abrupt changes in the measured runoff. The distributed dualistic water cycle model WEP-N was used to simulate the water cycle process in the SRB. WEP-N is driven by multiple factors. A multifactor attribution analysis [27] was used to determine water use and climate change contribution rates to runoff variation in the SRB.

2.3.1. Principles of the WEP-N Model Hydrological Cycle

The Water and Energy Transfer Processes and Nitrogen Cycle Processes Model in cold regions (WEP-N) [28] was developed based on the distributed hydrological model in cold regions known as WEP-COR, which couples simulations of natural hydrological and water use processes [24]. The model considered the influences of meteorology, underlying surfaces, and human activity on the water cycle process. The social water cycle simulation included water storage, intake, delivery, use, consumption, and drainage. Water intake and drainage connect the natural and social water cycles.

The model was calculated using the contour bands inside sub-watersheds. Each contour band was divided according to the land use of the underlying surface, namely, water body, impervious area, soil vegetation, irrigated farmland, or non-irrigated farmland [24,29]. The runoff was calculated based on the proportions of the underlying surface area for each group. Each unit included seven vertical layers. From top to bottom, they were: vegetation canopy or building interception layer, surface depression storage layer, root zone comprised of three layers, transition zone layer, and groundwater layer. The soil below the ground was divided into 11 layers for the soil water and heat coupling calculations. Simulation of the natural water cycle process in the watershed included runoff generation and confluence and comprised the water cycle processes in the surface soil, soil, groundwater, slope ditch, and river channel. The water cycle in the surface soil included precipitation, snow melting, evaporation, infiltration, and surface runoff. The snow melting process was calculated by the temperature-index method [30]. The evapotranspiration values of the surface soil, soil, and water were calculated using the Penman formula [31]. Evaporation from the vegetation canopy was calculated using the Penman–Monteith formula [31]. Infiltration was calculated on the basis of the rainfall intensity and using the Green–Ampt model or the Richards equation [32]. The surface runoff was calculated by the Hortonian or saturation overland flow theory when the precipitation intensity exceeded the infiltration capacity [33]. The soil hydrothermal cycle included heat and water transport during the freezing and thawing periods. Soil heat transport was determined using the basic one-dimensional vertical heat flow movement equation (Equation (1)). Soil water transport was determined using the one-dimensional vertical water flow equation (Equation (2)). The movement of liquid water in the soil was driven by the soil water potential, including pressure, gravity, temperature, and solute potentials (Equations (4)–(6)). The relationship between water and heat transport in frozen soil was characterized as the dynamic balance between the moisture content of the unfrozen water and the negative temperature of the soil. Groundwater movement was calculated with the Boussinesq equation [34]. The exchange between river and groundwater was calculated by Darcy's law and was based on the differences in the water level and the characteristics of the riverbed material [35]. Soil freezing would hinder the exchange between the groundwater and river channels when the temperature exceeds a certain critical value during the freeze-thaw period. The overland confluence was calculated by the kinematic wave method and from the uppermost to the lowermost contour zone of each sub-watershed [36]. Each river channel confluence was calculated using one-dimensional motion waves from upstream to downstream according to the elevation, slope, and Manning roughness coefficient of each contour zone [24]. The influences of temperature on ice formation and river melting were considered.

The temperature difference between the atmosphere and the surface was the heat conduction source. The surface temperature was determined, and the heat flux and temperature of each layer were calculated as follows [37,38]:

$$\frac{\partial}{\partial z}\left[\lambda_s \frac{\partial T}{\partial z}\right] = C_v \frac{\partial T}{\partial t} - L_i \rho_i \frac{\partial \theta_i}{\partial t} \tag{1}$$

where z is the soil layer thickness (m), T is the temperature of each soil layer (°C), λ_s and C_v are the soil thermal conductivity (W/[m·°C]) and volumetric heat capacity (J/[m^3·°C]), respectively, t is the time (s), and L_i, ρ_i, and θ_i are the latent heat of ice melting (3.35×10^5 J/kg), the ice density (920 kg/m^3), and the volumetric ice content (m^3/m^3) in the soil, respectively.

The soil liquid water movement was calculated using the Richards equation [39].

$$\frac{\partial \theta}{\partial t} = -\frac{\rho_i}{\rho_l}\frac{\partial \theta_i}{\partial t} - \frac{\partial}{\partial z}\left[-K(\theta)\frac{\partial H}{\partial z}\right] \tag{2}$$

where ρ_l and θ are the soil density (kg/m^3) and volumetric water content (m^3/m^3), respectively, $K(\theta)$ is hydraulic conductivity of the unsaturated soil (m/s), and H is the soil water potential (m).

Temperature drives the changes in the water phase. Hydrothermal coupling was calculated as follows:

$$\theta_l = \theta_m(t) \tag{3}$$

where $\theta_m(t)$ is the maximum liquid water content corresponding to a negative soil temperature.

The soil water potential H was calculated as follows [40]:

$$H = z + h_m + h_s + h_t \tag{4}$$

where h_m is the pressure potential (m), h_s is the solute potential (m), and h_t is the temperature potential (m).

The solute potential is a function of temperature and was calculated as follows [40]:

$$h_s = -\frac{cR}{\mu} T_K \tag{5}$$

where c is the mass of a solute in the unit volume of the solution (kg/m^3), R is the molar gas constant (8.3145 J/[mol·k]), μ is the molar mass of the solute (g/mol), and T_k is the thermodynamic temperature (K).

The temperature potential was calculated as follows [41]:

$$h_t = h_m G_{WT} \frac{1}{\gamma_0} \frac{d\gamma}{dt} \tag{6}$$

where G_{WT} is the gain factor (dimensionless), γ_0 (γ_0 = 71.89 g/s^2) is the surface tension of the soil at 25 °C (g/s^2), and γ is the surface tension of the soil (g/s^2). Note that $\gamma = 75.6 - 0.1425T_k - 2.38 \times 10^{-4}{T_k}^2$.

2.3.2. Multifactor Attribution Analysis

Multifactor attribution analysis decomposed the impact contribution according to the fixing-changing method and was calculated as follows [27,42].

$$\Delta X_j = \frac{1}{2^{n-1}} \sum_{i=1}^{2^n} \alpha_{i,j} \times S_i \quad j = 1, \cdots, n \tag{7}$$

$$A = \sum_{j=1}^{n} \Delta X_j = \frac{1}{2^{n-1}} \sum_{i=1}^{2^n} \beta_n^i \times S_i \tag{8}$$

$$\beta_n^i = \sum_{j=1}^{n} \alpha_{i,j} \tag{9}$$

where ΔX_j is the influence contribution of the jth factor, $\alpha_{i,j}$ is the weight coefficient of the jth factor corresponding to the ith scenario, S_i is the simulated result corresponding to scenario i, n is the number of factors considered, A is the sum of the contributions of all factors, and β_n^i is the sum of the weight coefficients of all factors in the ith scenario under the premise of considering n influencing factors.

The contribution rate of each factor to the change in the water cycle elements was calculated with the following equation:

$$\eta_i = \frac{\Delta X_i}{\sum_{j=1}^{n} \Delta X_j} \quad i = 1, \ldots, n \tag{10}$$

3. Results and Discussion

3.1. Model Calibration and Validation

The WEP-N model in the SRB was constructed according to a previously published method [28]. The SRB was partitioned into 9544 sub-basins and 29,488 contour zones based on DEM data and river observations. WEP-COR, the predecessor of the WEP-N model, was verified using the stratified soil temperature and liquid water content of the Qianguo Irrigation District and the monthly mean discharge between 1956 and 2000 at the Jiangqiao,

Fuyu, Jiamusi Stations in the main streams of the Songhua River [24]. The WEP-N model was validated using the stratified soil temperature and liquid water content, the daily discharge based on the hydrothermal coupling experiment, and the river flow monitoring experiment during two freeze-thaw periods (2017–2018 and 2018–2019) in the Heidingzi River Basin [28]. The WEP-N model was calibrated and validated using the discharge measured monthly between 1956 and 2018 at the Jiangqiao, Fuyu, and Jiamusi Stations in the main streams of the Songhua River. Data from the Jiangqiao, Fuyu, and Jiamusi Stations were divided into two parts. The data for the period 1956–1990 were used for calibration, and those for 1991–2018 were used for validation. The results of the monthly mean discharge simulations at the Jiangqiao, Fuyu, and Jiamusi Stations are shown in Figure 2. In general, the WEP-N model performed satisfactorily for the SRB and achieved efficiency coefficients of NSE > 0.75 and RE < 5 % for the validation period (Table 1). The simulated flow was suitable for application in the subsequent analyses.

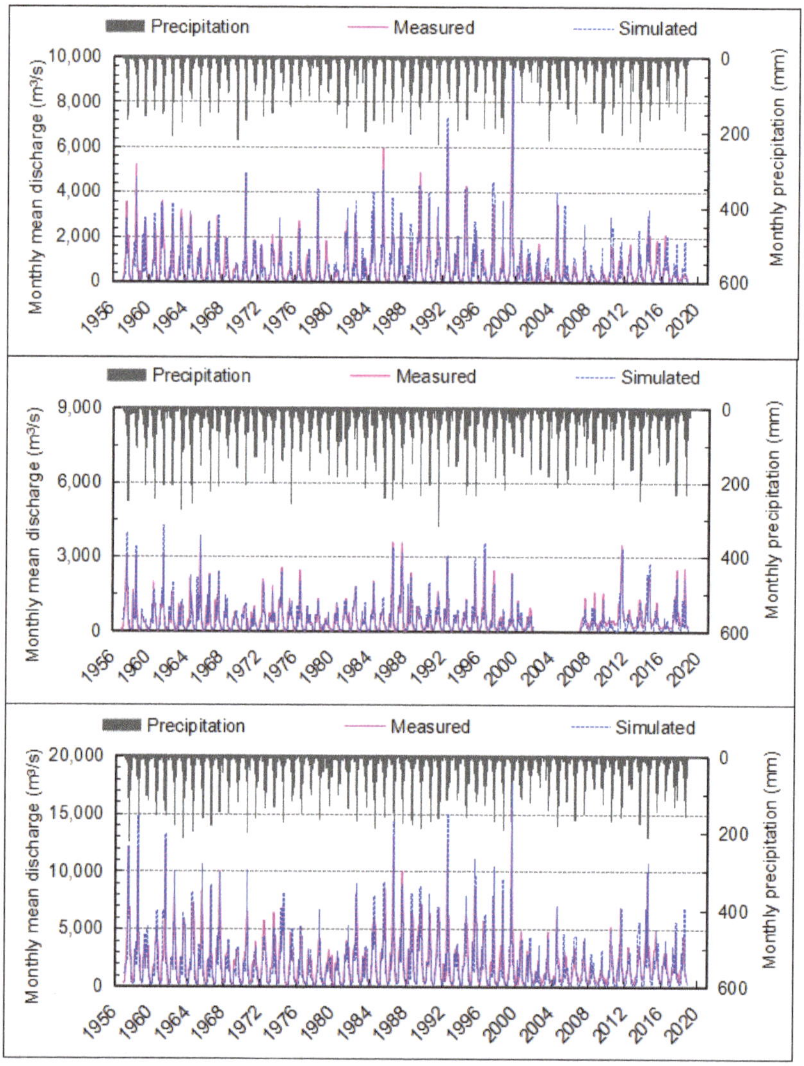

Figure 2. Validation of the WEP-N model at the Jiangqiao, Fuyu, and Jiamusi Stations.

Table 1. Validation of the WEP-N model at the Jiangqiao, Fuyu, and Jiamusi Stations.

Hydrological Site	N_{SE}		$R_E/\%$	
	1956–1990	1991–2018	1956–1990	1991–2018
Jiangqiao	0.80	0.77	4.98	4.85
Fuyu	0.86	0.73	4.26	−0.41
Jiamusi	0.81	0.76	4.51	0.83

3.2. Influence of Climate Change and Water Use on the Annual Runoff Variation in the SRB

The trends in measured annual runoff in the SRB were analyzed by the Mann–Kendall test method and abrupt changes were detected by the Pettitt test method (Figure 3). The measured annual runoff decreased by 20.3 billion in 63 years A significant abrupt change in the measured annual runoff appeared ca. 1998 ($p < 0.01$). To analyze the impact of climate change and water use on the water cycle, the data series were segregated into two periods using the abrupt change in measured annual runoff ca. 1998 as the dividing line. The data for the period 1956–1998 served as the base period, while those for 1999–2018 served as the change period.

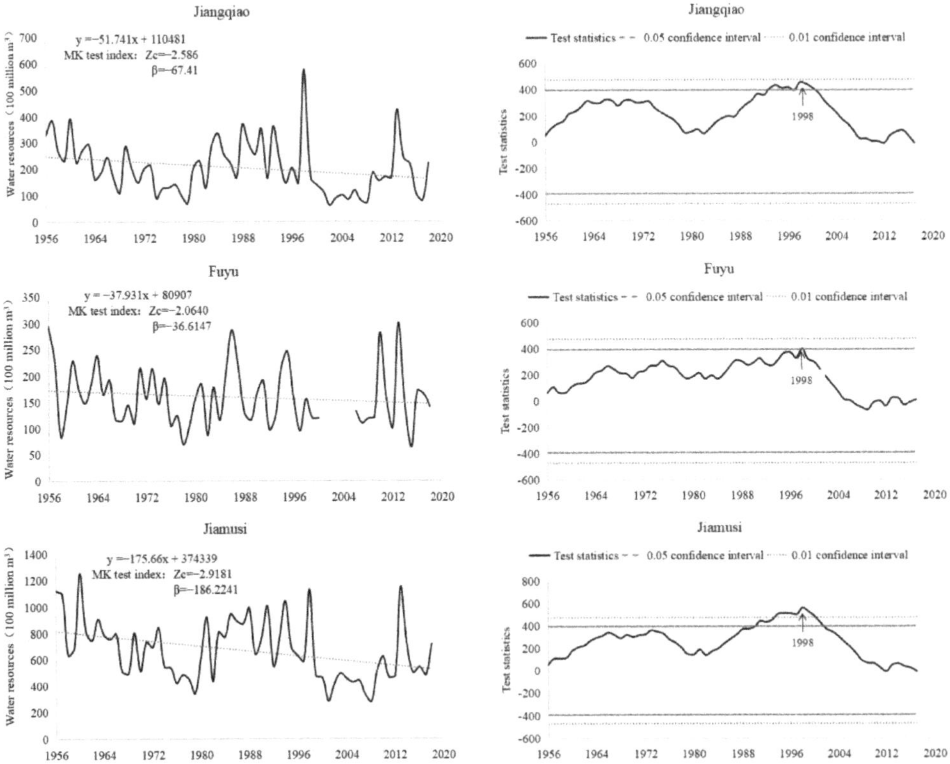

Figure 3. Trend and mutation analyses of measured annual runoff in the SRB.

Climate change and water use were the main factors contributing to runoff reduction in the SRB. The influences of climate change and water use on the annual runoff variation were then analyzed. A base scenario (BS) was created to represent the pre-1998 configuration. The meteorological and water use data for the base period were replaced with those of the change period. The other inputs remained unchanged, and four different scenarios were modeled (Table 2).

Table 2. Settings of scenarios for the multifactor attribution analysis.

Scenario	Description
BS	Base scenario
BSW	Change base period water use data to that of the change period
BSM	Change base period meteorological data to that of the change period
BSWM	Change base period water use and meteorological data to those of the change period

Table 3 shows relative changes in temperature, precipitation, and water use. The annual runoff decreased in response to climate change and water use. Compared with the BS scenario, the water use was 8.5 billion m^3 higher, the precipitation was 26.0 mm lower, and the temperature was 1.3 °C higher in the BSWM scenario.

Table 3. Changes in each impact factor.

Period	Temperature (°C)	Precipitation (mm)	Water Use (Billion m^3)
Base	2.5	540.2	19.2
Change	3.8	514.2	27.7
Variation	1.3	26.0	8.5

For the BS and BSWM scenarios, the annual runoff volumes in the SRB were 73.7 billion m^3 and 52.9 billion m^3, respectively (Table 4). The rate of change in the annual runoff was −28.2% in the BSWM scenario relative to that of the BS scenario. The contribution rates of water use and climate change to the annual runoff reduction in the SRB were 23.0% and 77.0%, respectively. These values were determined by multifactor attribution analysis, revealing that climate change was the dominant factor attenuating annual runoff.

Table 4. Contributions of various factors to annual runoff reduction in the SRB.

Item		Annual Runoff
BS		73.7 billion m^3
BSWM		52.9 billion m^3
BSWM-BS		−20.7 billion m^3
Rate of change		−28.2%
Contribution rate	Water use	23.0%
	Climate change	77.0%

We analyzed the influence of climate change on the reduction in the annual production flow components. Only climate change sets the scenarios for the comparative analyses based on the BS (Table 5).

Table 5. Scenario settings for multifactor attribution analysis.

Scenario	Description
BS	Base scenario
BST	Change base period air temperature to that of the change period
BSP	Change base period precipitation data to that of the change period
BSTP	Change base period air temperature and precipitation data to those of the change period

The annual average temperature increased by 1.3 °C while the precipitation decreased by 26.0 mm. Hence, there was a 26.9-mm decrease in annual production flow (Tables 3 and 6).

Table 6. Annual production flow influenced by climate change.

Item	Annual Production Flow			
	Production Flow	Surface Flow	Soil Flow	Base Flow
BS	160.2 mm	131.2 mm	2.8 mm	26.2 mm
BSTP	133.3 mm	114.5 mm	2.2 mm	16.6 mm
BSTP-BS	−26.9 mm	−16.7 mm	−0.6 mm	−9.6 mm
Rate of change	−16.8%	−12.7%	−21.4%	−36.6%
$\frac{\text{Annual production flow component}}{\text{Total annual production flow}}$	100.0%	62.1%	2.2%	35.7%

We then analyzed the annual production flow components (Table 6). Annual surface, soil, and base flow decreased by 16.7 mm, 0.6 mm, and 9.6 mm, respectively, in the BSTP scenario relative to the BS scenario. The rate of change in the annual surface flow in the BSTP scenario was −12.7% to that of the BS scenario, which was the minimal rate of change in the annual surface, soil, and base flows. Nevertheless, the annual surface flow reduction accounted for 62.1% of the annual production flow reduction. The rate of change in the annual base flow was −36.6%, but the annual base flow reduction accounted for 35.7%. The change in annual soil flow was minimal and accounted for only 2.2%. The observed decrease in total annual production flow in the SRB was caused mainly by the decrease in the annual surface flow. The decrease in annual base flow accounted for a relatively small proportion.

3.3. *Effect of Climate Change on Production Flow Variation during Different Periods in the SRB*

3.3.1. Production Flow Variation during Different Periods

The non-freeze-thaw was the main annual production flow reduction period. Production flow reduction during the non-freeze-thaw period accounted for 80.7% of the annual total under the influence of climate change (Tables 7 and 8). The production flow reduction during the thawing period accounted for 20.4% of the annual total. The production flow during the freezing period slightly increased and accounted for −1.1% of the annual total.

Table 7. Changes in meteorological factors.

Period	Meteorological Factor	Whole Year	Freezing Period	Thawing Period	Non-Freeze-Thaw Period
Base	Precipitation (mm)	540.1	23.6	71.5	445.0
	Temperature (°C)	2.5	−14.9	4.2	15.5
Change	Precipitation (mm)	519.3	29.5	86.5	403.3
	Temperature (°C)	3.9	−13.8	5.8	16.9
Variation	Precipitation (mm)	−20.8	6.0	15.0	−41.7
	Temperature (°C)	1.3	1.1	1.6	1.4

Table 8. Production flow variation during different periods.

Item	Production Flow during Different Periods of Year			
	Whole Year	Freezing Period	Thawing Period	Non-Freeze-Thaw Period
BS	160.2 mm	7.4 mm	37.2 mm	115.6 mm
BSTP	133.3 mm	7.7 mm	31.8 mm	93.8 mm
BSTP-BS	−27.0 mm	0.3 mm	−5.5 mm	−21.8 mm
$\frac{\text{(BSTP−BS) during different periods}}{\text{Annual (BSTP−BS)}}$	100.0%	−1.1%	20.4%	80.7%

3.3.2. Production Flow Component Variation during Freezing Period

The freezing period temperature/precipitation data series for the base period were replaced with those for the change period to study the influence of climate change on runoff component variation during the freezing period. The temperature and precipitation increased by 1.1 °C and 5.5 mm, respectively (Table 7). Hence, there was a 1-mm increase in the total production flow during the freezing period (Table 9). Relative to the BS scenario, the rate of change in the total production flow during the freezing period was 13.9% in the BSTP scenario.

Table 9. Influences of temperature and precipitation on the runoff component variation during the freezing period.

Item	Production Flow during Freezing Period			
	Total Production Flow	**Surface Flow**	**Soil Flow**	**Base Flow**
BS	7.4 mm	0.4 mm	0.1 mm	6.9 mm
BSTP	8.4 mm	0.8 mm	0.1 mm	7.5 mm
BSTP-BS	1.0 mm	0.4 mm	0.0 mm	0.6 mm
Rate of change	13.9%	99.3%	31.2%	8.5%
$\frac{\text{Variation in production flow component}}{\text{Variation in total production flow}}$	100.0%	40.0%	0%	60%

The increase in the base flow accounted for most of the increase in the total production flow. The rate of change in the base flow during the freezing period was 8.5%. Nevertheless, the increase in base flow explained 60.0% of the increase in total production flow during the freezing period. The increase in surface flow accounted for 40.0% of the increase in total production flow during the freezing period. However, the rate of change in the surface flow was 99.3% during the freezing period. There were minimal changes in the soil flow, and increases in it explained 0% of the increase in total production flow during the freezing period.

3.3.3. Production Flow Component Variation during the Thawing Period

The thawing period temperature/precipitation data series for the base period were replaced with those for the change period to study the influence of climate change on runoff component variation during the thawing period. Increases of 1.57 °C and 10.3 mm precipitation caused a 3.2-mm decrease in runoff during the thawing period (Table 10). Relative to the BS scenario, the rate of change in the total production flow during the thawing period was −8.4% in the BSTP scenario.

Table 10. Influences of temperature and precipitation on runoff component variation during the thawing period.

Item	Production Flow during Thawing Period			
	Total Production Flow	**Surface Flow**	**Soil Flow**	**BASE flow**
BS	37.2 mm	31.7 mm	1.0 mm	4.6 mm
BSTP	31.8 mm	28.1 mm	1.1 mm	4.9 mm
BSTP-BS	−3.2 mm	−3.6 mm	0.1 mm	0.3 mm
Rate of change	−8.4%	−11.4%	10.0%	6.5%
$\frac{\text{Variation in production flow component}}{\text{Variation in total production flow}}$	100.0%	112.5%	−3.1%	−9.4%

The rate of change in the surface flow was −11.4%, but the reduction in the surface flow accounted for 112.5% of the total reduction in the production flow during the thawing period. The reduction in the surface flow explained most of the reduction in the total production flow during the thawing period. The soil and base flow increased during the thawing period. The increases in soil and base flow accounted for −3.1% and −9.4% of the change in the total production flow during the thawing period, respectively.

3.3.4. Variations in Production Flow Components during the Non-Freeze-Thaw Period

The temperature/precipitation data series for the non-freeze-thaw period were replaced with those for the change period to study the influence of climate change on runoff component variation during the non-freeze-thaw period. The temperature increased by 1.4 °C while the precipitation decreased by 41.7 mm. Thus, there was a decrease of 17.7 mm in the total production flow during the non-freeze-thaw period (Table 11). Relative to the

BS scenario, the rate of change in the total production flow during the non-freeze-thaw period was −15.3% in the BSTP scenario.

Table 11. Influences of temperature and precipitation on the variation in runoff component during the non-freeze-thaw period.

Item	Production Flow during Non-Freeze-Thaw Period			
	Total Production Flow	Surface Flow	Soil Flow	Base Flow
BS	115.6 mm	99.1 mm	1.7 mm	14.8 mm
BSTP	97.9 mm	87.6 mm	1.5 mm	8.8 mm
BSTP-BS	−17.7 mm	−11.5 mm	−0.2 mm	−6.0 mm
Rate of change	−15.3%	−11.6%	−11.8%	−40.5%
$\frac{\text{Variation in production flow component}}{\text{Variation in total production flow}}$	100.0%	65.0%	1.1%	33.9%

The rate of change in the surface flow was −11.6%, but the reduction in surface flow accounted for 65.0% of the reduction in the total production flow during the non-freeze-thaw period. The rate of change in the base flow was −40.5%, and the reduction in the base flow accounted for 33.9%. The changes in the soil flow were minimal, and the reduction in the soil flow accounted for 1.1% of the total production flow during the non-freeze-thaw period.

3.4. *Effects of Climate Change on Groundwater Recharge*

The foregoing analysis demonstrated that under climate change, the surface flow caused most of the reduction in production flow in the SRB. In contrast, the reduction in base flow accounted for a relatively small proportion of the reduction in production flow in the SRB. The base flow increased during the freezing and thawing periods. Relative to the BS scenario, the rates of change in the groundwater recharge during the freezing and thawing periods increased by 9.2% and 4.1%, respectively, in the BSTP scenario (Table 12).

Table 12. Influences of temperature and precipitation on groundwater recharge variation.

Item	Groundwater Recharge during Different Periods		
	Freezing Period	Thawing Period	Non-Freeze-Thaw Period
BS	6.1 mm	12.2 mm	37.1 mm
BSTP	6.7 mm	12.7 mm	27.9 mm
BSTP-BS	0.6 mm	0.5 mm	−9.2 mm
Rate of change	9.2%	4.1%	−24.8%

The water use was 27.5 billion m^3, the groundwater exploitation was 9.4 billion m^3, and the groundwater exploitation accounted for 34.3 % of water use and 29.1% of groundwater resources in the SRB between 1980 and 2018. The utilization rate of groundwater resources is far below the red line for development and utilization of 40%, the internationally recognized alarm line, which shows potential for development [43]. The groundwater exploitation was much less in the SRB compared with the Yellow River Basin and Haihe River Basin in northern China. The attenuation of the production flow aggravates water resource shortages. Appropriate attention should be given to groundwater utilization in areas with relatively less groundwater exploitation.

4. Conclusions

The WEP-N model was used to simulate the SRB's hydrological cycle, and its overall performance was acceptable. The flow simulation was accurate, NSE > 0.75 and RE < 5 % for three hydrological stations and close to the actual measurements.

Climate change and water use were the main factors influencing the SRB's reduction in the annual production flow. Compared with the BS scenario, the rate of change in the annual production flow was −28.2% under the BSWM scenario. According to a multifactor attribution analysis, the rates of the contribution of climate change and water use to the reduction in annual production flow were 77.0% and 23.0%, respectively. Thus, climate change was the dominant factor attenuating runoff. The decrease in annual surface flow caused a 62.1% reduction in the annual production flow in the SRB. By contrast, the decrease in annual base flow accounted for only 35.7% of the reduction in the annual production flow in the SRB.

The change in annual production flow occurred mainly during the non-freeze-thaw period. The reductions in production flow during the non-freeze-thaw and thawing periods accounted for 80.7% and 20.4% of the annual reduction in the production flow, respectively. The production flow slightly increased during the freezing period. The change in the production flow occurred mainly during the non-freeze-thaw period. The increases in surface, soil, and base flow accounted for 60.0%, 0%, and 40.0% of the total increase in the production flow under the influences of increasing temperature and precipitation during the freezing period. The variations in surface, soil, and base flow accounted for 112.5%, −3.1%, and −9.4% of the total reduction in production flow during the thawing period. The reductions in surface, soil, and base flow accounted for 65.0%, 1.1%, and 33.9% of the total reduction in production flow during the non-freeze-thaw period. The foregoing analysis showed that surface flow caused the reduction in production flow in the SRB, where reduction in base flow accounted for a relatively small proportion under climate change. The base flow increased during the freezing and thawing periods.

Relative to the BS scenario, the rates of change in groundwater recharge during the freezing and thawing periods increased by 9.2% and 4.1%, respectively, in the BSTP scenario. The attenuation of the production flow aggravated the water resource shortage. Attention should be directed towards certain areas of SRB with less groundwater exploitation and similar areas in northern Eurasia and northern North America.

Author Contributions: S.L. performed the model programming and simulations. Z.Z., J.L. (Jiajia Liu), P.W., C.L. and J.L. (Jia Li) contributed to the model programming. S.L. and Z.Z. performed the writing. X.X., Y.J. and H.W. also contributed to the writing of the paper. All authors have read and agreed to the published version of the manuscript.

Funding: This research was funded by the National Natural Science Foundation of China (51679257, 51779270) and the National Key Research and Development Program of China (2016YFC0402405).

Institutional Review Board Statement: Not applicable.

Informed Consent Statement: Not applicable.

Data Availability Statement: The datasets and model codes relevant to the current study are available from the corresponding author upon reasonable request.

Acknowledgments: The authors are grateful to the editors and reviewers; the comments and suggestions have contributed significantly to the improvement of the manuscript.

Conflicts of Interest: The authors have no relevant financial or non-financial interests to disclose.

References

1. Oyler, J.W.; Dobrowski, S.Z.; Ballantyne, A.P.; Klene, A.E.; Running, S.W. Artificial amplification of warming trends across the mountains of the western United States. *Geophys. Res. Lett.* **2015**, *42*, 153–161. [CrossRef]
2. Gao, H.; Li, H.; Duan, Z.; Ren, Z.; Meng, X.; Pan, X. Modelling glacier variation and its impact on water resource in the Urumqi Glacier No. 1 in Central Asia. *Sci. Total Environ.* **2018**, *644*, 1160–1170. [CrossRef]
3. Yang, D.; Ye, B.; Kane, D.L. Streamflow changes over Siberian Yenisei River Basin. *J. Hydrol.* **2004**, *296*, 59–80. [CrossRef]
4. Yang, W.; Jin, F.; Si, Y.; Li, Z. Runoff change controlled by combined effects of multiple environmental factors in a headwater catchment with cold and arid climate in northwest China. *Sci. Total Environ.* **2021**, *756*, 143995. [CrossRef]
5. Li, B.; Chen, Y.; Chen, Z.; Li, W. Trends in runoff versus climate change in typical rivers in the arid region of northwest China. *Quat. Int.* **2012**, *282*, 87–95. [CrossRef]
6. Pan, C.G.; Kamp, U.; Munkhjargal, M.; Halvorson, S.J.; Dashtseren, A.; Walther, M. An estimated contribution of glacier runoff to Mongolia's Upper Khovd River Basin in the Altai Mountains. *Mt. Res. Dev.* **2019**, *39*, R12–R20. [CrossRef]

7. Tang, Q.; Oki, T. Historical and future changes in streamflow and continental runoff. In *Terrestrial Water Cycle and Climate Change: Natural and Human-Induced Impacts*; John Wiley & Sons: Hoboken, NJ, USA, 2016; Volume 221, pp. 17–37. [CrossRef]
8. Zhang, Y.; Liu, S.; Xu, J.; Shangguan, D. Glacier change and glacier runoff variation in the Tuotuo River basin, the source region of Yangtze River in western China. *Environ. Geol.* **2008**, *56*, 59–68. [CrossRef]
9. Shiyin, L.; Yong, Z.; Yingsong, Z.; Yongjian, D. Estimation of glacier runoff and future trends in the Yangtze River source region, China. *J. Glaciol.* **2009**, *55*, 353–362. [CrossRef]
10. Yang, Y.; Weng, B.; Man, Z.; Yu, Z.; Zhao, J. Analyzing the contributions of climate change and human activities on runoff in the Northeast Tibet Plateau. *J. Hydrol. Reg. Stud.* **2020**, *27*, 100639. [CrossRef]
11. Rakhimova, M.; Liu, T.; Bissenbayeva, S.; Mukanov, Y.; Gafforov, K.S.; Bekpergenova, Z.; Gulakhmadov, A. Assessment of the impacts of climate change and human activities on runoff using climate elasticity method and general circulation model (GCM) in the Buqtyrma River Basin, Kazakhstan. *Sustainability* **2020**, *12*, 4968. [CrossRef]
12. Qiu, L.; Peng, D.; Xu, Z.; Liu, W. Identification of the impacts of climate changes and human activities on runoff in the upper and middle reaches of the Heihe River basin, China. *J. Water Clim. Chang.* **2016**, *7*, 251–262. [CrossRef]
13. Thompson, L.G.; Brecher, H.H.; Mosley-Thompson, E.; Hardy, D.R.; Mark, B.G. Glacier loss on Kilimanjaro continues unabated. *Proc. Natl. Acad. Sci. USA* **2009**, *106*, 19770–19775. [CrossRef]
14. Otte, I.; Detsch, F.; Mwangomo, E.; Hemp, A.; Appelhans, T.; Nauss, T. Multidecadal trends and interannual variability of rainfall as observed from five lowland stations at Mt. Kilimanjaro, Tanzania. *J. Hydrol. Meteorol.* **2017**, *18*, 349–361. [CrossRef]
15. Hemp, A. Climate change-driven forest fires marginalize the impact of ice cap wasting on Kilimanjaro. *Glob. Chang. Biol.* **2005**, *11*, 1013–1023. [CrossRef]
16. Bozkurt, D.; Rojas, M.; Boisier, J.P.; Valdivieso, J. Climate change impacts on hydroclimatic regimes and extremes over Andean basins in central Chile. *Hydrol. Earth Syst. Sci. Discuss.* **2017**, 1–29. [CrossRef]
17. Crespo, P.J.; Feyen, J.; Buytaert, W.; Bücker, A.; Breuer, L.; Frede, H.; Ramírez, M. Identifying controls of the rainfall–runoff response of small catchments in the tropical Andes (Ecuador). *J. Hydrol.* **2011**, *407*, 164–174. [CrossRef]
18. St. Jacques, J.M.; Sauchyn, D.J.; Zhao, Y. Northern Rocky Mountain streamflow records: Global warming trends, human impacts or natural variability? *Geophys. Res. Lett.* **2010**, *37*, 1–5. [CrossRef]
19. Arrigoni, A.S.; Greenwood, M.C.; Moore, J.N. Relative impact of anthropogenic modifications versus climate change on the natural flow regimes of rivers in the Northern Rocky Mountains, United States. *Water Resour. Res.* **2010**, *46*, 1–16. [CrossRef]
20. Miao, C.; Yang, L.; Liu, B.; Gao, Y.; Li, S. Streamflow changes and its influencing factors in the mainstream of the Songhua River basin, Northeast China over the past 50 years. *Environ. Earth Sci.* **2011**, *63*, 489–499. [CrossRef]
21. Chu, H.; Wei, J.; Qiu, J.; Li, Q.; Wang, G. Identification of the impact of climate change and human activities on rainfall-runoff relationship variation in the Three-River Headwaters region. *Ecol. Indic.* **2019**, *106*, 105516. [CrossRef]
22. Meng, F.; Su, F.; Yang, D.; Tong, K.; Hao, Z. Impacts of recent climate change on the hydrology in the source region of the Yellow River Basin. *J. Hydrol. Reg. Stud.* **2016**, *6*, 66–81. [CrossRef]
23. Song, C.; Wang, G.; Sun, X.; Hu, Z. River runoff components change variably and respond differently to climate change in the Eurasian Arctic and Qinghai-Tibet Plateau permafrost regions. *J. Hydrol.* **2021**, *601*, 126653. [CrossRef]
24. Li, J.; Zhou, Z.; Wang, H.; Liu, J.; Jia, Y.; Hu, P.; Xu, C. Development of WEP-COR model to simulate land surface water and energy budgets in a cold region. *Hydrol. Res.* **2019**, *50*, 99–116. [CrossRef]
25. Burn, D.H.; Elnur, M. Detection of hydrologic trends and variability. *J. Hydrol.* **2002**, *255*, 107–122. [CrossRef]
26. Pettitt, A.N. A Non-Parametric Approach to the Change-Point Problem. *Appl. Statist.* **1979**, *28*, 126–135. [CrossRef]
27. Liu, J.J.; Zhou, Z.H.; Jia, Y.W.; Wang, H. A new method to quantitatively separate the effects of multi-factors on the water cycle evolution. *J. Hydraul. Eng.* **2014**, *45*, 658–665. [CrossRef]
28. Liu, S.; Zhou, Z.; Liu, J.; Wang, K.; Li, J.; Wang, P.; Xie, X.; Jia, Y.; Wang, H. Simulation of water and nitrogen movement mechanism in cold regions during freeze-thaw period based on a distributed nonpoint source pollution model closely coupled water, heat, and nitrogen processes at the watershed scale. *Environ. Sci. Pollut. Res. Int.* 2022. [CrossRef]
29. Flerchinger, G.N.; Saxton, K.E. Simultaneous heat and water model of a freezing snow-residue-soil system I. Theory and development. *Trans. ASAE.* **1989**, *32*, 573–576. [CrossRef]
30. Hock, R. A distributed temperature-index ice-and snowmelt model including potential direct solar radiation. *J. Glaciol.* **1999**, *45*, 101–111. [CrossRef]
31. Monteith, J.L. *Principles of Environmental Physics*; Edward Arnold: London, UK, 1973; Volume 214.
32. Jia, Y.; Tamai, N. Integrated analysis of water and heat balances in Tokyo metropolis with a distributed model. *J. Jpn. Soc. Hydrol. Water Resour.* **1998**, *11*, 150–163. [CrossRef]
33. Jia, Y.; Wang, H.; Zhou, Z.H.; Qiu, Y.Q.; Luo, X.Y.; Wang, J.H.; Yan, D.H.; Qin, D.Y. Development of the WEP-L distributed hydrological model and dynamic assessment of water resources in the Yellow River basin. *J. Hydrol.* **2006**, *331*, 606–629. [CrossRef]
34. Zaradny, H. *Groundwater Flow in Saturated and Unsaturated Soil*; A. A. Balkema Uitgevers: Rotterdam, The Netherlands, 1993.
35. Hubbert, M.K. Darcy's law and the field equations of the flow of underground fluids. *Trans. AIME.* **1956**, *207*, 222–239. [CrossRef]
36. Ponce, V.M. Diffusion wave modeling of catchment dynamics. *J. Hydraul. Eng.* **1986**, *112*, 716–727. [CrossRef]
37. Shang, S. Numerical simulation of soil moisture and thermal regime in winter. *Irrig. Drain.* **1997**, *16*, 12–17. (In Chinese)
38. Wang, S.; Prasher, S.O.; Patel, R.M.; Yang, C.; Kim, S.; Madani, A.; Macdonald, P.M.; Robertson, S.D. Fate and transport of nitrogen compounds in a cold region soil using DRAINMOD. *Comput. Electron. Agric.* **2006**, *53*, 113–121. [CrossRef]

39. Wang, A.W.; Xie, Z.H.; Feng, X.; Tian, X.; Qin, P. A soil water and heat transfer model including changes in soil frost and thaw fronts. *Sci. China Earth Sci.* **2014**, *57*, 1325–1339. [CrossRef]
40. Lei, Z.D.; Yang, S.X.; Xie, C.S. *Soil Hydrodynamics*; Tsinghua University Press: Beijing, China, 1988; pp. 8–12. (In Chinese)
41. Hansson, K.; Šimůnek, J.; Mizoguchi, M.; Lundin, L.C.; Van Genuchten, M.T. Water flow and heat transport in frozen soil: Numerical solution and freeze-thaw applications. *Vadose Zone J.* **2004**, *3*, 692–704. [CrossRef]
42. Sun, Y.; Tian, F.; Yang, L.; Hu, H. Exploring the spatial variability of contributions from climate variation and change in catchment properties to streamflow decrease in a mesoscale basin by three different methods. *J. Hydrol.* **2014**, *508*, 170–180. [CrossRef]
43. Falkenmark, M.; Widstrand, C. Population and water resources: A delicate balance. *Popul. Bull.* **1992**, *47*, 1–36. [PubMed]

water

MDPI

Article

Carbon Dioxide Emission Equivalent Analysis of Water Resource Behaviors: Determination and Application of CEEA Function Table

Qiting Zuo [1,2], Zhizhuo Zhang [1,*], Junxia Ma [1,2], Chenguang Zhao [1] and Xi Qin [1]

1 School of Water Conservancy Engineering, Zhengzhou University, Zhengzhou 450001, China
2 Henan International Joint Laboratory of Water Cycle Simulation and Environmental Protection, Zhengzhou 450001, China
* Correspondence: zhangzhizhuozzu@163.com

Abstract: To achieve the global temperature control target under the background of climate warming, it is necessary to establish a systematic carbon dioxide (CO_2) emission accounting method system in the field of water resources as soon as possible. In this study, the carbon dioxide emission equivalent analysis (CEEA) method for different water resource behaviors (WRBs) is proposed from four dimensions of development, allocation, utilization, and protection, and a function table of CEEA (FT-CEEA) for WRBs is constructed. The FT-CEEA includes CEEA formulae for 16 aspects in four categories of water resource development, allocation, utilization, and protection. The CEEA method is applied to 31 provinces in China. The results reveal that: (1) There are significant spatial differences in the carbon dioxide emission equivalent (CEE) of WRBs in different provinces of China under the influence of various factors such as water supply structure and natural conditions. (2) Reservoir storage, tap water allocation, and wastewater treatment are the main contributors to CEE in the categories of water resource development, allocation, and protection behaviors, respectively. (3) The water resource utilization behavior category has the most significant CO_2 emission and absorption effects, and industrial and domestic water utilization behaviors are the main sources of emission effects. (4) The overall CO_2 emission effect of WRBs is greater than the absorption effect. Measures such as increasing the proportion of hydroelectric power generation, improving ecological water security capacity, and strengthening the level of wastewater treatment and reclaimed water reuse are effective ways to promote the goal of carbon neutrality in the field of water resources.

Keywords: water resource behaviors (WRBs); carbon dioxide emission equivalent (CEE); equivalent analysis; carbon dioxide emission equivalent analysis (CEEA); function table of carbon dioxide emission equivalent analysis (FT-CEEA)

Citation: Zuo, Q.; Zhang, Z.; Ma, J.; Zhao, C.; Qin, X. Carbon Dioxide Emission Equivalent Analysis of Water Resource Behaviors: Determination and Application of CEEA Function Table. *Water* **2023**, *15*, 431. https://doi.org/10.3390/w15030431

Academic Editor: Songhao Shang

Received: 10 January 2023
Revised: 18 January 2023
Accepted: 18 January 2023
Published: 20 January 2023

1. Introduction

1.1. Motivation

Since the industrial civilization, under the combined influence of human activities and natural factors, the global warming trend has become increasingly significant. How to deal with the challenges posed by climate change to sustainable development has become a major scientific issue facing mankind [1]. Building a low-carbon future development mode has gradually become a global consensus. The United Nations Framework Convention on Climate Change (UNFCCC), as the world's first international convention to control carbon dioxide (CO_2) emissions, provides a basic framework for international cooperation on climate change [2]. The 21st United Nations Climate Change Conference (UNFCCC COP21) held in 2015 formally adopted the Paris Agreement, which sets the global average temperature increase within 2 °C as an explicit goal [3]. However, with the current trend of CO_2 emissions, the temperature control targets of the Paris Agreement will be difficult to achieve. Deep CO_2 reductions in the coming years are key to achieving that goal [4,5].

Currently, 133 countries have made carbon neutral commitments. China has adopted "carbon neutrality" as a long-term national strategy to address climate change. Compared with developed countries that have achieved carbon peak, developing countries are currently facing the dual pressures of low-carbon transformation and economic development [6,7].

Water resources are the material basis for human survival and the key support for social development and ecological protection. The field of water resources is an important area for implementing the goal of "carbon neutrality" and supporting sustainable development. The "2030 Carbon Peak Action Plan" issued by the Chinese government in 2021 regards hydropower generation, water ecological protection, and efficient utilization of water resources as important ways to promote carbon neutrality [8]. In addition, improving the carbon emission accounting mechanism in different fields and carrying out research on CO_2 emission accounting methods are also important contents of the action plan. Therefore, it is of great significance to explore the CO_2 emission equivalent analysis (CEEA) method for different water resource behaviors (WRBs), and to find a reference "ruler" for the accounting of CO_2 emission equivalent (CEE) in the field of water resources. This "ruler" also has certain positive significance for the global control of CO_2 emissions.

1.2. Literature Review

The identification of source-sink relationships and the assessment of emission intensity of CO_2 are the basis for scientific research in the field of climate change. Accounting for carbon emission and sink effects has been a popular research topic in this field. In terms of carbon emission assessment, most studies have focused on the carbon emission intensity of human activities and carbon footprint accounting in different fields. Carbon footprint can be simply defined as the total amount of greenhouse gases (GHG), mainly carbon dioxide, released directly or indirectly from human activities [9]. Carbon footprint accounting can be divided into macro and micro levels. The methods involved are mainly input-output analysis (IOA) [10] and life cycle assessment (LCA) [11]. In recent years, many scholars have carried out multidimensional accounting of carbon footprints at the mesoscale and macroscale, such as countries [12], cities [13], and industries [14]. Carbon footprint accounting research on the microscale such as enterprise [15], product [16], and technology [17] is also ongoing. For example, Chai et al. [17] used a life-cycle approach to compare the carbon footprints of three mainstream wastewater treatment technologies in China. Accounting for CO_2 emissions caused by land use change [18] is also a popular research topic. In addition, some studies have explored the carbon emission effects of lake wetland [19], reservoir [20], farmland [21], and other ecosystems. For example, Keller et al. conducted a study on carbon emissions from reservoir fallout zones and concluded that reservoirs are a source rather than a sink of carbon in the global carbon cycle [20].

In terms of carbon sink effect assessment, relevant studies have mostly focused on carbon sink effects in terrestrial ecosystems such as forest, grassland, and wetland. The research methods include ground investigation, eddy covariance carbon flux observation [22], ecosystem process model simulation [23], etc. It is worth mentioning that in 2019, the IPCC added the "top-down" atmospheric inversion methodological system to the basic methodological framework for future global GHG accounting [24]. Atmospheric inversion methods have received more attention in recent years in the study of ecosystem carbon sinks. For example, Fernández-Martínez et al. [25] analyzed the trend of global carbon sinks based on atmospheric inversion and vegetation models and explored the relationship between CO_2 emissions and temperature. The research on the carbon sink effects of terrestrial ecosystems has formed a sound theoretical and methodological system. However, compared to terrestrial ecosystems, research on carbon sinks in marine ecosystems is still in the developmental stage.

In particular, studies on energy consumption and CO_2 emission accounting in the field of water resources have been carried out by relevant institutions. In 2005, California released a report on California's water–energy nexus [26], which systematically studied the energy consumption of water supply, water transmission, water utilization, and water treatment in California. Further, the River Network, a U.S. research organization, released The Carbon Footprint of Water [27] in 2009 comprehensively assessed the various water-related carbon footprints in the United States. In addition, the issue of energy consumption and carbon emissions in the field of water resources has been actively discussed by scholars from different countries. LCA and IOA are still the most mainstream research methods under this research topic. The research results basically cover all aspects of social water cycle and urban water system. Although more research has been done in urban water systems, the scale of research has involved countries [28], regions [29], cities [30], schools [31], etc. For example, Wakeel et al. [28] analyzed the energy consumption of different countries in various segments of the social water cycle and compared different methods for measuring energy consumption in the water sector. Using energy consumption as a bridge to quantitatively assess the relationship between water and carbon emissions, Rothausen and Conway [29] systematically explored the GHG emissions in the water sector in different countries and regions. At the urban scale, Valek et al. [32] quantified the CO_2 emissions associated with the water system in Mexico City based on survey data. Based on the LCA method, Friedrich et al. [30] assessed the carbon footprint of different parts of the urban water system (storage, treatment, distribution, collection, and wastewater treatment) in Durban, South Africa. Similarly, Sambito and Freni [33] used the LCA method to quantify the carbon footprint of a metropolitan water system in Italy. In addition, Li et al. [31] quantified the water–energy–carbon relationship on a campus in northern China and explored the spatial distribution pattern of carbon sources/sinks at a small scale.

In addition to the overall study of the carbon emissions from the social water cycle and urban water system, some scholars have carried out targeted discussions on different links of the water system (water production and supply, desalination, water utilization, wastewater treatment, etc.). In terms of water production and supply, relevant studies mainly focus on carbon footprint accounting of water supply system and water distribution system. For example, Fang and Newell [34] used the LCA method to assess the carbon footprint of Southern California's water supply system, arguing that the carbon footprint of local reclaimed water is much lower than that of long-distance water supply. Boulos and Bros [35] proposed a WNEE (Water network energy efficiency) method for measuring the carbon footprint of energy consumption in a water distribution system, which was applied in a European city. Moreover, Heihsel and Lenzen [36] constructed a multi-regional input-output model (MIOA) for measuring GHG emissions from desalination in Australia, which provides a solution for the calculation of the carbon footprint of desalination at a macro-scale. In terms of water end-use, the studies mainly cover energy consumption and carbon emission measurement for domestic and agricultural water use. For example, Siddiqi and Fletcher [37] summarized the range of energy intensity of domestic water and agricultural water in the end-use process. Escriva-Bou et al. [38] simulated GHG emissions associated with domestic water use in California using probability distribution models and emission factors. Wang et al. [39] evaluated the carbon footprint of agricultural groundwater use in 31 provinces of China based on statistical survey data. In terms of wastewater collection and treatment, the carbon emission effects and measurement methods of wastewater treatment plants and municipal wastewater sectors in different countries such as China [40], the United States [41], and Italy [42] have been deeply discussed. Further, the carbon emission effects of different wastewater treatment technologies and options have been studied in comparison [43]. It is worth mentioning that research on carbon emissions accounting for water saving behavior has also been carried out, covering different scales such as city [44] and campus [45]. In addition, Wang et al. [46] explored the water footprint and carbon footprint in hydropower stations in China and made recommendations for carbon emission reduction of hydropower stations. Some of the studies addressing the carbon emission effects in the water resources sector are summarized in Table 1.

Table 1. Selected representative literature of carbon emission effect studies in the field of water resources.

Author(s)	Region(s)	Water-Related Activities	Methodology
Griffiths-Sattenspiel et al. [27]	United States	Water Supply and Conveyance Water Treatment Water Distribution Water End-Uses Wastewater Collection and Treatment Wastewater Discharge	Carbon emission estimation based on statistical survey data and emission factors
Friedrich et al. [30]	Durban, South Africa	Water Impoundment Water Treatment Water Distribution Water Collection Wastewater Treatment Water Recycling Bottled Water	Carbon footprint analysis based on LCA method
Zhang et al. [47]	All cities in Guangdong Province, China	Water Extraction and Conveyance Water Purification and Supply Water Distribution Wastewater Treatment	Accounting for CO_2 emissions based on energy intensity and emission factors
Venkatesh et al. [48]	Nantes (France), Oslo (Norway), Turin (Italy), Toronto (Canada)	Water Supply Water Treatment Water Distribution Wastewater Collection Wastewater Treatment	System analysis method
Bakhshi and Demonsabert [49]	Loudoun, United States	Raw Water Extraction and Treatment Water Distribution Wastewater Collection Wastewater Treatment	Carbon emission estimation based on survey data and Geographic information system models
Stokes and Horvath [50]	Southern California, United States	Imported Water Desalinated Ocean Water (Conventional pretreatment) Desalinated Ocean Water (Membrane pretreatment) Desalinated Brackish Groundwater Recycled Water	Carbon emission measurement of water supply system based on hybrid LCA method
Valek et al. [32]	México City, México	Water Supply Water Treatment System	CO_2 equivalent analysis based on statistical survey data and emission factors
Sambito and Freni [33]	Sicily, Italy	Water Supply and Treatment System Distribution of Water and Sewer System Wastewater Treatment Plant	Carbon footprint analysis based on LCA approach
Presura and Robescu [51]	Constanta, Romania	Potable Water Treatment Wastewater Treatment	Carbon footprint analysis based on energy intensity and emission factors
Heihsel and Lenzen [36]	Australia	Seawater Desalination	Carbon footprint analysis based on multi-regional input-output model
Wang et al. [39]	China	Groundwater Use for Agriculture	Carbon footprint analysis based on energy intensity and emission factors
Wu et al. [43]	Australia	Wastewater Treatment (Direct emission) Wastewater Treatment (Indirect emission) Wastewater Treatment (Value chain emission)	Carbon footprint analysis based on emission factors

In general, relevant research results provide important reference value for the quantitative identification of water-carbon relationship and carbon neutrality in the field of water resources. However, some of the studies are too targeted, difficult to obtain data, and the experimental methods are not easily reproducible to meet the demand for systematic research on CO_2 emission accounting in the field of water resources. In addition, carbon dioxide emissions related to water resource behaviors involve many links and are not limited to the scope discussed above. How to make a comprehensive and feasible "ruler" to provide convenience and reference for the estimation of water-related CO_2 emissions is still a problem to be further explored. To facilitate the discussion of CO_2 emission or absorption effects in the field of water resources, this paper is devoted to the study of water resource behaviors, that is, a series of activities related to the development, allocation, utilization, and protection of water resources. Different links of the water cycle or water resources system can be understood as different WRBs. Researching the methodologies for quantifying the CO_2 emission effects of different WRBs is a further refinement and extension of the carbon source/sink effects accounting in the field of water resources.

1.3. Contribution and Objectives

Based on the literature review, this study proposes a carbon dioxide emission equivalent analysis (CEEA) method for several common water resource behaviors (WRBs) from four dimensions: water resources development, water resources allocation, water resources utilization, and water resources protection. The function table of CO_2 emission equivalent analysis (FT-CEEA) of WRBs is constructed for the first time, which provides a method set for researchers in different regions and industries to evaluate the CO_2 emission equivalent (CEE) of WRBs. Compared to existing studies, the contributions of this study are: (1) The CEEA method is proposed to realize the quantitative calculation of CEE for different WRBs; (2) the FT-CEEA is developed to provide a convenient and feasible "ruler" for the measurement of CEE in the field of water resources; (3) based on the FT-CEEA, the spatial distribution characteristics of CO_2 emission or absorption effects of WRBs in 31 provinces in China are clarified.

This paper is organized as follows: Section 2 is the introduction of the CEEA method and FT-CEEA; Section 3 is the study area and data description, as well as results analysis and discussion; Section 4 is the main conclusion and research prospect.

2. Methodology

2.1. CEEA Method Framework of Water Resource Behaviors

Water resource behavior (WRB) is a collective term for a range of activities related to the development, allocation, utilization, and protection of water resources. The carbon dioxide emission equivalent (CEE) of water resource behaviors refers to the CO_2 emission or absorption effects directly or indirectly caused by water resource behaviors. In this study, the method to quantify the CEE generated by WRBs is called the carbon dioxide emission equivalent analysis method (CEEA) of WRBs. Most WRBs do not emit CO_2 themselves and are not explicitly linked to CO_2. However, WRBs are often accompanied by energy consumption, which in turn leads to CO_2 emissions. Therefore, compared to "carbon dioxide emission", "carbon dioxide emission equivalent" is more accurate to represent the CO_2 emission or absorption effects of WRBs.

This study proposes the CEEA method and develops the FT-CEEA (the function table of carbon dioxide emission equivalent analysis), aiming to find a reference "ruler" to provide methodological reference and technical support for the accounting of CEE related to WRBs. The general idea of the CEEA method is to develop diversified CEE functions for WRBs in different dimensions by direct reference, refinement, and innovation, and finally integrate them into a unified calculation platform to form a relatively complete "ruler", namely FT-CEEA. The general idea diagram of the CEEA method is shown in Figure 1.

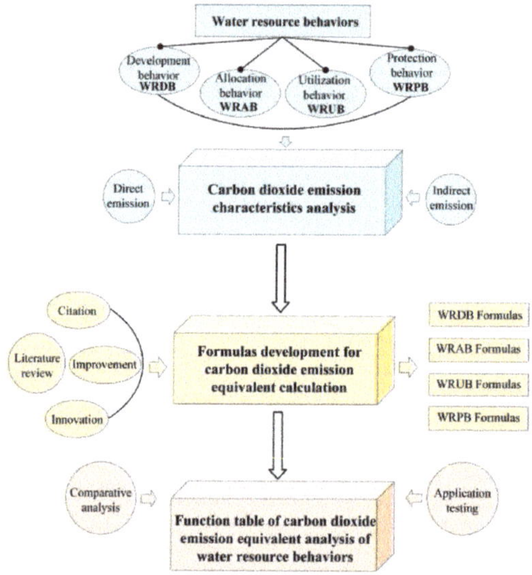

Figure 1. The general idea of the CEEA method.

The WRBs involve a wide range of fields and factors, and the CEE accounting of WRBs should have a clear system boundary to avoid the infinite extension of indirect calculations. The principle of this study for system boundary formulation is to focus on CEE directly caused by WRBs, with appropriate consideration of indirect CEE that are closely related to such WRBs. Based on the definition of WRBs, the system boundary of CEE accounting is determined, as shown in Figure 2. The categories of WRBs can be roughly divided into water resource development behaviors (WRDBs), water resource allocation behaviors (WRABs), water resource utilization behaviors (WRUBs), and water resource protection behaviors (WRPBs). Each category contains a variety of typical WRBs, each WRB has a corresponding CEEA method.

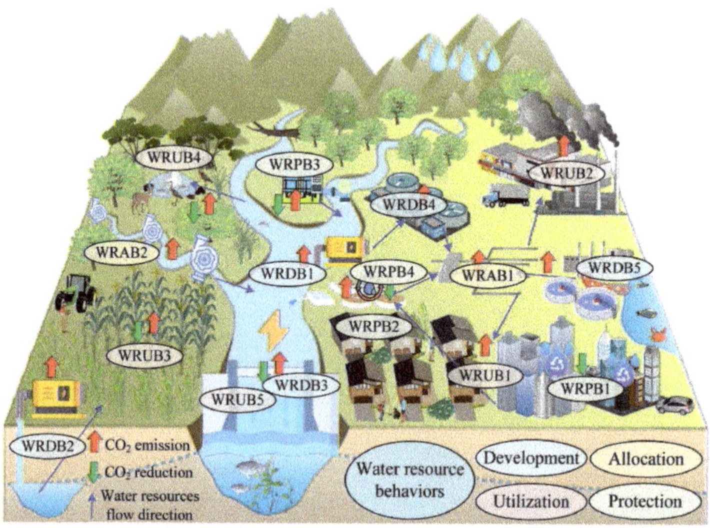

Figure 2. Schematic diagram of CEE accounting for WRBs.

2.2. CEEA Method to Water Resource Development Behaviors

Water resource development behaviors (WRDBs) refer to a series of activities related to water resources development. In this study, WRDBs are preliminarily defined as surface water lifting (WRDB1), groundwater extraction (WRDB2), reservoir storage (WRDB3), raw water treatment (WRDB4), and seawater desalination (WRDB5).

(1) Surface water lifting (WRDB1): Surface water resources are extracted from natural rivers or lakes to higher elevations using water extraction projects to achieve centralized treatment and unified distribution of "raw freshwater". The electric energy consumed by the water lifting project is converted into the mechanical energy needed for water resources lifting, so the CO_2 emissions of this behavior are mainly concentrated in the energy consumption link of the water lifting project. The CEEA formula of WRDB1 based on emission factor [52] is as follows:

$$E_1 = Q_1 \times EI_1 \times EF \tag{1}$$

$$EI_1 = \frac{\rho \times g \times h_1}{3.6 \times 10^6 \times \eta} \tag{2}$$

$$EF = \frac{\sum_i \left(FC_{i,y} \times NCV_{i,y} \times EF_{CO_2,i,y}\right)}{EG_y} \tag{3}$$

where E_1 is the carbon dioxide emission equivalent of surface water lifting behavior, kg; Q_1 is the amount of surface water lifting, m^3; EI_1 is the energy intensity of WRDB1 (the amount of electricity required to lift per unit of surface water) kWh/m^3; ρ is the density of surface water (typically 1000), kg/m^3; g is the acceleration of gravity (typically 9.8), m/s^2; h_1 is the surface water lifting head, m; η is the efficiency of the water lifting project; EI_1 is the power system CO_2 emission factor (the amount of CO_2 emitted per unit of electricity consumed), kg/kWh; EG is the total net power generation during the calculation period of the power system, kWh; FC is the consumption of fuel by the generator set during the calculation period, in mass or volume units; NCV is the average low-level heat content of the fuel during the calculation period, in GJ/mass or volume units; EF_{CO_2} is the CO_2 emission factor of the fuel (amount of CO_2 emitted per unit of energy) during the calculation period, $kgCO_2/GJ$; i is the type of fossil fuels consumed to generate electricity; and y is the year. Power-related departments in different countries will regularly release the EF of the power system. For example, the Ministry of Ecology and Environment of China has issued EF reference values for different provinces in China. In China, EI_1 is mainly related to the water head, which can be 0.2 kWh/m^3 [53,54] on average. The global level can refer to the value range given by relevant research: 0.0002–1.74 kWh/m^3 [55].

(2) Groundwater extraction (WRDB2): Similar to the principle of WRDB1, groundwater extraction behavior also needs to convert the electrical energy of pumping equipment into the mechanical energy required for groundwater rise. CO_2 emissions are mainly concentrated in the energy consumption of pumping equipment:

$$E_2 = Q_2 \times EI_2 \times EF \tag{4}$$

$$EI_2 = \frac{9.8 \times \rho \times h_2}{3.6 \times 10^6 \times \eta} \tag{5}$$

where E_2 is the CEE of WRDB2, kg; Q_2 is the amount of groundwater extraction, m^3; EI_2 is the energy intensity of WRDB2 (the amount of electricity required to extract per unit of groundwater) kWh/m^3; h_2 is the groundwater depth, m. Other variables have the same meaning as above. Unlike surface water, the energy intensity of WRDB2 varies considerably with different groundwater burial depths. The value of EI_2 can be obtained according to the actual situation of the study area, and EI_2 in different regions of China can also refer to Table 4 [39]. In addition, the EI_2 of different countries is available in studies: 0.18–0.49 kWh/m^3 (USA) [56], 0.48–0.53 kWh/m^3 (Australia) [57], and 0.37–1.44 kWh/m^3 (Global) [55].

(3) Reservoir storage (WRDB3): The energy consumption of WRDB3 mainly comes from the daily operation and management of water storage infrastructure [54], such as gate control, lighting, and monitoring equipment operation. This process also produces CO_2 emissions. The calculation formula is as follows:

$$E_3 = Q_3 \times EI_3 \times EF \quad (6)$$

where E_3 is the CEE of WRDB3, kg; Q_3 is the actual volume of water stored in the reservoir, m^3; EI_3 is the energy intensity of WRDB3 (the amount of electricity required to store per unit of water in the reservoir) kWh/m^3. EI_3 varies due to differences in reservoir conditions in different regions. Field visits to reservoirs can be conducted to obtain the value of EI_3. EI_3 can also refer to existing research. Studies have shown that the energy intensity range of WRDB3 in China is [0.07,0.2] kWh/m^3 [54], and 0.14 kWh/m^3 can be used to study the average state of China [58].

(4) Raw water treatment (WRDB4): After taking raw water from the water source, it needs to be treated by the waterworks, including coagulation, sedimentation, filtration, and disinfection [41]. Each process relies mainly on electricity to maintain the normal operation of the processing equipment, so WRDB4 also produces CO_2 emissions [51]:

$$E_4 = Q_4 \times EI_4 \times EF \quad (7)$$

where E_4 is the CEE of WRDB4, kg; Q_4 is the volume of raw water treatment, m^3; EI_4 is the energy intensity of WRDB4 (the amount of electricity required to treat per unit of raw water) kWh/m^3. EI_4 can be determined by the statistical calculation of energy consumption data of each link of WRDB4. According to the yearbook of Chinese urban water supply, the national average is 0.31 kWh/m^3 [54,59], which can be used for reference. Existing studies have also given the values of different countries for reference: 0.371–0.392 kWh/m^3 (USA) [56]; 0.1–0.6 kWh/m^3 (Australia) [60]; 0.38–1.44 kWh/m^3 (Canada) [61]; 0.11–1.5 kWh/m^3 (Spain) [62]; 0.15–0.44 kWh/m^3 (New Zealand) [63].

(5) Seawater Desalination (WRDB5): Nine coastal provinces in China have large-scale seawater desalination capacity. Although the industrialization process of desalination in China is slow, seawater desalination is an important behavior in the process of sustainable development of water resources in the future [51].

$$E_5 = Q_5 \times EI_5 \times EF \quad (8)$$

where E_5 is the CEE of WRDB5, kg; Q_5 is the volume of seawater desalination, m^3; EI_5 is the energy intensity of WRDB5 (the amount of electricity required to treat per unit of seawater) kWh/m^3; EI_5 should be obtained based on the survey data of desalination plants, and can also refer to existing studies: 5.9 kWh/m^3 (China) [64–66]; 4 kWh/m^3 (Australia) [57]; 2.4–8.5 kWh/m^3 (Global) [55,67].

2.3. *CEEA Method to Water Resource Allocation Behaviors*

Water resource allocation behaviors (WRABs) refer to a series of activities related to water resource transportation and distribution. Representative WRABs include urban-rural tap water allocation (WRAB1) and inter-regional water transfer (WRAB2).

(1) Tap water allocation (WRAB1): The treated water from the waterworks is distributed to individual water users through the urban and rural water distribution system. The energy consumption of WRAB1 is mainly the head loss in the water transmission and distribution process, and the CEE is focused on the power consumption in the pressurization process [68]:

$$E_6 = Q_6 \times EI_6 \times EF \quad (9)$$

$$EI_6 = \frac{9.8 \times \rho \times (h_f + h_j)}{3.6 \times 10^6 \times \eta} \quad (10)$$

$$h_f = \lambda \frac{l}{4R} \frac{v^2}{2g} \quad (11)$$

$$h_j = \zeta \frac{v^2}{2g} \quad (12)$$

where E_6 is the CEE of WRAB1, kg; Q_6 is the amount of urban and rural tap water allocation, m^3; EI_6 is the energy intensity of WRAB1 (the amount of electricity required to distribute per unit of tap water) kWh/m^3; h_f is head loss along the path, λ is the drag coefficient along the path, l is the length of tap water allocation, R is the hydraulic radius, m; v is the average velocity of tap water transmission and distribution, m^3/s; h_j is local head loss, m; ζ is local drag coefficient; η is the efficiency of the pressurized pump station. Head loss can be calculated by the Darcy formula. EI_6 can be obtained according to the investigation and statistics of unit water distribution power consumption data of water supply company. The energy intensity of tap water companies in different regions of China is quite different [69]. Combined with the China Urban Water Supply Yearbook and related research [68–70], the recommended value is 0.2 kWh/m^3 for reference. Reference values of EI_6 in other countries: 0.2–0.32 kWh/m^3 (California, USA) [26]; 0.12–0.22 kWh/m^3 (Spain) [71]; 0.1 kWh/m^3 (South Africa) [72].

(2) Inter-regional water transfer (WRAB2): Most of the inter-regional water transfer projects require pumping stations for pressurized delivery to overcome the energy loss from head loss. The CEE calculation principle of WRAB2 is similar to that of WRAB1. The difference is that the urban and rural tap water allocation system is mostly pressure pipe flow, while the inter-regional water transfer is mostly open channel constant flow:

$$E_7 = Q_7 \times EI_7 \times EF \quad (13)$$

$$EI_7 = \frac{\rho \times g \times (h_f + h_j)}{3.6 \times 10^6 \times \eta} \quad (14)$$

where E_7 is the CEE of WRAB2, kg; Q_7 is the amount of water transferred across regions, m^3; EI_7 is the energy intensity of WRAB2 (the amount of electricity required to transfer per unit water resources across regions) kWh/m^3; h_f and h_j are the head loss along the open channel and local head loss, m; the specific calculation can be referred to the relevant formula of open channel hydraulics [73]. In the absence of the necessary investigation conditions, EI_7 can refer to the value of 0.815 kWh/m^3 (China) taken in existing studies [54,74].

2.4. CEEA Method to Water Resource Utilization Behaviors

Water resource utilization behaviors (WRUBs) refer to a series of activities related to water use. WRUBs include domestic water utilization (WRUB1), industrial water utilization (WRUB2), agricultural water utilization (WRUB3), ecological water utilization (WRUB4), and hydroelectric power generation (WRUB5). Many carbon emission studies based on LCA methods do not consider the end-use process, because the emission effects caused by end-use are not part of the life cycle [29]. However, some indirect emission effects closely related to WRBs are generated or caused by these behaviors, and end-use often results in a high proportion of CEE [27]. Therefore, based on the definition of WRBs, this study also includes CEE in the end-use process of water resources in the calculation range.

(1) Domestic water utilization (WRUB1): WRUB1 does not include public domestic water because the end-use purpose of public domestic water is so broad that it is difficult to achieve a relatively accurate quantification. The main source of CO_2 emissions from WRUB1 is the energy consumption of the heating process [27]. Combined with the actual domestic water consumption in China, CO_2 emissions in the energy-consuming process of cooking and bath heating can be taken as the CEE of WRUB1, and its CEEA method is as follows:

$$E_8 = Q_8 \times EI_8 \times EF \quad (15)$$

$$EI_8 = \rho \times R_{household} \times (R_{heat1} + R_{heat2}) \times C_w \times \Delta T \times 1/\eta \quad (16)$$

where E_8 is the CEE of WRUB1, kg; Q_8 is the total amount of domestic water consumption, m^3; ρ is the density of surface water (typically 1000), kg/m^3; C_w is the heat capacity of the water (generally 1.162×10^{-3} kWh/(kg·°C) [74]); ΔT is the temperature difference before and after heating, °C; η is the efficiency of the heating equipment (generally 95% [74]). $R_{household}$ is the proportion of residential household domestic water consumption in total domestic water consumption; R_{heat} is the proportion of water used for heating in residential household domestic water consumption, where R_{heat1} is the proportion of cooking and drinking water, and R_{heat2} is the proportion of bathing water. Depending on different research needs, $R_{household}$ can be obtained according to the actual investigation, or according to the proportion in the water resources bulletin. In addition, studies have examined the energy intensity (EI_8) of household water use in different regions for reference: 7.43 kWh/m^3 (China) [75], 24.6 kWh/m^3 (Ontario, Canada) [61].

(2) Industrial water utilization (WRUB2): China has a wide range of industrial sectors, and the water use processes in different sectors have different CO_2 emission characteristics. The energy consumption of WRUB2 is mainly concentrated in the link of water cooling and water heating [59], which is also the main source of CO_2 emission. There are two ideas for calculating the CEE of WRUB2:

$$E_9 = Q_9 \times EI_9 \times EF \tag{17}$$

$$E_9 = C_{industry} \times R_{water} \times EF \tag{18}$$

where E_9 is the CEE of WRUB2, kg; Q_9 is the total amount of industrial water consumption, m^3; EI_9 is the energy intensity of WRUB2 (energy consumption per unit of industrial water) kWh/m^3. EI_9 can be determined from field surveys, and relevant studies have concluded that the energy intensity of industrial water use in a typical Chinese city is 5.033 kWh/m^3 [76]. Another idea is to calculate CEE by determining the power consumption of WRUB2 through the power consumption structure of the industrial sector [59]. A study suggests that water-related electricity consumption in the industrial sector in typical Chinese cities accounts for about 10% [59]. $C_{industry}$ is total industrial electricity consumption, kWh; R_{water} is the ratio of water cooling and water heating power consumption to total power consumption in the industrial sector, %.

(3) Agricultural water utilization (WRUB3): Unlike domestic and industrial water, CO_2 emissions from agricultural water utilization are mainly concentrated in the irrigation process. There are five main sources of carbon emissions from farmland ecosystems: chemical fertilizers, pesticides, agricultural films, agricultural machinery, and agricultural irrigation [77]. In this study, CO_2 emissions from agricultural irrigation are used as the CEE of WRUB3. In addition, the carbon sink effect occurs on farmland due to photosynthesis during crop growth [78]. Therefore, the CO_2 absorption effect of WRUB3 should be considered [79]. The three elements of crop growth are: sunlight, water, and fertilizer, and the carbon sink effect in farmland is the result of the joint action of these three elements. Obviously, it is not appropriate to consider the entire amount of CO_2 absorbed by the farmland as the CO_2 absorption effect of WRUB3. Therefore, the CO_2 absorption effect of WRUB3 is separated from the overall CO_2 absorption effect of farmland by setting weights. Assuming that the three elements of sunlight, water, and fertilizer are equally important for the crop growth process [80], the contribution of these three elements to the carbon sink effect can be distributed by equal weight method. Of course, the weight distribution scheme can be discussed and adjusted according to the actual situation of crop planting. The CEE calculation method of WRUB3 is as follows:

$$E_{10} = E_{10emission} - E_{10absorption} \tag{19}$$

$$E_{10emission} = A \times \delta_e \times \frac{44}{12} \tag{20}$$

$$E_{10absorption} = \omega \times A \times \delta_a \times \frac{44}{12} \tag{21}$$

where E_{10} is the CEE of WRUB3, kg; $E_{10emission}$ is the total CO_2 emissions of WRUB3, kg; $E_{10absorption}$ is the amount of CO_2 absorbed by agricultural water utilization; A is the actual agricultural irrigation area, ha; δ_e and δ_a are CO_2 emission and absorption coefficient per unit irrigated area, t/ha; ω is the weight, which is initially set to 1/3.

(4) Ecological water utilization (WRUB4): Water resources are the foundation and core of ecosystem functions. The function of ecosystems such as woodlands, grasslands, wetlands, and watersheds cannot be performed without the maintenance of ecological water [81]. WRUB4 refers to artificial ecological water, that is, urban environmental water and rivers, lakes, and wetland replenishment water supplied by human measures [82]. Different from domestic and production water utilization behaviors, the CEE of WRUB4 cannot be directly quantified by energy as a medium. Therefore, in this study, the CO_2 absorbed by four land types closely related to ecological water use, namely, urban garden, urban green space (excluding garden area), water area within the jurisdiction, and wetland within the jurisdiction, is roughly taken as the CEE of WRUB4. Of course, the actual process of CO_2 absorption from WRUB4 is far more complicated than described.

$$E_{11} = -\sum_{i}^{n} A_i \times \delta_i \times \frac{44}{12} \tag{22}$$

where E_{11} is the CEE of WRUB4, kg; A is the area of ecological water land type, ha; δ is the CO_2 absorption coefficient of ecological water land type (the amount of CO_2 absorbed per unit area of ecological water land), t/ha. i is the type of land. δ can be obtained by field measurements in the study area, or by referring to existing studies [78].

(5) Hydroelectric power generation (WRUB5): CO_2 emissions from hydropower generation are much lower than those from thermal power [83]. Based on the UN CDM (United Nations' Clean Development Mechanism), GHG emissions from hydropower generation can be disregarded in the calculation of hydropower CDM projects [84]. Therefore, the relative carbon reduction effect of hydropower compared to thermal power is used in this study to quantify the CEE of WRUB5.

$$E_{12} = -G \times CPG \times EF_c \tag{23}$$

where E_{12} is the CEE of WRUB5, kg; G is the total amount of hydroelectric power, kWh; CPG is the standard coal consumption of power generation unit, tce/kWh; EF_c is the CO_2 emission coefficient of standard coal, kg/tce. CPG can be obtained from the investigation of the thermal power industry in the study area. Studies have shown that the average coal consumption of thermal power generating units in China is 3.7×10^{-4} tce/kWh [85]. EF_c can refer to IPCC guidelines for national greenhouse gas inventories [52] or existing studies [85].

2.5. CEEA Method to Water Resource Protection Behaviors

Water resource protection behaviors (WRPBs) refer to a series of activities related to water resources protection, including water saving (WRPB1), wastewater collection (WRPB2), wastewater treatment (WRPB3), and reclaimed water reuse (WRPB4).

(1) Water saving (WRPB1): Water saving behavior directly avoids part of the energy consumed in the development and allocation of water resources, so it can be regarded as a carbon reduction behavior [32,86]. Its CEEA method is as follows:

$$E_{13} = -Q_{13} \times (EP_{exploitation} + EP_{distribution}) \tag{24}$$

$$EP_{exploitation} = (E_1 + E_2)/(Q_1 + Q_2) \tag{25}$$

$$EP_{distribution} = E_7/Q_7 \tag{26}$$

where E_{13} is the CEE of WRPB1, kg; Q_{13} is the total amount of water saved, m^3; $EP_{exploitation}$ is the comprehensive CO_2 emission coefficient of water resource exploitation (CO_2 emissions per unit of water resource exploitation), kg/m^3; $EP_{distribution}$ is the comprehensive

CO_2 emission coefficient of water resource allocation (CO_2 emissions per unit of water resource allocation), kg/m^3. Other variables have the same meaning as above.

(2) Wastewater collection (WRPB2): Wastewater from different sources usually relies on gravity to converge to the wastewater network, and then is pressurized by the wastewater network pump to the wastewater treatment plant. Similar to WRAB1, the CEE of WRPB2 is mainly generated by energy consumption to overcome head loss [49]:

$$E_{14} = Q_{14} \times EI_{14} \times EF \tag{27}$$

$$EI_{14} = \frac{9.8 \times \rho \times (h_f + h_j)}{3.6 \times 10^6 \times \eta} \tag{28}$$

where E_{14} is the CEE of WRPB2, kg; Q_{14} is the total amount of wastewater collected, m^3; EI_{14} is the energy intensity of WRPB2 (electricity consumption by collecting unit of wastewater), kWh/m^3. EI_{14} should be obtained based on the investigation and statistics of the wastewater collection system in the study area, and can also refer to the values in related studies: 0.013 kWh/m^3 (China) [86].

(3) Wastewater treatment (WRPB3): The treatment methods of wastewater treatment plants in different countries are different, but generally include three stages: primary treatment, secondary treatment, and tertiary treatment. Each stage has different processes, and the energy consumption intensity of each process is different. The main CO_2 emissions are concentrated in the secondary and tertiary treatment stages [87]. On the other hand, untreated wastewater contains more pollutants such as *COD* and *BOD5*, which can produce large amounts of carbon emissions. WRPB3 has a positive CO_2 reduction effect by reducing the concentration of such pollutants [88]. In addition, wastewater treatment plants generally use sludge in wastewater for power generation [89], and its carbon reduction effect should also be considered. In this study, the CO_2 absorption effect of WRPB3 is considered based on the concentration difference of major carbon emission pollutants before and after wastewater treatment and the sludge power generation:

$$E_{15} = E_{15emission} - E_{15absorption} \tag{29}$$

$$E_{15emission} = Q_{15} \times EI_{15} \times EF - Q_{15} \times R_s \times P_s \times EF \tag{30}$$

$$EI_{15} = \sum_{i=1}^{3} \sum_{j} EI_{ij} \tag{31}$$

$$E_{15absorption} = Q_{15} \times \Delta R_{COD} \times EF_{COD} + Q_{15} \times \Delta R_{BOD5} \times EF_{BOD5} \tag{32}$$

where E_{15} is the CEE of wastewater treatment behavior, kg; Q_{15} is the total amount of wastewater treatment, m^3; EI_{15} is the energy intensity of WRPB3 (electricity consumption by treating unit of wastewater), kWh/m^3. EI_{ij} is the energy consumption intensity of the process j in stage i, kWh/m^3. The energy intensity or emission factor of unit wastewater treatment can be obtained by investigating the energy consumption and treatment capacity of the wastewater treatment plant [28,29]. EI_{15} from relevant studies are available for reference: 0.24 kWh/m^3 (China) [74]; 0.8–1.5 kWh/m^3 (Australia) [60]; 0.177–0.78 kWh/m^3 (USA) [56]; 0.41–0.61 kWh/m^3 (Spain) [71]; 0.44 kWh/m^3 (South Africa) [72]; 0.38–1.122 kWh/m^3 (Global) [55]. Rs is the sludge concentration in wastewater, generally 0.3–0.5% [90]; Ps is the power generation of unit sludge, and the coefficient in related research is 14.27 kWh/m^3 for reference [89]. ΔR_{COD} and ΔR_{BOD5} are the concentration differences of *COD* and *BOD5* before and after wastewater treatment, respectively. When the measurement conditions are available, the measurement results shall prevail. When conducting large-scale research, ΔR_{COD} and ΔR_{BOD5} can also be determined according to relevant emission standards. According to China's comprehensive wastewater discharge standard, the concentration difference between *COD* and *BOD5* before wastewater treatment (Level 3 standard) and after wastewater treatment (Level 1 standard) is 0.94 kg/m^3 and 0.58 kg/m^3. EF_{COD} and EF_{BOD5} are the amount of CO_2 reduced by removing unit *COD* and *BOD5*, and the units are $kgCO_2/kgCOD$ and $kgCO_2/kgBOD5$, respectively. According to the relevant emission factors released by IPCC [52], EF_{COD} and EF_{BOD5} are 0.69 and 1.65, respectively.

(4) Reclaimed water reuse (WRPB4): Reclaimed water reuse reduces the extraction of surface water and groundwater, and can therefore be considered as a WRB to reduce CO_2 emissions. The calculation formula of CEE is as follows:

$$E_{16} = -Q_{16} \times EP_{exploitation} \tag{33}$$

$$EP_{exploitation} = (E_1 + E_2)/(Q_1 + Q_2) \tag{34}$$

where E_{16} is the CEE of reclaimed water reuse behavior, kg; Q_{16} is the amount of reclaimed water reuse, m^3; EP is the comprehensive CO_2 emission coefficient of water resources exploitation (CO_2 emissions per unit of water resource exploitation), kg/m^3.

2.6. Function Table of CEEA for Water Resource Behaviors

The above methods and ideas are summarized and all the CEEA formulas are combined to form a table, which is the function table of CEEA (FT-CEEA) for WRBs (Table 2). In addition, in view of the large regional differences in the grid CO_2 emission factor and the energy intensity of groundwater extraction, the referenceable values (Tables 3 and 4) for different regions of China are given [54], which can be selected according to the actual situation of the study area. The instructions for using FT-CEEA are as follows.

Table 2. FT-CEEA for water resource behaviors.

WRBs	CEEA Formulas	Parameter Reference Values
WRDB1 (Surface water lifting)	$E_1 = Q_1 \times EI_1 \times EF$ $EI_1 = \frac{\rho \times g \times h_1}{3.6 \times 10^6 \times \eta}$ $EF = \frac{\sum_i \left(FC_{i,y} \times NCV_{i,y} \times EF_{CO_2,i,y}\right)}{EG_y}$	EI_1: 0.2 kWh/m^3 (China); 0.0002–1.74 kWh/m^3 (Global) EF: Table 3 (China)
WRDB2 (Groundwater extraction)	$E_2 = Q_2 \times EI_2 \times EF$ $EI_2 = \frac{9.8 \times \rho \times h_2}{3.6 \times 10^6 \times \eta}$	EI_2: Table 4 (China); 0.18–0.49 kWh/m^3 kWh/m^3 (USA); 0.48–0.53 kWh/m^3 (Australia); 0.37–1.44 kWh/m^3 (Global)
WRDB3 (Reservoir storage)	$E_3 = Q_3 \times EI_3 \times EF$	EI_3: 0.14 kWh/m^3 (China)
WRDB4 (Raw water treatment)	$E_4 = Q_4 \times EI_4 \times EF$	EI_4: 0.31 kWh/m^3 (China); 0.371–0.392 kWh/m^3 (USA); 0.1–0.6 kWh/m^3 (Australia); 0.38–1.44 kWh/m^3 (Canada); 0.11–1.5 kWh/m^3 (Spain); 0.15–0.44 kWh/m^3 (New Zealand)
WRDB5 (Seawater Desalination)	$E_5 = Q_5 \times EI_5 \times EF$	EI_5: 5.9 kWh/m^3 (China); 4 kWh/m^3 (Australia); 2.4–8.5 kWh/m^3 (Global)
WRAB1 (Tap water allocation)	$E_6 = Q_6 \times EI_6 \times EF$ $EI_6 = \frac{9.8 \times \rho \times (h_f + h_j)}{3.6 \times 10^6 \times \eta}$ $h_f = \lambda \frac{l}{4R} \frac{v^2}{2g}; h_j = \zeta \frac{v^2}{2g}$	EI_6: 0.2 kWh/m^3 (China); 0.2–0.32 kWh/m^3 (California, USA); 0.12–0.22 kWh/m^3 (Spain); 0.1 kWh/m^3 (South Africa)
WRAB2 (Inter–regional water transfer)	$E_7 = Q_7 \times EI_7 \times EF$ $EI_7 = \frac{\rho \times g \times (h_f + h_j)}{3.6 \times 10^6 \times \eta}$	EI_7: 0.815 kWh/m^3 (China)
WRUB1 (Domestic water utilization)	$E_8 = Q_8 \times EI_8 \times EF$ $EI_8 = \rho \times R_{household} \times (R_{heat1} + R_{heat2}) \times Cw \times \Delta T \times 1/\eta$	EI_8: 7.43 kWh/m^3 (China); 24.6 kWh/m^3 (Ontario, Canada)
WRUB2 (Industrial water utilization)	$E_9 = Q_9 \times EI_9 \times EF$ $E_9 = C_{industry} \times R_{water} \times EF$	EI_9: 5.033 kWh/m^3 (China) R_{water}: 10% (China)
WRUB3 (Agricultural water utilization)	$E_{10} = E_{10emission} - E_{10absorption}$ $E_{10emission} = A \times \delta_e \times \frac{44}{12}$ $E_{10absorption} = \omega \times A \times \delta_a \times \frac{44}{12}$	δ_e: 0.266 tC/ha (China) δ_a: 4.05 tC/ha (China) ω: 1/3

Table 2. *Cont.*

WRBs	CEEA Formulas	Parameter Reference Values
WRUB4 (Ecological water utilization)	$E_{11} = -\sum_i^n A_i \times \delta_i \times \frac{44}{12}$	δ: Garden 3.81 tC/ha; Green Space 0.948 tC/ha; Wetland 0.567 tC/ha; Water Area 0.567 tC/ha (China)
WRUB5 (Hydroelectric power generation)	$E_{12} = -G \times CPG \times EF_c$	CPG: 3.7 × 10^{-4} tce/kWh (China) EF_c: 670 kg/tce (China)
WRPB1 (Water saving)	$E_{13} = -Q_{13} \times (EP_{exploitation} + EP_{distribution})$ $EP_{exploitation} = (E_1 + E_2)/(Q_1 + Q_2)$ $EP_{distribution} = E_7/Q_7$	For the parameters of E_1, E_2, and E_7, see WRDB1, WRDB2, and WRAB2
WRPB2 (Wastewater collection)	$E_{14} = Q_{14} \times EI_{14} \times EF$ $EI_{14} = \frac{9.8 \times \rho \times (h_f + h_j)}{3.6 \times 10^6 \times \eta}$	EI_{14}: 0.013 kWh/m^3 (China)
WRPB3 (Wastewater treatment)	$E_{15} = E_{15emission} - E_{15absorption}$ $E_{15emission} = Q_{15} \times EI_{15} \times EF - Q_{15} \times R_s \times P_s \times EF$ $EI_{15} = \sum_{i=1}^{3} \sum_j EI_{ij}$ $E_{15absorption} = Q_{15} \times (\Delta R_{COD} \times EF_{COD} + \Delta R_{BOD5} \times EF_{BOD5})$	EI_{15}: 0.24 kWh/m^3 (China); 0.8–1.5 kWh/m^3 (Australia); 0.177–0.78 kWh/m^3 (USA); 0.41–0.61 kWh/m^3 (Spain); 0.44 kWh/m^3 (South Africa); 0.38–1.122 kWh/m^3 (Global) R_s: 0.3~0.5% (China) EF_{COD}: 0.69 $kgCO_2$/kgCOD (IPCC); EF_{BOD}: 1.65 $kgCO_2/kgBOD_5$ (IPCC)
WRPB4 (Reclaimed water reuse)	$E_{16} = -Q_{16} \times EP_{exploitation} EP_{exploitation} = (E_1 + E_2)/(Q_1 + Q_2)$	For the parameters of E_1 and E_2, see WRDB1 and WRDB2

Table 3. Average CO_2 emission factor of power grids in different regions of China ($kgCO_2$/kWh).

Provinces	*EF*	Provinces	*EF*
Beijing	0.8292	Henan	0.8444
Tianjin	0.8733	Hubei	0.3717
Hebei	0.9148	Hunan	0.5523
Shanxi	0.8798	Chongqing	0.6294
Inner Mongolia	0.8503	Sichuan	0.2891
Shandong	0.9236	Guangdong	0.6379
Liaoning	0.8357	Guangxi	0.4821
Jilin	0.6787	Guizhou	0.6556
Heilongjiang	0.8158	Yunnan	0.415
Shanghai	0.7934	Hainan	0.6463
Jiangsu	0.7356	Shaanxi	0.8696
Zhejiang	0.6822	Gansu	0.6124
Anhui	0.7913	Qinghai	0.2263
Fujian	0.5439	Ningxia	0.8184
Jiangxi	0.7635	Xinjiang	0.7636

Table 4. Energy intensity of unit groundwater extraction in different regions of China (kWh/m^3).

Provinces	EI_2	Provinces	EI_2
Beijing	0.44	Henan	0.3
Tianjin	0.66	Hubei	0.22
Hebei	0.53	Hunan	0.4
Shanxi	0.62	Chongqing	0.57
Inner Mongolia	0.3	Sichuan	0.3
Shandong	0.47	Guangdong	0.41
Liaoning	0.21	Guangxi	0.34
Jilin	0.35	Guizhou	0.36
Heilongjiang	0.43	Yunnan	0.45
Shanghai	0.39	Hainan	0.41
Jiangsu	0.36	Shaanxi	0.64
Zhejiang	0.43	Gansu	0.5
Anhui	0.32	Qinghai	0.52
Fujian	0.4	Ningxia	0.27
Jiangxi	0.37	Xinjiang	0.6

(1) FT-CEEA is a collection of formulas for estimating and cross-sectionally comparing the CEE of various WRBs. The CEEA formulas for different WRBs in FT-CEEA can be used selectively depending on the study purpose and study scale. The quantity, type, and calculation method of WRBs in FT-CEEA are not static and can be updated and improved according to the changing situation and new research progress.

(2) The results of each formula are not necessarily an absolute measurement of the emission or absorption effects of CO_2, but the idea of each formula is relatively reasonable. FT-CEEA is equivalent to setting up a "ruler" as a relative comparison of CEE generated by WRBs calculated by different researchers. FT-CEEA has no scale limitation and can be applied to different scales with limited accuracy requirements. However, the specific parameters need to be adjusted according to the actual situation of the research object.

(3) Most of the formulas in FT-CEEA need to be supported by relevant parameters, but in most cases, it is difficult to carry out field investigations and measurements of the parameters. Given this situation, some valuable reference values are provided in this table. Of course, some changes can be made in the selection of parameter reference values according to different research needs and actual conditions.

3. Case Study

3.1. Overview of the Study Area

China has a vast territory, and there are significant spatial differences in industrial structure, water use mode, and carbon emission intensity in different regions. In terms of CO_2 emissions in 2019, Shanxi (the province with the highest emission intensity) is 37 times higher than Qinghai (the province with the lowest emission intensity) under different development orientation [91]. In the past 20 years, under the background of rapid economic and social development, some provinces in China are facing many challenges such as the insufficient capacity for sustainable utilization of water resources and prominent conflict between carbon emission reduction and economic development [92]. Since the 1990s, China has been in a new period of rapid growth in carbon emissions, lagging behind developed countries in time. Although China's total carbon emissions ranked first in the world in recent years, China's per capita carbon emissions are still far lower than developed countries. Many traditional industries in China still maintain a production mode with high consumption and high emission. Promoting the low-carbon transformation of traditional industries has become an urgent bottleneck to achieving China's carbon neutrality goal [7].

In this study, 31 provincial administrative regions in mainland of China are divided into 8 regions [93]. The regional division, elevation distribution, water supply structure, and CO_2 emission intensity of the study area are shown in Figure 3.

3.2. Data source and Description

In addition to the important parameters in FT-CEEA, the data used in the case study are mainly the data of indicators involved in different WRBs of 31 provinces in China in 2020. The data involved in WRABs include tap water allocation and inter-regional water transfer. The data involved in WRUBs include domestic water consumption, industrial water consumption, actual agricultural irrigation area, land area of four kinds of artificial ecological water utilization, and hydroelectric power generation. The data involved in WRPBs include water saving, wastewater treatment, and reclaimed water reuse.

The sources of the above data include China Water Resources Bulletin 2020, China Seawater Utilization Bulletin 2020, Water Resources Bulletin of 31 provinces in 2020, China Statistical Yearbook 2021, China Water Statistical Yearbook 2021, China Energy Statistical Yearbook 2021, China Environmental Statistical Yearbook 2021, and China Urban Construction Statistical Yearbook 2021.

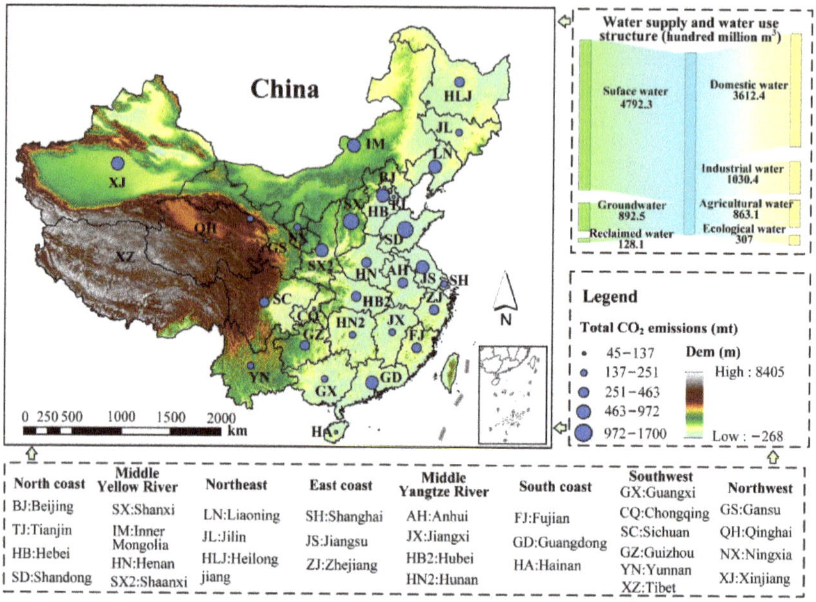

Figure 3. The study area.

3.3. *Results and Discussion*

3.3.1. Carbon Dioxide Emission Equivalent Analysis of WRDBs

Based on FT-CEEA and the above data, the CEE of WRBs in 31 provinces and 8 regions of China in 2020 was calculated. The calculation results of the eight regions are obtained by summing the included provinces.

The CEEA results of WRDBs are presented in Table 5. In 2020, the surface water lifting behavior (WRDB1) in eight regions of China generated 63.52 million tons of CEE, accounting for 29.8% of the total CEE produced by WRDBs. Among them, the WRDB1 in middle Yangtze River and east coast provinces produced higher CEE of 12.63 million tons and 11.9 million tons, respectively. The three provinces of Shanghai, Jiangsu, and Zhejiang in the east coast region are dominated by surface water utilization. The surface water supply of Jiangsu Province in 2020 is 55.6 billion cubic meters, resulting in the CEE generated by WRDB1 ranking first among 31 provinces (8.18 million tons). The region with the smallest CEE of WRDB1 is the north coast region (4.64 million tons). On the other hand, WRDB2 in the north coast region produced the most CEE (8.2 million tons). In contrast, groundwater extraction in the east coast region produced only 0.12 million tons of CEE in 2020. The spatial distribution characteristics of CEEA results of WRDB1 and WRDB2 are closely related to the water supply structure in different regions. Compared with the southern provinces of China, the northern provinces have a higher degree of groundwater exploitation and a larger proportion of groundwater utilization, which is also a manifestation of the uneven spatial distribution of water resources in China [94]. In addition, Xinjiang is the province with the most CEE generated by WRDB2 in 31 provinces (5.69 million tons). The reason is that Xinjiang has a large amount of groundwater supply. In 2020, the groundwater supply in Xinjiang is 12.43 billion cubic meters, second only to Heilongjiang (12.94 billion cubic meters). Another important factor is that Xinjiang's higher altitude means it takes much more energy to extract per unit of groundwater than the eastern provinces [39].

Table 5. CEE of WRDBs and WRABs in eight regions of China in 2020 (10,000 tons).

Regions	WRDB1	WRDB2	WRDB3	WRDB4	WRDB5	WRAB1	WRAB2
North coast	464.39	819.75	595.18	408.86	193.61	275.50	1062.86
Middle Yellow River	550.14	799.23	943.75	425.88	0.00	268.61	447.79
Northeast	528.18	628.60	983.30	243.08	20.66	157.29	0.00
East coast	1190.44	12.27	468.52	1061.67	61.60	725.56	34.30
Middle Yangtze River	1262.68	103.99	1650.82	793.48	0.00	530.31	40.53
South coast	746.55	39.34	663.31	516.33	16.56	336.86	90.46
Southwest	720.44	44.23	1807.78	385.87	0.00	260.99	6.60
Northwest	889.61	660.88	477.18	119.54	0.00	80.88	12.48
Total	6352.42	3108.29	7589.83	3954.71	292.43	2636.02	1695.02

WRDB3 is the behavior that produces the most CEE in WRDBs, generating 75.9 million tons of CEE in 2020, accounting for 35.6% of the total CEE produced by WRDBs. Among them, the CEE produced in middle Yangtze River and southwest regions was significantly higher than that in other regions, and the CEE produced by WRDB3 in the east coast region was less (4.69 million tons). Raw water treatment behavior (WRDB4) produced 39.55 million tons of CEE in 2020. Due to the high proportion of domestic and industrial water, the east coast, the middle Yangtze River and the southern coastal provinces have become the main contributors to the CEE generated by WRDB4. In 2020, the CEE generated by seawater desalination behavior (WRDB5) was 2.92 million tons, accounting for 1.4% of the total CEE generated by WRDBs. China's desalination plants are mainly concentrated in 9 coastal provinces [64], which are Shandong, Hebei, Zhejiang, Tianjin, Liaoning, Guangdong, Fujian, Hainan, and Jiangsu in descending order according to CEE. The proportion of CEE generated by WRDB5 in the north and east coast provinces exceeded 87%.

3.3.2. Carbon Dioxide Emission Equivalent Analysis of WRABs

The CEEA results of water resource allocation behaviors (WRABs) are shown in Table 5. WRAB1 produced 26.36 million tons of CEE in 2020, accounting for 60.9% of the total CEE produced by WRABs. The CEE of WRAB1 is similar to WRDB4 in spatial distribution. The difference in water resources utilization structure in different regions of China can explain the distribution characteristics to some extent. Compared with the eastern provinces of China, the northwest provinces have a higher proportion of agricultural water and a lower proportion of industrial and domestic water [95]. Tap water supply is mainly concentrated in industrial and domestic water. Therefore, the water use structure dominated by agricultural water has led to the CO_2 emission effect of WRAB1 in the northwest region being much lower than that in the eastern region. Cross-regional water transfer behavior produced 16.95 million tons of CEE in 2020. Due to the existence of large-scale water diversion projects such as the South-to-North Water Diversion Project and the Luanhe River Diversion Project, the CEE generated by WRAB2 in the north coast and the middle Yellow River provinces accounted for up to 89%. This spatial distribution feature is similar to the research results of Xiang and Jia [54].

3.3.3. Carbon Dioxide Emission Equivalent Analysis of WRUBs

The CEEA results of WRUBs are shown in Table 6. Among the five kinds of WRUBs, the CEE value of domestic water utilization and industrial water utilization is positive, resulting in the CO_2 emission effect. The CEE value of agricultural water utilization, ecological water utilization, and hydroelectric power generation is negative, resulting in the CO_2 absorption effect. Among them, the CEE calculation of WRUB2 is based on the first calculation scheme (energy intensity scheme).

Table 6. CEE of WRUBs in eight regions of China in 2020 (10,000 tons).

Regions	WRUB1	WRUB2	WRUB3	WRUB3 Emission	WRUB3 Absorption	WRUB4	WRUB5
North coast	4684.22	2643.82	−3600.23	885.44	4485.67	−873.39	−87.73
Middle Yellow River	4514.91	3112.55	−3901.06	959.42	4860.49	−1514.52	−923.25
Northeast	2547.60	1812.01	−2873.56	706.72	3580.28	−1629.72	−452.12
East coast	5229.36	12,308.51	−2085.14	512.82	2597.96	−1188.59	−598.08
Middle Yangtze River	5225.80	8202.69	−4165.40	1024.44	5189.84	−1616.46	−6028.88
South coast	5044.67	3755.17	−1065.50	262.05	1327.55	−858.01	−1472.20
Southwest	4103.96	2716.85	−2855.96	702.39	3558.36	−3286.00	−20,571.24
Northwest	1337.90	802.65	−2635.73	648.23	3283.96	−3439.65	−3461.95
Total	32,688.42	35,354.25	−23,182.59	5701.51	28,884.10	−14,406.34	−33,595.46

In 2020, the CEE of WRUB1 (326.88 million tons) and WRUB2 (353.54 million tons) in 31 provinces of China are not very different in total, but there are large differences between regions. The CEE generated by WRUB2 in the east coast and middle Yangtze River provinces is higher than that generated by WRUB1, especially in the east coast provinces. The outstanding proportion of industrial and domestic water in Jiangsu Province leads to the highest CEE generated by WRUB2. The difference in water use structure is the main reason for the difference in CEE of industrial and domestic water in different regions [96]. The absorption effect of WRUB3 (288.84 million tons) is greater than the emission effect (57.02 million tons), so the CEE of agricultural water utilization behavior is negative in total (−231.83 million tons). The WRUB3 of the northwest provinces has produced a considerable CO_2 emission effect (6.48 million tons), which is consistent with the local water resource utilization structure [97]. The middle Yangtze River provinces have more agricultural irrigation area, and the CO_2 absorption effect produced by WRUB3 is also the highest among the eight regions (51.9 million tons).

Ecological water utilization behavior (WRUB4) produced −144.06 million tons of CEE in 2020. The CEE of WRUB4 in southwest and northwest provinces was nearly half of the total CEE produced by WRUB4. The main reason is that the wetland and water area of Sichuan, Tibet, Qinghai, Xinjiang, and other provinces is much higher than other regions. The strong guarantee of ecological water use in the above-mentioned provinces has played an important role in maintaining the carbon sink function of wetland and water ecosystem [98]. In 2020, the hydroelectric power generation behavior (WRUB5) in eight regions of China produced a total of −335.95 million tons of CEE with significant spatial differences. Southwest provinces have the most abundant hydropower resources [99], while the proportion of hydropower in the energy structure of the north coast provinces is very small. The distribution of hydropower resources in China is the main reason for the CEE spatial difference of WRUB5.

3.3.4. Carbon Dioxide Emission Equivalent Analysis of WRPBs

The CEEA results of WRPBs are shown in Table 7. Among the four WRPBs, only the CEE value of wastewater collection behavior (WRPB2) is positive, resulting in CO_2 emission effect. The CEE values of the other three WRPBs are negative, resulting in the CO_2 absorption effect. Water saving behavior (WRPB1) can undoubtedly provide a positive impact on reducing CO_2 emissions [100]. If only the energy saving effect of WRPB1 on water resources development and allocation is considered, the CEE of WRPB1 in 2020 is −2.05 million tons. In general, Shanghai, Guangdong, Zhejiang, Jiangsu, and Beijing are at the forefront of the construction of water-saving society [101], and there is still a large room for improvement in the capacity of water-saving and emission reduction in northwest provinces. The CEE of WRPB2 is the smallest among all WRBs in FT-CEEA (0.5 million tons). The CO_2 absorption effect produced by wastewater treatment behavior (WRPB3) is significantly greater than the emission effect. The spatial distribution characteristics of CEE of WRPB3 are directly related to the wastewater treatment capacity of different regions.

The east coast and south coast provinces have a large amount of wastewater discharge and a strong wastewater treatment capacity [102], which correspondingly brings a higher carbon dioxide emission and absorption effect. If only the energy saving effect of reclaimed water reuse behavior on water resources development is considered, the CEE generated by WRPB4 in 2020 is −2.68 million tons. The CEEA results of WRPB4 are closely related to regional water resource endowment and water supply structure. Compared with the southern provinces, Beijing, Hebei, Shandong, Henan, and other northern provinces are relatively short of water, so the reuse of reclaimed water has become an effective means to alleviate the contradiction between local water supply and demand [103]. As a result, the amount of reclaimed water supplied by these provinces is much higher than that of other provinces, and correspondingly, more CO_2 absorption effect is generated.

Table 7. CEE of WRPBs in eight regions of China in 2020 (10,000 tons).

Regions	WRPB1	WRPB2	WRPB3	WRPB3 Emission	WRPB3 Absorption	WRPB4
North coast	−24.69	9.33	−1158.76	131.27	1290.03	−109.55
Middle Yellow River	−24.16	5.42	−703.43	76.30	779.73	−67.26
Northeast	−12.51	5.73	−809.33	80.63	889.96	−17.91
East coast	−57.17	9.61	−1489.48	135.17	1624.65	−22.95
Middle Yangtze River	−38.73	6.29	−1255.82	88.55	1344.37	−16.36
South coast	−33.13	8.07	−1482.34	113.53	1595.87	−7.12
Southwest	−10.26	4.51	−1153.53	63.40	1216.93	−12.19
Northwest	−4.62	1.35	−231.80	19.06	250.86	−14.82
Total	−205.27	50.31	−8284.47	707.91	8992.38	−268.15

4. Conclusions

In this study, the carbon dioxide emission equivalent analysis (CEEA) method of water resource behaviors (WRBs) was developed, and a function table of carbon dioxide emission equivalent (FT-CEEA) was constructed. Based on the FT-CEEA, the CEE of different WRBs in 31 provinces of China in 2020 was analyzed. Some valuable conclusions are as follows:

(1) Four categories of WRBs in 31 provinces of China produced a total of 0.137 billion tons of CEE in 2020, of which the emission effect was 1.001 billion tons and the absorption effect was 0.864 billion tons. There is significant spatial variability in CEE of WRBs in eight regions of China, and the spatial distribution characteristics of CEE produced by different WRBs are also different. Water supply/utilization structure, energy consumption structure, water resources endowment, physical geographic characteristics, hydropower resources distribution are important reasons for the spatial differences of CEE.

(2) The WRDBs and WRABs produced a total of 0.256 billion tons of CEE. Among the WRDBs, reservoir storage and surface water lifting have the most CO_2 emission effect. Among the WRABs, the CEE from inter-regional water transfer is smaller than that from tap water allocation. Water resource protection behaviors produced −87 million tons of CEE. The absorption effect of wastewater treatment behavior is the main contributor to CEE, followed by reclaimed water reuse behavior and water saving behavior.

(3) The CO_2 emission and absorption effects of WRUBs are most significant among four categories. Domestic water and industrial water utilization are the two main sources of emission effects, hydroelectric power generation behavior produced the greatest absorption effect. There is still a certain distance to achieve carbon neutrality in the field of water resources.

Based on the above conclusions, some targeted measures and suggestions are discussed for the carbon neutrality goal in the field of water resources. Increasing the proportion of hydropower generation, improving the capacity of ecological water security, strengthening wastewater treatment and reclaimed water reuse, and promoting the construction of water-saving society can be considered as effective ways to promote carbon neutrality in this field.

However, there are still some limitations. The consideration of water resource behavior categories may not be comprehensive. In this study, the water resource behaviors were divided into four categories: development, allocation, utilization, and protection. However, water resource behaviors are not limited to the four categories, and the number of WRBs is far more than 16. Therefore, FT-CEEA is dynamic rather than static, and needs to be constantly updated. In addition, many CEE calculations of WRB are completed by using energy as an intermediate medium, which is the quantitative scheme adopted by most related studies. Although the energy consumption is the major factor in the generation of CEE by those WRBs, it cannot be excluded that there may be other potential factors contributing to carbon emissions. When these potential factors reach a certain scale, the resulting CEE also needs to be considered. Moreover, for some WRBs, the CEEA method may not be considered perfect. For example, the CO_2 absorbed by the four types of land closely related to ecological water utilization was roughly used as the CEE of WRUB3. In fact, the CO_2 absorbed by the lands is due to many factors, including ecological water utilization. How to separate the CEE of ecological water and CEE produced by other factors? Further exploration and refinement are still needed.

Author Contributions: Conceptualization, formal analysis, funding acquisition, project administration, Q.Z.; data curation, methodology, software, visualization, writing—original draft, Z.Z., C.Z. and X.Q.; investigation, J.M., Z.Z., C.Z., and X.Q.; supervision, Q.Z. and J.M.; resources, Q.Z. and Z.Z.; writing—review and editing, Q.Z., Z.Z., J.M., C.Z. and X.Q. All authors have read and agreed to the published version of the manuscript.

Funding: This research was funded by the Natural Science Foundation of China (no. 52279027), the National Key Research and Development Program of China (no. 2021YFC3200201), and the Major Science and Technology Projects for Public Welfare of Henan Province (no. 201300311500).

Institutional Review Board Statement: Not applicable.

Informed Consent Statement: Not applicable.

Data Availability Statement: Publicly available datasets were analyzed in this study. These data can be found and downloaded on relevant websites.

Conflicts of Interest: The authors declare no conflict of interest.

References

1. Michalak, A.M. Study role of climate change in extreme threats to water quality. *Nature* **2016**, *535*, 349–350. [CrossRef] [PubMed]
2. Hermwille, L.; Obergassel, W.; Ott, H.E.; Beuermann, C. UNFCCC before and after Paris–what's necessary for an effective climate regime? *Clim. Policy* **2017**, *17*, 150–170. [CrossRef]
3. Christoff, P. The promissory note: COP 21 and the Paris Climate Agreement. *Environ. Politics* **2016**, *25*, 765–787. [CrossRef]
4. Azar, C.; Johansson, D.J.A. IPCC and the effectiveness of carbon sinks. *Environ. Res. Lett.* **2022**, *17*, 041004. [CrossRef]
5. Salvia, M.; Reckien, D.; Pietrapertosa, F.; Eckersley, P.; Spyridaki, N.-A.; Krook-Riekkola, A.; Olazabal, M.; De Gregorio Hurtado, S.; Simoes, S.G.; Geneletti, D.; et al. Will climate mitigation ambitions lead to carbon neutrality? An analysis of the local-level plans of 327 cities in the EU. *Renew. Sustain. Energy Rev.* **2021**, *135*, 110253. [CrossRef]
6. Mallapaty, S. How China could be carbon neutral by mid-century. *Nature* **2020**, *586*, 482–483. [CrossRef]
7. Liu, Z.; Deng, Z.; He, G.; Wang, H.; Zhang, X.; Lin, J.; Qi, Y.; Liang, X. Challenges and opportunities for carbon neutrality in China. *Nat. Rev. Earth Environ.* **2022**, *3*, 141–155. [CrossRef]
8. Wu, Y.; Xu, B. When will China's carbon emissions peak? Evidence from judgment criteria and emissions reduction paths. *Energy Rep.* **2022**, *8*, 8722–8735. [CrossRef]
9. Mancini, M.S.; Galli, A.; Niccolucci, V.; Lin, D.; Bastianoni, S.; Wackernagel, M.; Marchettini, N. Ecological Footprint: Refining the carbon Footprint calculation. *Ecol. Indic.* **2016**, *61*, 390–403. [CrossRef]
10. Wiedmann, T.; Minx, J.; Barrett, J.; Wackernagel, M. Allocating ecological footprints to final consumption categories with input–output analysis. *Ecol. Econ.* **2006**, *56*, 28–48. [CrossRef]
11. Dong, H.; Geng, Y.; Xi, F.; Fujita, T. Carbon footprint evaluation at industrial park level: A hybrid life cycle assessment approach. *Energy Policy* **2013**, *57*, 298–307. [CrossRef]
12. Lenzen, M.; Sun, Y.-Y.; Faturay, F.; Ting, Y.-P.; Geschke, A.; Malik, A. The carbon footprint of global tourism. *Nat. Clim. Chang.* **2018**, *8*, 522–528. [CrossRef]

13. Lombardi, M.; Laiola, E.; Tricase, C.; Rana, R. Assessing the urban carbon footprint: An overview. *Environ. Impact Assess. Rev.* **2017**, *66*, 43–52. [CrossRef]
14. Onat, N.C.; Kucukvar, M. Carbon footprint of construction industry: A global review and supply chain analysis. *Renew. Sustain. Energy Rev.* **2020**, *124*, 109783. [CrossRef]
15. Radonjič, G.; Tompa, S. Carbon footprint calculation in telecommunications companies—The importance and relevance of scope 3 greenhouse gases emissions. *Renew. Sustain. Energy Rev.* **2018**, *98*, 361–375. [CrossRef]
16. Henriksson, P.J.G.; Heijungs, R.; Dao, H.M.; Phan, L.T.; De Snoo, G.R.; Guinée, J. Product Carbon Footprints and Their Uncertainties in Comparative Decision Contexts. *PLoS ONE* **2015**, *10*, e0121221. [CrossRef]
17. Chai, C.; Zhang, D.; Yu, Y.; Feng, Y.; Wong, M.S. Carbon Footprint Analyses of Mainstream Wastewater Treatment Technologies under Different Sludge Treatment Scenarios in China. *Water* **2015**, *7*, 918–938. [CrossRef]
18. Zhou, Y.; Chen, M.; Tang, Z.; Mei, Z. Urbanization, land use change, and carbon emissions: Quantitative assessments for city-level carbon emissions in Beijing-Tianjin-Hebei region. *Sustain. Cities Soc.* **2021**, *66*, 102701. [CrossRef]
19. Raymond, P.A.; Hartmann, J.; Lauerwald, R.; Sobek, S.; McDonald, C.; Hoover, M.; Butman, D.; Striegl, R.; Mayorga, E.; Humborg, C.; et al. Global carbon dioxide emissions from inland waters. *Nature* **2013**, *503*, 355–359. [CrossRef]
20. Keller, P.S.; Marcé, R.; Obrador, B.; Koschorreck, M. Global carbon budget of reservoirs is overturned by the quantification of drawdown areas. *Nat. Geosci.* **2021**, *14*, 402–408. [CrossRef]
21. Guotong, Q.; Fei, C.; Na, W.; Dandan, Z. Inter-annual variation patterns in the carbon footprint of farmland ecosystems in Guangdong Province, China. *Sci. Rep.* **2022**, *12*, 14134. [CrossRef] [PubMed]
22. Ran, Y.; Li, X.; Sun, R.; Kljun, N.; Zhang, L.; Wang, X.; Zhu, G. Spatial representativeness and uncertainty of eddy covariance carbon flux measurements for upscaling net ecosystem productivity to the grid scale. *Agric. For. Meteorol.* **2016**, *230–231*, 114–127. [CrossRef]
23. Feng, X.; Fu, B.; Lu, N.; Zeng, Y.; Wu, B. How ecological restoration alters ecosystem services: An analysis of carbon sequestration in China's Loess Plateau. *Sci. Rep.* **2013**, *3*, 2846. [CrossRef] [PubMed]
24. Thompson, R.L.; Lassaletta, L.; Patra, P.K.; Wilson, C.; Wells, K.C.; Gressent, A.; Koffi, E.N.; Chipperfield, M.P.; Winiwarter, W.; Davidson, E.A.; et al. Acceleration of global N2O emissions seen from two decades of atmospheric inversion. *Nat. Clim. Chang.* **2019**, *9*, 993–998. [CrossRef]
25. Fernández-Martínez, M.; Sardans, J.; Chevallier, F.; Ciais, P.; Obersteiner, M.; Vicca, S.; Canadell, J.G.; Bastos, A.; Friedlingstein, P.; Sitch, S.; et al. Global trends in carbon sinks and their relationships with CO_2 and temperature. *Nat. Clim. Chang.* **2018**, *9*, 73–79. [CrossRef]
26. Trask, M. *Water-Energy Relationship*; California Energy Commission: Sacramento, CA, USA, 2005.
27. Griffiths-Sattenspiel, B.; Wilson, W. *The Carbon Footprint of Water*; River Network: Portland, OR, USA, 2009.
28. Wakeel, M.; Chen, B.; Hayat, T.; Alsaedi, A.; Ahmad, B. Energy consumption for water use cycles in different countries: A review. *Appl. Energy* **2016**, *178*, 868–885. [CrossRef]
29. Rothausen SG, S.A.; Conway, D. Greenhouse-gas emissions from energy use in the water sector. *Nat. Clim. Change* **2011**, *1*, 210–219. [CrossRef]
30. Friedrich, E.; Pillay, S.; Buckley, C. Carbon footprint analysis for increasing water supply and sanitation in South Africa: A case study. *J. Clean. Prod.* **2009**, *17*, 1–12. [CrossRef]
31. Li, R.; Zhao, R.; Xie, Z.; Xiao, L.; Chuai, X.; Feng, M.; Zhang, H.; Luo, H. Water–energy–carbon nexus at campus scale: Case of North China University of Water Resources and Electric Power. *Energy Policy* **2022**, *166*, 113001. [CrossRef]
32. Valek, A.M.; Sušnik, J.; Grafakos, S. Quantification of the urban water-energy nexus in México City, México, with an assessment of water-system related carbon emissions. *Sci. Total Environ.* **2017**, *590*, 258–268. [CrossRef]
33. Sambito, M.; Freni, G. LCA Methodology for the Quantification of the Carbon Footprint of the Integrated Urban Water System. *Water* **2017**, *9*, 395. [CrossRef]
34. Fang, A.; Newell, J.P.; Cousins, J. The energy and emissions footprint of water supply for Southern California. *Environ. Res. Lett.* **2015**, *10*, 114002. [CrossRef]
35. Boulos, P.F.; Bros, C.M. Assessing the carbon footprint of water supply and distribution systems. *J. Am. Water Work. Assoc.* **2010**, *102*, 47–54. [CrossRef]
36. Heihsel, M.; Lenzen, M.; Malik, A.; Geschke, A. The carbon footprint of desalination: An input-output analysis of seawater re-verse osmosis desalination in Australia for 2005–2015. *Desalination* **2019**, *454*, 71–81. [CrossRef]
37. Siddiqi, A.; Fletcher, S. Energy intensity of water end-uses. *Curr. Sustain./Renew. Energy Rep.* **2015**, *2*, 25–31. [CrossRef]
38. Escriva-Bou, A.; Lund, J.R.; Pulido-Velazquez, M. Modeling residential water and related energy, carbon footprint and costs in California. *Environ. Sci. Policy* **2015**, *50*, 270–281. [CrossRef]
39. Wang, J.; Rothausen, S.G.; Conway, D.; Zhang, L.; Xiong, W.; Holman, I.P.; Li, Y. China's water–energy nexus: Greenhouse-gas emissions from ground-water use for agriculture. *Environ. Res. Lett.* **2012**, *7*, 014035. [CrossRef]
40. Zeng, S.; Chen, X.; Dong, X.; Liu, Y. Efficiency assessment of urban wastewater treatment plants in China: Considering greenhouse gas emissions. *Resour. Conserv. Recycl.* **2017**, *120*, 157–165. [CrossRef]
41. Zib, I.I.I.L.; Byrne, D.M.; Marston, L.T. Operational carbon footprint of the US water and wastewater sector's en-ergy consumption. *J. Clean. Prod.* **2021**, *321*, 128815. [CrossRef]

42. Marinelli, E.; Radini, S.; Foglia, A.; Lancioni, N.; Piasentin, A.; Eusebi, A.L.; Fatone, F. Validation of an evidence-based methodology to support regional carbon foot-print assessment and decarbonisation of wastewater treatment service in Italy. *Water Res.* **2021**, *207*, 117831. [CrossRef]
43. Wu, Z.; Duan, H.; Li, K.; Ye, L. A comprehensive carbon footprint analysis of different wastewater treatment plant configurations. *Environ. Res.* **2022**, *214*, 113818. [CrossRef]
44. Zhou, Y.; Zhang, B.; Wang, H.; Bi, J. Drops of Energy: Conserving Urban Water to Reduce Greenhouse Gas Emissions. *Environ. Sci. Technol.* **2013**, *47*, 10753–10761. [CrossRef] [PubMed]
45. Parece, T.E.; Grossman, L.; Geller, E.S. Reducing Carbon Footprint of Water Consumption: A Case Study of Water Conservation at a University Campus. In *The Handbook of Environmental Chemistry*; Younos, T., Grady, C., Eds.; Springer: Berlin/Heidelberg, Germany, 2013; pp. 199–218. [CrossRef]
46. Wang, J.; Chen, X.; Liu, Z.; Frans, V.F.; Xu, Z.; Qiu, X.; Xu, F.; Li, Y. Assessing the water and carbon footprint of hydropower stations at a national scale. *Sci. Total. Environ.* **2019**, *676*, 595–612. [CrossRef] [PubMed]
47. Zhang, L.; Chen, S. Carbon peaks of water systems in Chinese cities under varying water demand dynamics and energy transition pathways. *J. Clean. Prod.* **2022**, *379*, 134695. [CrossRef]
48. Venkatesh, G.; Chan, A.; Brattebø, H. Understanding the water-energy-carbon nexus in urban water utilities: Comparison of four city case studies and the relevant influencing factors. *Energy* **2014**, *75*, 153–166. [CrossRef]
49. Bakhshi, A.A.; Demonsabert, S.M. Estimating the carbon footprint of the municipal water cycle. *J. Am. Water Work. Assoc.* **2012**, *104*, E337–E347. [CrossRef]
50. Stokes, J.R.; Horvath, A. Energy and Air Emission Effects of Water Supply. *Environ. Sci. Technol.* **2009**, *43*, 2680–2687. [CrossRef]
51. Presura, E.; Robescu, L.D. Energy use and carbon footprint for potable water and wastewater treatment. *Proceedings of the International Conference on Business Excellence*. 2017, Volume 11, pp. 191–198. Available online: https://sciendo.com/article/10.1515/picbe-2017-0020 (accessed on 17 January 2023).
52. Eggleston, H.S.; Buendia, L.; Miwa, K.; Ngara, T.; Tanabe, K. *2006 IPCC Guidelines for National Greenhouse Gas Inventories*; Institute for Global Environmental Strategies: Hayama, Japan, 2006.
53. Hu, G.; Ou, X.; Zhang, Q.; Karplus, V.J. Analysis on energy–water nexus by Sankey diagram: The case of Beijing. *Desalination Water Treat.* **2013**, *51*, 4183–4193. [CrossRef]
54. Xiang, X.; Jia, S. China's water-energy nexus: Assessment of water-related energy use. *Resour. Conserv. Recycl.* **2019**, *144*, 32–38. [CrossRef]
55. Plappally, A.K.; Lienhard, V.J.H. Energy requirements for water production, treatment, end use, reclamation, and disposal. *Renew. Sustain. Energy Rev.* **2012**, *16*, 4818–4848. [CrossRef]
56. Appelbaum, B. Water and sustainability: US electricity consumption for water supply and treatment—The next half century. *Water Supply.* **2002**, *4*, 93.
57. Rocheta, E.; Peirson, W. *Urban Water Supply in a Carbon Constrained Australia*; UNSW Water Research Centre: Kensington, Australia, 2011.
58. Li, X.; Liu, J.; Zheng, C.; Han, G.; Hoff, H. Energy for water utilization in China and policy implications for integrated planning. *Int. J. Water Resour. Dev.* **2016**, *32*, 477–494. [CrossRef]
59. He, G.; Zhao, Y.; Wang, J.; Zhu, Y.; Jiang, S.; Li, H.; Wang, Q. The effects of urban water cycle on energy consumption in Beijing, China. *J. Geogr. Sci.* **2019**, *29*, 959–970. [CrossRef]
60. Marsh, D. The Water–Energy Nexus: A Comprehensive Analysis in the Context of New South Wales. Ph.D. Dissertation, Faculty of Engineering and Information Technology, University of Technology, Sydney, Australia, 2008.
61. Maas, C. *Ontario's Water-Energy Nexus: Will We Find Ourselves in Hot Water or Tap into Opportunity?* POLIS Project on Ecological Governance; University of Victoria: Victoria, BC, Canada, 2010.
62. Muñoz, I.; Milà-i-Canals, L.; Fernández-Alba, A.R. Life cycle assessment of water supply plans in Mediterranean Spain: The Ebro river transfer versus the AGUA Programme. *J. Ind. Ecol.* **2010**, *14*, 902–918. [CrossRef]
63. Kneppers, B.; Birchfield, D.; Lawton, M. Energy-water relationships in reticulated water infrastructure systems. *Water Supply* **2009**, *76*, 1–27.
64. Lin, S.; Zhao, H.; Zhu, L.; He, T.; Chen, S.; Gao, C.; Zhang, L. Seawater desalination technology and engineering in China: A review. *Desalination* **2020**, *498*, 114728. [CrossRef]
65. Liu, S.; Wang, Z.; Han, M.; Wang, G.; Hayat, T.; Chen, G. Energy-water nexus in seawater desalination project: A typical water production system in China. *J. Clean. Prod.* **2020**, *279*, 123412. [CrossRef]
66. Buonomenna, M.G. Membrane processes for a sustainable industrial growth. *RSC Adv.* **2012**, *3*, 5694–5740. [CrossRef]
67. von Medeazza, G.M. "Direct" and socially-induced environmental impacts of desalination. *Desalination* **2005**, *185*, 57–70. [CrossRef]
68. Sharif, M.N.; Haider, H.; Farahat, A.; Hewage, K.; Sadiq, R. Water–energy nexus for water distribution systems: A literature review. *Environ. Rev.* **2019**, *27*, 519–544. [CrossRef]
69. Smith, K.; Liu, S.; Liu, Y.; Guo, S. Can China reduce energy for water? A review of energy for urban water supply and wastewater treatment and suggestions for change. *Renew. Sustain. Energy Rev.* **2018**, *91*, 41–58. [CrossRef]
70. He, G.; Zhao, Y.; Wang, J.; Li, H.; Zhu, Y.; Jiang, S. The water–energy nexus: Energy use for water supply in China. *Int. J. Water Resour. Dev.* **2018**, *35*, 587–604. [CrossRef]
71. Corominas, J. Agua y energía en el riego, en la época de la sostenibilidad. *Ing. Del Agua* **2010**, *17*, 219–233. [CrossRef]

72. Buckley, C.; Friedrich, E.; von Blottnitz, H. Life-cycle assessments in the South African water sector: A review and future challenges. *Water SA* **2011**, *37*, 719–726. [CrossRef]
73. Sturm, T.W. *Open Channel Hydraulics*; McGraw-Hill: New York, NY, USA, 2001.
74. Sousa, V.; Meireles, I. Dynamic simulation of the energy consumption and carbon emissions for domestic hot water production in a touristic region. *J. Clean. Prod.* **2022**, *355*, 131828. [CrossRef]
75. Qiu, G.Y.; Zou, Z.; Li, W.; Li, L.; Yan, C. A quantitative study on the water-related energy use in the urban water system of Shenzhen. *Sustain. Cities Soc.* **2022**, *80*, 103786. [CrossRef]
76. Zhao, R.; Yu, J.; Xiao, L.; Sun, J.; Luo, H.; Yang, W.; Chuai, X.; Jiao, S. Carbon emissions of urban water system based on water-energy-carbon nexus. *Acta Geogr. Sin.* **2021**, *76*, 3119–3134.
77. Duan, H.P.; Zhang, Y.; Zhao, J.B.; Bian, X.M. Carbon footprint analysis of farmland ecosystem in China. *J. Soil Water Conserv.* **2011**, *25*, 203–208.
78. Feng, M.; Zhao, R.; Huang, H.; Xiao, L.; Xie, Z.; Zhang, L.; Sun, J.; Chuai, X. Water–energy–carbon nexus of different land use types: The case of Zhengzhou, China. *Ecol. Indic.* **2022**, *141*, 109073. [CrossRef]
79. Wu, H.; Guo, S.; Guo, P.; Shan, B.; Zhang, Y. Agricultural water and land resources allocation considering carbon sink/source and water scarcity/degradation footprint. *Sci. Total. Environ.* **2021**, *819*, 152058. [CrossRef]
80. van Diepen, C.A.; Wolf, J.; van Keulen, H.; Rappoldt, C. WOFOST: A simulation model of crop production. *Soil Use Manag.* **1989**, *5*, 16–24. [CrossRef]
81. Qiu, M.; Zuo, Q.; Wu, Q.; Yang, Z.; Zhang, J. Water ecological security assessment and spatial autocorrelation analysis of prefectural regions involved in the Yellow River Basin. *Sci. Rep.* **2022**, *12*, 5105. [CrossRef] [PubMed]
82. Zuo, Q.; Zhou, K.; Yang, L. Study on the quantity of water resources and the water quantity for ecosystem use in water resources programming. *Arid. Land Geogr.* **2002**, *4*, 296–301.
83. Xu, J.; Wang, F.; Lv, C.; Xie, H. Carbon emission reduction and reliable power supply equilibrium based daily scheduling towards hydro-thermal-wind generation system: A perspective from China. *Energy Convers. Manag.* **2018**, *164*, 1–14. [CrossRef]
84. Whittington, R. Hydro and the CDM: The role of hydroelectricity in meeting Kyoto obligations. *Refocus* **2007**, *8*, 54–56. [CrossRef]
85. Wu, B.; Chen, Y.; Zeng, Y.; Zhao, Y.; Yuan, C. Evaluation of carbon emission reduction in power generation and shipping of the Three Gorges Reservoir. *Resour. Environ. Yangtze Basin* **2011**, *20*, 257–261.
86. Meng, F.; Liu, G.; Liang, S.; Su, M.; Yang, Z. Critical review of the energy-water-carbon nexus in cities. *Energy* **2019**, *171*, 1017–1032. [CrossRef]
87. Racoviceanu, A.I.; Karney, B.W.; Kennedy, C.A.; Colombo, A.F. Life-Cycle Energy Use and Greenhouse Gas Emissions Inventory for Water Treatment Systems. *J. Infrastruct. Syst.* **2007**, *13*, 261–270. [CrossRef]
88. Qamar, M.A.; Javed, M.; Shahid, S.; Iqbal, S.; Abubshait, S.A.; Abubshait, H.A.; Ramay, S.M.; Mahmood, A.; Ghaithan, H.M. Designing of highly active g-C3N4/Co@ ZnO ternary nanocomposites for the disinfection of pathogens and degradation of the organic pollutants from wastewater under visible light. *J. Environ. Chem. Eng.* **2021**, *9*, 105534. [CrossRef]
89. Zhang, Q.; Sun, D.; Wang, M.; Yin, C. Analysis of Typical Energy Saving Technology in the Sewage Treatment Plant. *Energy Procedia* **2017**, *142*, 1230–1237. [CrossRef]
90. Peng, L.; Nairuo, Z.; Wei, X. COD and Carbon Emission Reduction in Sludge Deep Dewatering Treatment and Disposal. *Environ. Sanit. Eng.* **2012**, *20*, 9–12.
91. Guan, Y.; Shan, Y.; Huang, Q.; Chen, H.; Wang, D.; Hubacek, K. Assessment to China's recent emission pattern shifts. *Earth's Future* **2021**, *9*, e2021EF002241. [CrossRef]
92. Wu, Q.; Zuo, Q.; Ma, J.; Zhang, Z.; Jiang, L. Evolution analysis of water consumption and economic growth based on Decomposition-Decoupling Two-stage Method: A case study of Xinjiang Uygur Autonomous Region, China. *Sustain. Cities Soc.* **2021**, *75*, 103337. [CrossRef]
93. Li, D.; Zuo, Q.; Zhang, Z. A new assessment method of sustainable water resources utilization considering fair-ness-efficiency-security: A case study of 31 provinces and cities in China. *Sustain. Cities Soc.* **2022**, *81*, 103839. [CrossRef]
94. Wang, Y.; Wang, Y.; Su, X.; Qi, L.; Liu, M. Evaluation of the comprehensive carrying capacity of interprovincial water resources in China and the spatial effect. *J. Hydrol.* **2019**, *575*, 794–809. [CrossRef]
95. Liu, X.; Xu, Y.; Sun, S.; Zhao, X.; Wang, Y. Analysis of the Coupling Characteristics of Water Resources and Food Security: The Case of Northwest China. *Agriculture* **2022**, *12*, 1114. [CrossRef]
96. Zhou, Y.; Ma, M.; Gao, P.; Xu, Q.; Bi, J.; Naren, T. Managing water resources from the energy-water nexus perspective under a changing climate: A case study of Jiangsu province, China. *Energy Policy* **2018**, *126*, 380–390. [CrossRef]
97. Shi, Q.; Chen, S.; Shi, C.; Wang, Z.; Deng, X. The Impact of Industrial Transformation on Water Use Efficiency in Northwest Region of China. *Sustainability* **2014**, *7*, 56–74. [CrossRef]
98. Yang, Q.; Liu, G.; Casazza, M.; Hao, Y.; Giannetti, B.F. Emergy-based accounting method for aquatic ecosystem services valuation: A case of China. *J. Clean. Prod.* **2019**, *230*, 55–68. [CrossRef]
99. Li, X.-Z.; Chen, Z.-J.; Fan, X.-C.; Cheng, Z.-J. Hydropower development situation and prospects in China. *Renew. Sustain. Energy Rev.* **2018**, *82*, 232–239. [CrossRef]
100. Shimizu, Y.; Toyosada, K.; Yoshitaka, M.; Sakaue, K. Creation of Carbon Credits by Water Saving. *Water* **2012**, *4*, 533–544. [CrossRef]

101. Xu, Y.; Tian, Q.; Yu, Y.; Li, M.; Li, C. Water-Saving Efficiency and Inequality of Virtual Water Trade in China. *Water* **2021**, *13*, 2994. [CrossRef]
102. Chen, K.; Liu, X.; Ding, L.; Huang, G.; Li, Z. Spatial Characteristics and Driving Factors of Provincial Wastewater Discharge in China. *Int. J. Environ. Res. Public Health* **2016**, *13*, 1221. [CrossRef]
103. Yi, L.; Jiao, W.; Chen, X.; Chen, W. An overview of reclaimed water reuse in China. *J. Environ. Sci.* **2011**, *23*, 1585–1593. [CrossRef] [PubMed]

water

MDPI

Article

System Simulation and Prediction of the Green Development Level of the Chengdu-Chongqing City Group

Yuxin Liang [1], Liping Zhang [1,2,3,*], Mengsi Leng [1], Yi Xiao [1,2,3,*] and Jun Xia [1,2,3]

1 State Key Laboratory of Water Resources and Hydropower Engineering Science, Wuhan University, Wuhan 430072, China
2 Hubei Key Laboratory of Water System Science for Sponge City Construction, Wuhan University, Wuhan 430072, China
3 Institute of Water Security, Wuhan University, Wuhan 430072, China
* Correspondence: zhanglp@whu.edu.cn (L.Z.); 00011515@whu.edu.cn (Y.X.)

Abstract: Green development is a low-carbon, sustainable model for the achievement of the harmonious development of the economy and nature. Nowadays, the problems of resource scarcity and environmental pollution in the process of economic development are pressing, and the promotion of green development is the general trend. As one of the three growth poles of China's Yangtze River economic belt, the Chengdu-Chongqing City Group is an important platform to lead toward green development in the western region of China. Based on the understanding of the connotation of green development, this study established a green development-level evaluation system, including 19 indicators in three dimensions: target level, criterion level, and indicator level, and used the entropy weight method to measure the green development level of the Chengdu-Chongqing City Group. In view of the dynamic nature of the green development process, this study constructed a system dynamics model of the green development level of the Chengdu-Chongqing City Group and simulated and compared it between 2022 and 2050 under five shared socio-economic pathway (SSP) scenarios so as to provide a reference basis for future development. The results show that the overall green development level of the Chengdu-Chongqing City Group is on an upward trend, with the highest green development level under the SSP1 path and the lowest under the SSP3 path, and the lagging distance tends to increase further. In the next 30 years, the Chengdu-Chongqing City Group should initially follow SSP2 as the basis for development and then gradually perform a transition to SSP1 by 2035 to achieve real sustainable development, after which it should continue to develop according to the SSP1 path until 2050.

Keywords: urban agglomeration; green development; system dynamics; shared socio-economic pathways; simulation prediction

Citation: Liang, Y.; Zhang, L.; Leng, M.; Xiao, Y.; Xia, J. System Simulation and Prediction of the Green Development Level of the Chengdu-Chongqing City Group. *Water* **2022**, *14*, 3947. https://doi.org/10.3390/w14233947

Academic Editor: Bahram Gharabaghi

Received: 9 November 2022
Accepted: 30 November 2022
Published: 4 December 2022

1. Introduction

Green development theory originated from the green movement in the West. British scholars first noticed the constraint relationship between socio-economic development and environmental capacity. In 1662, William Petty recognized that value originates from labor, whose ability to create wealth is not infinite, being constrained by conditions such as land and technology [1]. Although the theory of green development has been deeply studied by international scholars for decades, there is still no uniform definition.

The research on green development theory in China is relatively late compared to that in Western countries; it focuses on reducing the use of resources and environmental damage while ensuring economic growth.

The representative one is the "people-oriented" concept of green development, proposed by Hu Angang of the National Research Center of Tsinghua University, which emphasizes the win–win situation of economic development and ecological environmental protection, i.e., the rational exploitation of resources and the reduction in pollution under the premise of sustained and stable economic growth [2]. In synthesis, green development is a model of economic development that does not entail damage to the ecological environment or over-exploitation and consumption of resources. It differs from the traditional crude development mode as it is environmentally friendly and entails resource-saving; moreover, it increases green wealth, allows the realization of a green life, and promotes the harmonious coexistence of humans and nature.

In recent years, green development, as a transformative development concept, has gradually achieved global consensus. With the maturity of the green development theory, scholars have started to conduct research related to the green development level, mainly focusing on the relationship between the economy, resources, and the environment [3]. More specifically, the related research focused mainly on three aspects: the measurement and evaluation of the green development level [3,4]; the identification of its influence factors [5,6]; and its simulation prediction [7,8]. These studies focused on both single cities and the whole country.

At present, the green development level is measured and evaluated mainly through the comprehensive evaluation index system method, the TOPSIS model, and the data envelopment analysis method. At the end of the 20th century, the comprehensive green development index system was developed and was further improved [9–13] as a comprehensive evaluation index system to measure and evaluate the green development level in specific regional and urban scopes. In 2009, the Economist Intelligence Unit (London, UK) constructed eight categories of green city indicators to analyze the environmental situation in 30 European cities [14]. In 2018, Carli et al. constructed a multi-objective integrated indicator system to analyze the sustainable development of urban energy, water, and environmental systems [15]. In 2019, Tian et al. adopted the entropy method to measure the greening of the Yangtze River Delta city agglomeration in terms of green growth, green welfare, green wealth, and green governance [16]. In 2021, Hu et al. established an urban green development evaluation index system for 108 cities in the Yangtze River Economic Zone to analyze the role of technological innovation in promoting green development [17].

To fully reflect the dynamics of the green development process and the interaction of its subsystems, as well as to improve the guidance and support for realistic decision making, scholars introduced the system dynamics approach to study green development from a system perspective. Founded by Professor Forrester at MIT, system dynamics was initially called "industrial dynamics" [18]; after the 1960s, he transformed system dynamics into urban dynamics when he studied the rise and fall of cities. This concept was then gradually improved and developed by Mass and Alfeld [19,20].

In the field of green development, the research on system dynamics is mostly focused on future simulation, and system dynamics models have been constructed to analyze the change trend of the future green development level by setting different simulation scenarios so as to obtain the optimal development path or countermeasure suggestions. Rudneva et al. constructed a "green" economic system dynamics model and analyzed the scenarios of regional "green" economic development [21]. Yang Guangming et al. used Chongqing city as an example to simulate and predict the sustainable development of the water resource carrying capacity, based on a system dynamics model [22]. Zhou studied the optimal path of urban agglomeration development by simulating the system dynamics of green development in the Beijing-Tianjin-Hebei urban agglomeration under different paths, putting forward relevant policy suggestions [23]. Wang et al. simulated and predicted the future scenarios of the ecological security index of the Beijing-Tianjin-Hebei city agglomeration by constructing a system dynamics model [24]. Li et al. considered three scenarios of economic development, technological innovation, and government investment to simulate and predict the industrial green development of the Beijing-Tianjin-Hebei urban agglomeration [25]. Jing et al. developed a system dynamics model of green low-carbon development for the Beijing-Tianjin-Hebei urban agglomeration considering four factors: economy, energy, environment, and urban aggregation, to explore the future system effects under different scenarios [26]. O'Keeffe et al. proposed a natural capital system dynamics model framework to predict and assess the natural capital performance of urban development in London [27]. Gudlaugsson, B et al. apply a system dynamics approach to investigate the complexity of energy transition in the Tees Valley region of the UK under a sustainable transition pathway and make policy recommendations [28]. In synthesis, a large number of studies have shown the general applicability of system dynamics to perform simulation and prediction studies in the field of green development.

Through a literature review, it can be found that the simulation and prediction research on green development is relatively mature, although there are still some areas for improvement:

1. In terms of spatial scope, the majority of the current research scholars on green development have focused on a single city, a single province, or a more developed socio-economic urban agglomeration, and there is almost no research on the green development of urban agglomerations in western China.
2. In terms of temporal span, the current research on urban green development has focused on a relatively short period, and the future forecast research period is generally up to 2030.
3. In terms of future simulation and prediction using system dynamics, most scholars have set up various scenarios for simulation based on the interplay of policies and indicators based on the construction of system dynamics models, and there is a lack of prediction and simulation under the different social system scenarios publicly released by the Intergovernmental Panel on Climate Change.

To address these shortcomings, this study selected the Chengdu-Chongqing City Group as the study area and divided the whole green development level system into four subsystems: social, economic, resource, and environmental subsystems. Next, a green development level system dynamics model was constructed, whose reliability and stability was verified through a historical test and a sensitivity test. On this basis, a simulation scheme of the system dynamics model of each city in the Chengdu-Chongqing City Group from 2019 to 2050 under five shared socio-economic pathways (SSPs) was set to analyze the change trend of the green development level, select the optimal pathway, and put forward corresponding suggestions and countermeasures.

2. Materials and Methods

2.1. Description of the Study Area

The Chengdu-Chongqing City Group, the Yangtze River midstream urban agglomeration, and the Yangtze River Delta urban agglomeration are the three major growth poles of China's Yangtze River economic belt, and the core areas for the promotion of urbanization and economic growth. The Chengdu-Chongqing City Group is located in the upper reaches of the Yangtze River; it spans between 28°24′ N~32°27′ N and 101°56′ E~110°11′ E, and it is a region with both development strength and potential in western China. As such, it is an important basis for the green development of the Yangtze River economic belt. The core of the Chengdu-Chongqing City Group is represented by the cities of Chengdu and Chongqing; it includes 27 districts (counties) in Chongqing and parts of Kaixian and Yunyang and 15 cities in Sichuan Province [29]. The location and extension of the Chengdu-Chongqing City Group is shown in Figure 1.

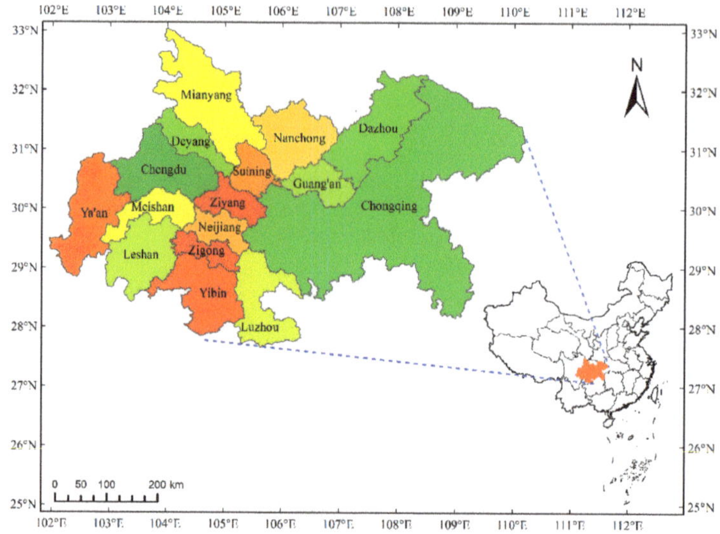

Figure 1. The Chengdu-Chongqing City Group scope diagram.

In terms of natural resources, the Chengdu-Chongqing City Group has abundant precipitation, with an average annual precipitation of about 900~1300 mm and abundant underground thermal energy and drinking water, such that the groundwater and transit water can basically meet the residents' water needs for living and production. At the same time, there are 12 main streams, including the Minjiang River and the Tuo River, and dozens of tributaries; hence, there are sufficient natural water resources in the Chengdu-Chongqing City Group area. According to the incomplete statistics alone, the animal and plant resources reach 11 classes, 200 families, 764 genera, and more than 3000 species, including rare animals such as giant pandas, red pandas, and golden snub-nosed monkeys, which are protected by the state. Thus, it can be seen that the Chengdu-Chongqing City Group is rich in natural resources such as water resources, mineral resources, land resources, and biological resources.

In terms of socio-economics, in 2018 the Chengdu-Chongqing City Group had a resident population of 95 million, accounting for 6.8% of the total population of the country. This area is rich in human resources and is a multi-ethnic area with more than 50 ethnic minorities, including Tibetans, Hui, Yi, and Tujia, in addition to the Han Chinese. Moreover, it is one of the regions with the best economic foundation and the strongest comprehensive strength in western China. Chengdu was awarded the first batch of national culture and tourism consumption demonstration cities in 2020; Chongqing has formed the world's largest electronic information industry cluster and China's largest domestic automotive industry cluster; the Chengdu and Chongqing region has strong strength in electronic information, equipment manufacturing, finance, and other industries; it has a better environment for innovation and entrepreneurship, and the open economic system is gradually taking shape.

In terms of spatial structure, the Chengdu-Chongqing City Group is located at the intersection of the horizontal axis along the Yangtze River corridor and the vertical axis of the Baokun corridor, with the Chengdu metropolitan area and the Chongqing metropolitan area as the main development axis, complemented by the urban belt along the river and the cities of Chengdu, Deyang, Mianyang, and Leshan, forming a spatial development pattern characterized by "one axis, two belts, two cores and three districts" [30]. As such, this region is key for the rapid development of central and western China and the creation of new space for economic growth [31].

2.2. Data Sources

The statistical data used in this article are mainly from the 1995–2018 "China Statistical Yearbook" (http://www.stats.gov.cn/tjsj./ndsj/) (accessed on 17 February 2022), released by the National Bureau of Statistics of China; the "Sichuan Statistical Yearbook" (http://tjj.sc.gov.cn/scstjj/c105855/nj.shtml) (accessed on 17 February 2022); the statistical yearbooks of 16 cities in the Chengdu-Chongqing City Group; the statistical bulletin of national economic and social development; the water resources bulletin; and other documents published in recent years. The interpolation method was employed to fill in the data gaps for single years.

2.3. The Entropy Weight Method

The entropy weight method is one of the most widely used objective weighting methods. This method determines the weights through a certain mathematical operation according to the relationships among the original data. It can avoid the non-objectivity and bias of the measurement results to a certain extent and has obvious advantages in calculating the level of green development [32]. The calculation steps are as follows:

Step 1: Data standardization processing

Because of the need to calculate the logarithm of the standardized index values in the calculation process of the entropy weighting method, in this study the index standardization method of non-zero transformation was adopted for data standardization [33]. The calculation formula is as follows:

Positive indicators:

$$x'_{ij} = \frac{(x_{ij} - x_{j\min})}{(x_{j\max} - x_{j\min})} \times 0.99 + 0.01 \tag{1}$$

Negative indicator:

$$x'_{ij} = \frac{(x_{j\max} - x_{ij})}{(x_{j\max} - x_{j\min})} \times 0.99 + 0.01 \tag{2}$$

where x'_{ij} is the standardized value of the j-th indicator in the i-th year; x_{ij} is the original value of the j-th indicator in the i-th year; $x_{j\min}$ is the minimum value of the j-th indicator; and $x_{j\max}$ is the maximum value of the j-th indicator.

Step 2: Calculation of the specific gravity, as follows:

$$Y_{ij} = \frac{x'_{ij}}{\sum_{i=1}^{m} x'_{ij}} \tag{3}$$

where Y_{ij} is the proportion of the j-th indicator in the i-th year, and m is the number of years of the study.

Step 3: Calculation of the information entropy, as follows:

$$e_j = \frac{1}{-\ln m} \sum_{i=1}^{m} (Y_{ij} \times \ln Y_{ij}) \tag{4}$$

where e_j is the information entropy of the j-th index.

Step 4: Calculation of the information utility value, as follows:

$$d_j = 1 - e_j \tag{5}$$

where w_j is the weight of the j-th indicator, and n is the number of indicators.

Step 5: Calculation of the level of green development

The subsystem of the green development level was calculated as follows:

$$w_j = \frac{d_j}{\sum_{j=1}^{n} d_j} \tag{6}$$

where w_j is the weight of the j-th indicator, and n is the number of indicators.

Step 6: Calculation of the level of green development

The subsystem of the green development was calculated as follows:

$$GDL = \sum_{j=1}^{n} x'_{ij} w_j \tag{7}$$

where GDL is the overall green development level of the system, with a value range of (0,1). The larger the value, the higher the overall green development level of the system, and vice versa.

2.4. System Dynamics Model Theory

System dynamics (SD) is based on the feedback control theory. Through the qualitative analysis of the research problem, real data and information are used to quantitatively deduce the coupling relationship between variables, while computer simulation technology is employed to study the behavior of dynamic social systems [34], ensuring the integrated application of natural and social sciences [35]. The establishment of an SD model mainly includes the drawing of a causal loop diagram, a system flow chart, a model equation, and parameter alignment; the stability of the system is ensured through the historical test and the sensitivity test [36].

The commonly used software for modeling SD includes Vensim, Ithink, DYNAMO, and Stella [37]. Vensim is easy to operate, has a friendly interface, uses arrows to connect variables when building an SD model, and the relationship between the variables is written in the form of model equations, which allows users to easily modify the content of the model. Therefore, the Vensim PLE software was selected in this study to model the SD of the green development level of the Chengdu-Chongqing City Group in the Yangtze River economic belt.

Model Variables and Equations

The SD model mainly includes four types of variables: level variables, rate variables, auxiliary variables, and constant variables. The state variables and the rate variables are the key variables [38].

The four main types of model equations for SD are the state equation, the rate equation, the auxiliary equation, and the constant equation. A table function can be used when the nonlinear relationships between the variables are difficult to present through mathematical operations [39]. The following equations represent the basic mathematical expressions used by Vensim.

1. State equation:

$$LvS(t) = St_0 + \int_{t_0}^{t} rateS(t)dt \tag{8}$$

where $LvS(t)$: is the value of the state variable at time t; St_0 is the value of the state variable at initial time t_0; and $rateS(t)$ is the rate of change of the state variable. The state equation can also be written in a discrete form as follows:

$$LEVEL.K = LEVEL.J + DT \times (INFLOW.JK - OUTFLOW.JK) \tag{9}$$

where $LEVEL.K$ is the value of the state variable at time K; $LEVEL.J$ is the value of the state variable value at time J; $INFLOW$ is the input rate; $OUTFLOW$ is the output rate; and DT is the temporal interval from time J to time K.

2. Rate equation:

$$rateS(t) = g[LvS(t), aux(t), const] \quad (10)$$

where $rateS(t)$: is the rate of change of the state variable; $LvS(t)$: is the value of the state variable at time t; and $aux(t)$ is the value of the auxiliary variable at time t.

3. Table Functions:

The table function is a function that describes the relationship between the independent and dependent variables by means of a list. The table function can be edited in Vensim PLE software using the "With lookup" function; its mathematical description is as follows:

$$lookupname([(X_{\min}, X_{\max}) - (Y_{\min}, Y_{\max})], (X_1, Y_1), (X_2, Y_2), \ldots (X_n, Y_n)) \quad (11)$$

where $(X_1, Y_1), (X_2, Y_2), \ldots (X_n, Y_n)$ are the values of the independent and dependent variables that have been assigned.

2.5. Shared Socio-Economic Pathways (SSPs)

2.5.1. Overview of Shared Socio-Economic Pathways (SSPs)

SSPs are a powerful tool to describe global socio-economic development scenarios. This tool was launched in 2010 by the IPCC on the basis of Representative Concentration Pathways (RCP)s, taking into account factors such as population growth, economic development, environmental conditions, and technological progress [40–42]. They describe the possible future development of society without the impact of climate change or climate policy [43].

Through a comprehensive analysis of existing global development frameworks, the IPCC has identified five typical SSPs from the perspectives of both climate challenge adaptation and climate challenge mitigation, namely SSP1 (sustainable development pathway), SSP2 (intermediate pathway), SSP3 (regional competitive pathway), SSP4 (uneven pathway), and SSP5 (fossil fuel-based development pathway), as shown in Figure 2. The mitigation challenge in Figure 2 refers to slowing down the pace of climate change by multiple means, including carbon dioxide emissions reduction; the adaptation challenge refers to the adoption of measures to adapt to the various impacts of climate change and avoid catastrophic consequences.

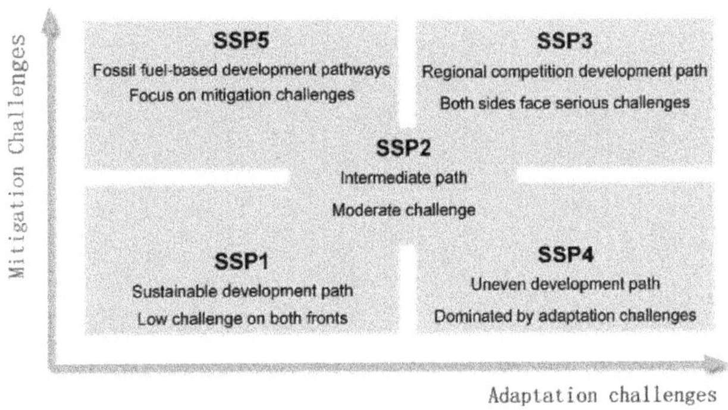

Figure 2. Shared social economy path diagram.

2.5.2. Simulation Scheme of the Future Green Development Level of the Chengdu-Chongqing City Group under the SSPs

The green development of each city in the Chengdu-Chongqing City Group entails a process of continuous change; accordingly, the indicators that characterize the level of urban green development are also continuously changing. In order to ensure the validity of the SD model, it was necessary to carry out a historical statistical description of the model parameters of each city in the Chengdu-Chongqing City Group [44]. In this study, the maximum, minimum, and average values of the six rate variables in each city from 2004 to 2018 were assessed to provide a reference interval for the setting of the model parameters for each city. The results for Chengdu and Chongqing are detailed in Appendix A.

The SSPs introduced by the IPCC contain almost all the possible future development trends. A research team led by Jiang Tong conducted a study on the future population and economy of China under the five SSPs, based on domestic and international population and economic forecasting studies [45–49]. At present, the research of Jiang Tong's team on China's population and economic future prediction has reached a temporal scale up to 2100 and has proved to be very mature [45,50]. Therefore, in this study, when setting the parameters of the model for the prediction of the future green development level of each city in the Chengdu-Chongqing City Group, the population growth rate and the GDP growth rate referred to the research results of Jiang Tong's team. The rate of the increase in the total energy consumption, arable land area, public green space area, and total water supply were assumed to be either high, medium-high, medium, or low according to the characteristics of the five SSPs considered. The details are shown in Table 1.

Table 1. Hypothetical scenarios of model parameters under the shared social economy path.

Path	Increase Rate of Total Energy Consumption	Increase Rate of Cultivated Land	Increase Rate of Public Green Space	Increase Rate of Total Water Supply
SSP 1	Low	High	High	Low
SSP 2	Medium	Medium	Medium	Medium
SSP 3	High	Low	Low	High
SSP 4	Medium-high	Medium	Medium	Medium-high
SSP 5	High	Low	High	High

In synthesis, the following principles were followed to formulate the future simulation plan of each city in the Chengdu-Chongqing City Group based on the five SSPs. The model parameters of each city in the Chengdu-Chongqing City Group under the SSP2 path were mainly based on the development trend of the statistical data in the historical periods and cities. Moreover, the values of the model parameters under the SSP1, SSP3, SSP4, and SSP5 paths were either higher or lower than a certain level of the SSP2 path, according to the parameter assumptions and historical data fluctuations.

3. Results and Discussion

3.1. Construction of the Green Development Level Indicator System and Indicator Weights

Based on the understanding of the connotation of green development, this study constructed an indicator system to evaluate the level of green development of the urban agglomerations by referring to the evaluation index system used in previous studies on green development and the 56 green development indicators suggested by the National Development and Reform Commission of China.

The developed evaluation index system consists of three levels: the target level, which is the green development level of the urban agglomerations; the guideline level, which includes the socio-economic, resource, and environment levels; and the quality-of-life level. It is composed of 19 indicators at the indicator level. The results of the calculation of the indicator weights are shown in the Table 2.

Table 2. Evaluation indicator system of green development level of the Chengdu-Chongqing City Group.

Target Layer	Guideline Layer	Indicator Layer	Unit	Attribute	w_1	w_2
Green Development Level	Socio-economic level	Real GDP per capita	Yuan	+	0.0949	0.2273
		Growth rate of regional GDP	%	+	0.0301	0.0721
		Value added of the secondary industry as a proportion of GDP	%	-	0.0551	0.1321
		Value added of tertiary industry as a proportion of GDP	%	+	0.0594	0.1423
		Total imports and exports	Billions of dollars	+	0.1307	0.3131
		Per capita disposable income ratio of urban and rural households	/	–	0.0162	0.0389
		Engel's coefficient of consumption of urban residents	%	–	0.0310	0.0742
	Resource environment level	Arable land area at the end of the year	hm^2	+	0.0504	0.1952
		Forest coverage rate	%	+	0.0451	0.1747
		Total annual water supply	Million tons	–	0.0248	0.0963
		Decrease in water consumption of CNY 10,000 GDP	%	+	0.0322	0.1250
		Decrease in energy consumption per CNY 10,000 GDP	%	+	0.0224	0.0866
		Emission of wastewater per unit of industrial value added	Tons/million yuan	–	0.0291	0.1127
		Industrial solid waste utilization rate	%	+	0.0541	0.2094
	Quality-of life level	Public green space per capita	m^2	+	0.0592	0.1823
		Ratio of good air quality days	%	+	0.0329	0.1015
		Number of medical beds per 10,000 people		+	0.1258	0.3875
		Road area per capita	m^2	+	0.0576	0.1775
		Sewage treatment rate	%	+	0.0491	0.1512

w_1: weight of the indicator layer relative to the target layer; w_2: weight of the indicator layer relative to the guideline layer.

3.2. Green Development Level SD Model Construction

3.2.1. Boundary Definition and Systematic Analysis

The SD model constructed in this study considered the administrative boundaries of 16 cities in the Chengdu-Chongqing City Group as the spatial boundaries. Considering China's national construction goals and the reliability of the historical data, the temporal limit of the system was set as covering the period 2004–2050; 2004–2018 was set as the historical data period, while 2019–2050 was set as the model simulation period, and the time step was set as one year.

The systematic analysis of the green development level of the Chengdu-Chongqing City Group allowed the identification of the key factors affecting the green development level. The complex green development level system was divided into four subsystems: society, economy, resources, and environment. The indicator types of each subsystem are shown in Figure 3.

After the analysis of the four subsystems, an analysis framework of the subsystem of the green development level of the urban agglomeration was constructed, and the mutual influence relationship among the subsystems was clarified, as shown in Figure 4.

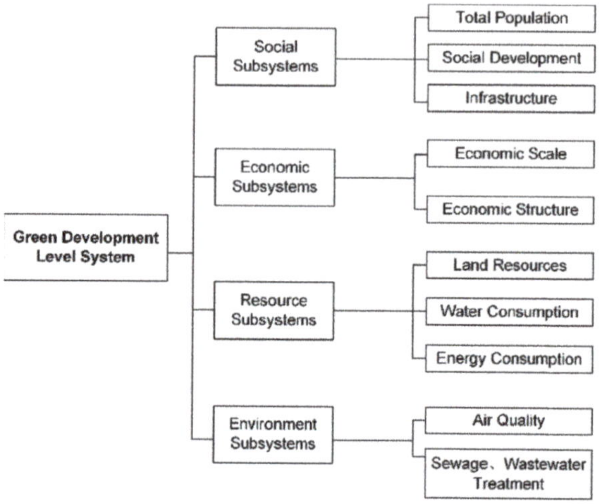

Figure 3. Systematic analysis of SD model of the green development level in the Chengdu-Chongqing City Group.

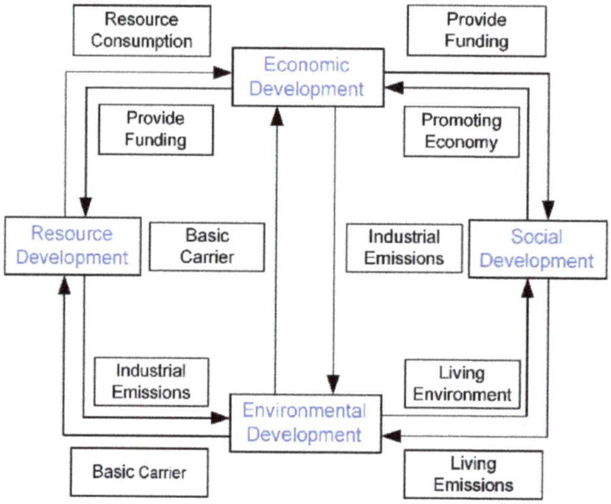

Figure 4. Analytical framework of SD model of the green development level in the Chengdu-Chongqing City Group.

3.2.2. Causal Feedback Relation

A system can be divided into an open system and a feedback system [51]. The SD model of the green development level of the Chengdu-Chongqing City Group belongs to the feedback system; accordingly, the feedback loop of this system and the causal relationship between the variables could be clarified through the causality diagram. Combining the model index analysis and the green development level system analysis framework, the causal relationship between the subsystems and the variables was assessed, and the causal loop diagram of the SD model of the green development level of the Chengdu-Chongqing City Group in the Yangtze economic belt was drawn using the SD software Vensim PLE, as shown in Figure 5.

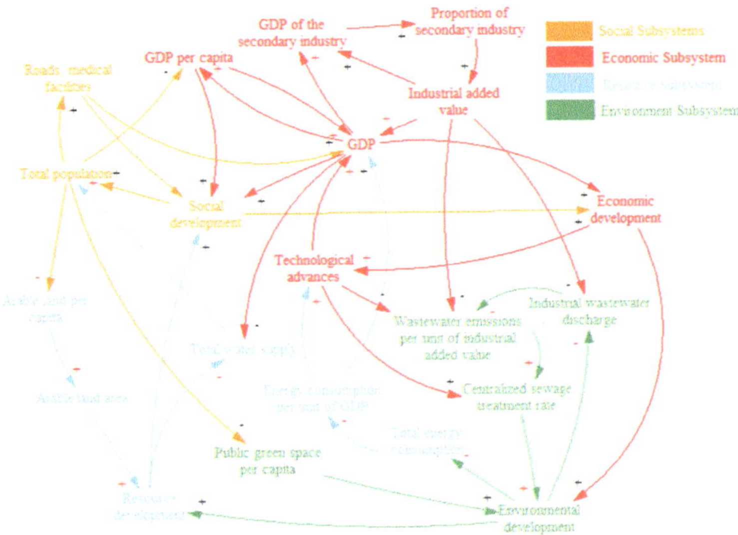

Figure 5. The causal loop of the SD model of green development level in the Chengdu-Chongqing City Group.

3.2.3. System Flow Chart

The system flow chart is the core of an SD model. Based on the causal loop diagram of the green development level of the Chengdu-Chongqing City Group, the system flow chart was further drawn. It includes 48 variables, including 6 state variables (i.e., total population, total GDP, total energy consumption, cultivated land area, public green area, and total water supply), 6 rate variables (i.e., population growth rate, GDP growth rate, increase rate of total energy consumption, increase rate of cultivated land area, increase rate of public green space, and increase rate of total water supply), and 36 auxiliary variables, as shown in Figure 6.

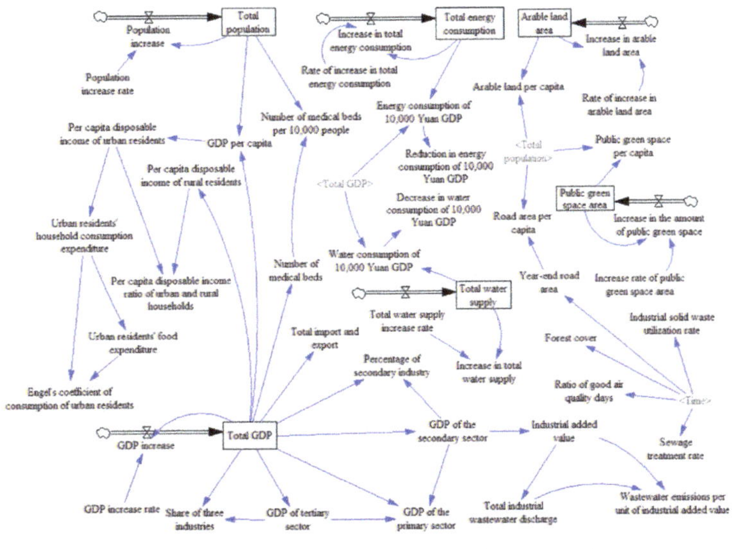

Figure 6. System flow chart of SD model of the green development level in the Chengdu-Chongqing City Group.

3.2.4. Establishment of Model Parameters and Equations

The variables of the SD model were divided into four main types: state variables, rate variables, auxiliary variables, and constants. The variables used in this model and their properties are shown in Table 3.

Table 3. SD model variables of the green development level in the Chengdu-Chongqing City Group.

Serial Number	Variable Name	Type	Serial Number	Variable Name	Type
1	Total population	L	25	Total industrial wastewater discharge	A
2	Population increase	R	26	Industrial value added	A
3	Population growth rate	C	27	Wastewater discharge per unit of industrial value added	A
4	Total GDP	L	28	Public green space area	L
5	GDP growth	R	29	Increase in public green space area	R
6	GDP growth rate	C	30	Increase rate of public green space	C
7	GDP per capita	A	31	Public green space per capita	A
8	Total Import and Export	A	32	Arable land area	L
9	Per capita disposable income of urban residents	A	33	Increase in arable land area	R
10	Per capita disposable income of rural residents	A	34	Rate of increase in arable land area	C
11	Per capita disposable income ratio of urban and rural households	A	35	Arable land area per capita	A
12	Urban residents' household consumption expenditure	A	36	Number of medical beds	A
13	Urban residents' food expenditure	A	37	Number of medical beds per 10,000 people	A
14	Engel coefficient of urban residents' consumption	A	38	Road area at the end of the year	A
15	Output value of primary industry	A	39	Road area per capita	A
16	Output value of secondary industry	A	40	Industrial solid waste utilization rate	A
17	Output value of tertiary industry	A	41	Forest coverage rate	A
18	Secondary industry output value as a proportion of GDP	A	42	Ratio of good air quality days	A
19	The proportion of the output value of the tertiary industry to GDP	A	43	Sewage treatment rate	A
20	Total energy consumption	L	44	Total water supply	L
21	Increase in total energy consumption	R	45	Increase in total water supply	R
22	Rate of increase in total energy consumption	C	46	Rate of increase in total water supply	C
23	Energy consumption of CNY 10,000 GDP	A	47	Water consumption of CNY 10,000 GDP	A
24	Reduction in energy consumption of CNY 10,000 GDP	A	48	Decrease in water consumption of CNY 10,000 GDP	A

L: state variables; R: rate variables; A: auxiliary variables; C: constants.

The model equations were established in the following three ways: for variables with clear mathematical relationships, they were established directly; for variables with random distributions and no clear mathematical relationships, the dataset was fitted using fitted regression analysis; for variables without functional relationships, or with functional relationships that were or difficult to establish, the table functions were created with the help of the "With lookup" function in the Vensim software. The model equations are shown in Table 4.

Table 4. Model equations for explicit mathematical relationships.

Subsystems	Variables	Variable Equation
Social Subsystems	Total population	INTEG (annual population growth, initial population)
	Population increase	Total population × Population growth rate
	Number of medical beds per 10,000 people	Number of medical beds/Total population
	Road area per capita	Road area/Total population
	Per capita disposable income ratio of urban and rural households	Per capita disposable income of urban households/Per capita disposable income of rural households
Economy Subsystems	Total GDP	INTEG (amount of GDP growth, initial GDP)
	GDP Growth	Total GDP × GDP growth rate
	GDP per capita	Total GDP/Total population
	Engel coefficient of urban residents' consumption	Urban residents' food expenditure/urban residents' household consumption expenditure
	GDP share of output value of secondary industry	Output value of secondary industry/Total GDP
	The proportion of the output value of the tertiary industry to GDP	Output value of tertiary industry/Total GDP
	Output value of primary industry	Total GDP—Output value of secondary industry—Output value of tertiary industry
Resources Subsystems	Total energy consumption	INTEG (total increase in energy consumption, total initial energy consumption)
	Increase in total energy consumption	Total energy consumption × Total energy consumption increase rate
	Arable land area	INTEG (increase in arable land area, initial arable land area)
	Increase in arable land area	Arable land area × Arable land area growth rate
	Arable land area per capita	Arable land area/Total population
	Energy consumption of CNY 10,000 GDP	Total energy consumption/Total GDP
	Total water supply	INTEG (increase in total water supply, initial total water supply)
	Increase in total water supply	Total water supply × Total water supply growth rate
	Water consumption of CNY 10,000 GDP	Total water supply/Total GDP
Environment Subsystems	Wastewater emissions per unit of industrial value added	Total industrial wastewater discharge/Industrial value added
	Public green space area	INTEG (increase in public green space area, initial public green space area)
	Increase in public green space area	Public green space area × Public green space area growth rate
	Public green space per capita	Public green space area/Total population

The parameters in the SD model generally include the constant values and the initial values of the state variables in the model. The constant values in the SD model of the green development level of the Chengdu-Chongqing City Group refer to six rate variables. Considering that the various cities were at different stages of development during the period of 2004–2018, the rate variables were assigned fixed values that were not in line with the actual situation. Therefore, a time-continuous function was adopted in this study to represent these variables; the specific functional relationship is shown in Table 5. The model also includes six state variables, whose initial values were taken from the statistical data of each city in 2003, as detailed in Appendix A.

Table 5. Chengdu and Chongqing rate variable time-continuous functions.

City	Rate Variables	Time-Continuous Function
Chengdu	Population growth rate	$-0.07843\ln(t)+5.9833$
	GDP growth rate	$-8.3395\ln(t)+63.5984$
	Rate of increase in total energy consumption	$27.81\ln(t)-211.51$
	Rate of increase in arable land area	$-0.593\ln(t)+4.519$
	Growth rate of public green space	$0.00264t^2-10.6277t+10698.4652$
	Rate of increase in total water supply	$-0.000407t+0.867970$
Chongqing	Population growth rate	$-0.40583\ln(t)+3.09324$
	GDP growth rate	$-15.4244\ln(t)+117.4724$
	Rate of increase in total energy consumption	$-9.8597\ln(t)+75.0677$
	Rate of increase in arable land area	$-2.29896\ln(t)+17.4902$
	Growth rate of public green space	$-51.03124\ln(t)+388.2950$
	Rate of increase in total water supply	$12.197\ln(t)-92.722$

3.3. *Model Validity Test*

3.3.1. Historical Test

The historical test aims to verify whether the relative error between the model simulation results and the historical values is within an acceptable range [52]; the formula employed in this study is as follows:

$$D_t = \frac{X_t' - X_t}{X_t} \times 100\% \tag{12}$$

where D_t is the relative error value at time t; X_t' is the model simulation value at time t; and X_t is the historically true value at time t.

Six core indicators in the SD model of the green development level of the Chengdu-Chongqing City Group were selected to perform the historical test: total population, total GDP, total energy consumption, total water supply, arable land area, and public green space area. The passing rate of the historical test of each city in the Chengdu-Chongqing City Group is shown in Appendix A. The results of the historical test show that the SD model established in this study can fit the historical data well.

3.3.2. Sensitivity Check

In a stable SD model, the change in its simulation values when the model parameters change is relatively low, i.e., the SD model has low sensitivity [8,53]. To verify whether the SD model constructed in this study could operate effectively, six rate variables were selected to test its sensitivity. The sensitivity of the rate variables of each city in the Chengdu-Chongqing City Group from 2004 to 2018 was derived based on the sensitivity formula with a 10% increase and a 10% decrease. The sensitivity check formula employed is as follows:

$$S_{Lij} = \left| \frac{\Delta L_{it}}{L_{it}} \times \frac{X_{jt}}{\Delta X_{jt}} \right| \tag{13}$$

$$S_{Xj} = \frac{1}{N} \sum\nolimits_{i=1}^{N} S_{Lij} \tag{14}$$

where L_{it} is the value of the i-th state variable L at time t; ΔL_{it} is the change in the i-th state variable L at time t; X_{jt} is the value of the j-th parameter X at time t; ΔX_{jt} is the change in the j-th parameter X at time t; S_{Lij} is the sensitivity of the i-th state variable L to the j-th parameter X; and S_{Xj} is the sensitivity of the model to the j-th parameter X.

The results of the sensitivity test showed that the average sensitivity of the six rate variables to the SD model was lower than 0.01 in the two cases of the 10% increase and the 10% decrease. Hence, the structure of the SD model of the green development level of the

Chengdu-Chongqing City Group constructed in this study was stable and could be used for the purposes of simulation and prediction.

3.4. Estimation of the Future Green Development Level of the Chengdu-Chongqing City Group under the SSPs

3.4.1. Analysis of Changes in Green Development Levels in the Chengdu-Chongqing City Group

The abovementioned model parameters of each city in the Chengdu-Chongqing City Group were included in the SD model; the values of the various socio-economic indicators from 2019 to 2050 were simulated, and the entropy weight method was applied to calculate the green development level and criterion level of each city under the five SSPs considered, to further analyze the changes in the green development level in the Chengdu-Chongqing City Group. The simulation results are shown in Figure 7.

Under the five SSPs, the green development level and the guideline level of each city in the Chengdu-Chongqing City Group will increase overall. Specifically, the green development level of Chengdu and Chongqing will be significantly higher than that of other cities in the Chengdu-Chongqing City Group, and the gap will tend to further widen. Mianyang ranked third for green development level, showing a clear advantage over the other cities in 2022–2035. However, the gap between the green development level of Leshan and Mianyang is projected to gradually narrow, while that of Dazhou will be the lowest among all the cities investigated. In general, the green development level and the guideline level of each city will have different development growth rates and coordination levels under different pathways.

Under SSP1, the green development level of each city will have the fastest growth among all pathways considered; this growth is also expected to be coordinated and stable, although the gap between Dazhou and other cities will further expand.

Under SSP2, the cities were divided into four echelons according to their level of green development. Chengdu and Chongqing are in the first echelon and are characterized by similar levels of green development; the second echelon includes the cities of Mianyang and Leshan, the latter being expected to have almost the same level of green development as Mianyang by 2050. The fourth echelon includes only the city of Dazhou, and its gap with the other cities is projected to further expand. The other cities are in the third tier.

Under SSP3, the growth rate of the green development level will be the slowest among all the pathways considered; Chengdu and Chongqing will maintain the lead, with a large gap between them and the rest of the cities. Dazhou will have the lowest level of green development, although the gap between it and the other cities will first widen and then narrow.

Under SSP4, the growth rate of the green development level of each city will be stable overall, with a slowdown and a small plateau period appearing from 2030 to 2040, after which the growth will accelerate. Chengdu and Chongqing will have significantly higher levels of green development than the other cities, although this gap will first increase and then decrease. Mianyang and Leshan will follow in the ranking, and the gap between them will tend to shrink continuously until 2050, when it is expected to almost disappear.

Under SSP5, the green development level of each city will increase steadily, with Chengdu and Chongqing recording an almost equal level. Compared to other paths, the gap between Chengdu and Chongqing and the other cities will be the smallest, while the gap between Dazhou and the other cities will be the largest and will tend to expand continuously.

Looking at these results from the perspective of each city in the Chengdu-Chongqing City Group, it can be seen that the green development level of each city under SSP1 will be the highest and will have an overall steady growth trend; Chengdu will maintain the leading position, followed by Chongqing, and the gap between the remaining cities will tend to be stable. Under SSP3, all the cities in the Chengdu-Chongqing City Group will have the lowest level of green development and the slowest growth rate; moreover, the trend of growth will be uneven, with Chengdu and Chongqing expected to have a significantly higher growth trend than the other cities.

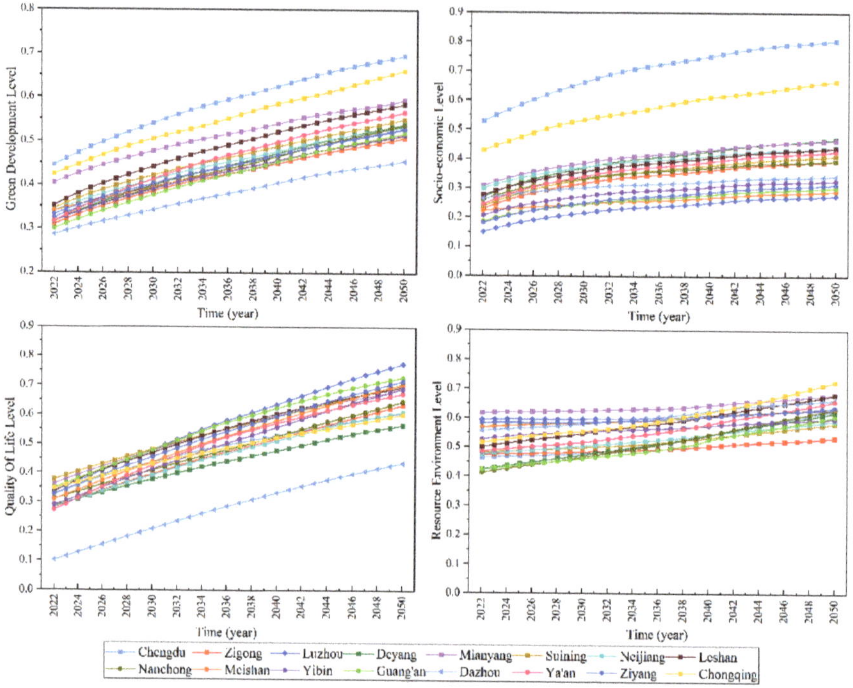

(**a**)

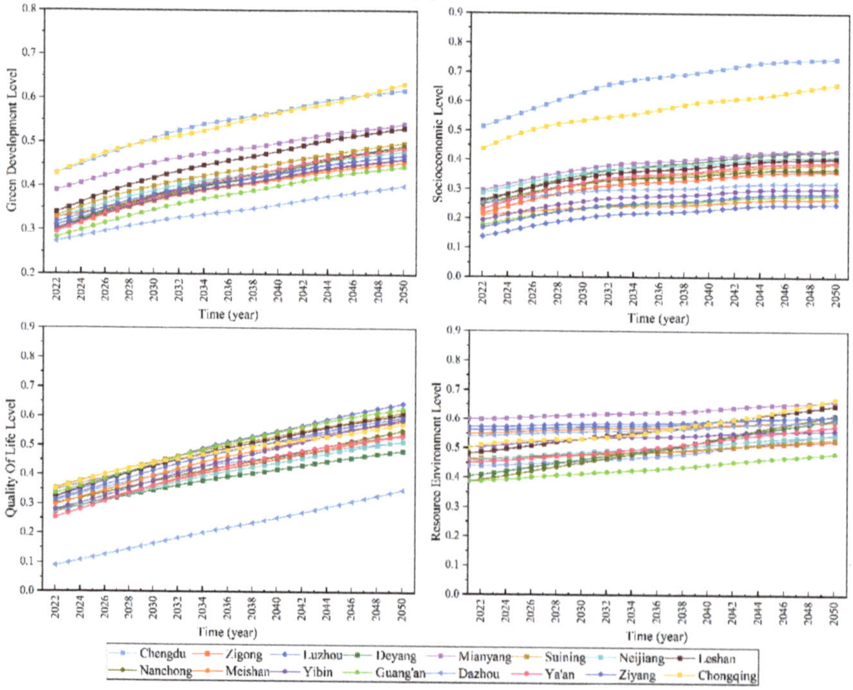

(**b**)

Figure 7. *Cont.*

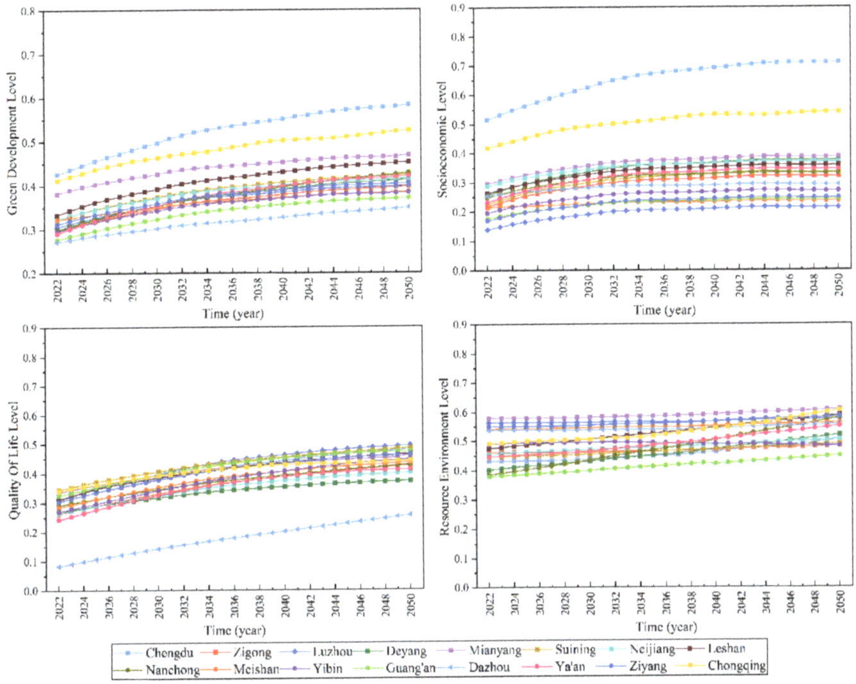

(**c**)

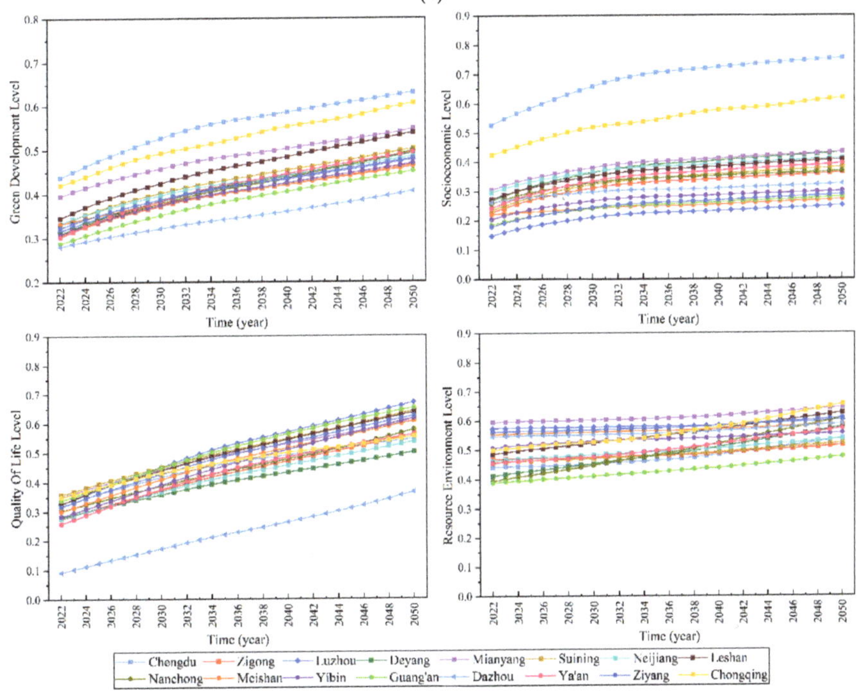

(**d**)

Figure 7. *Cont.*

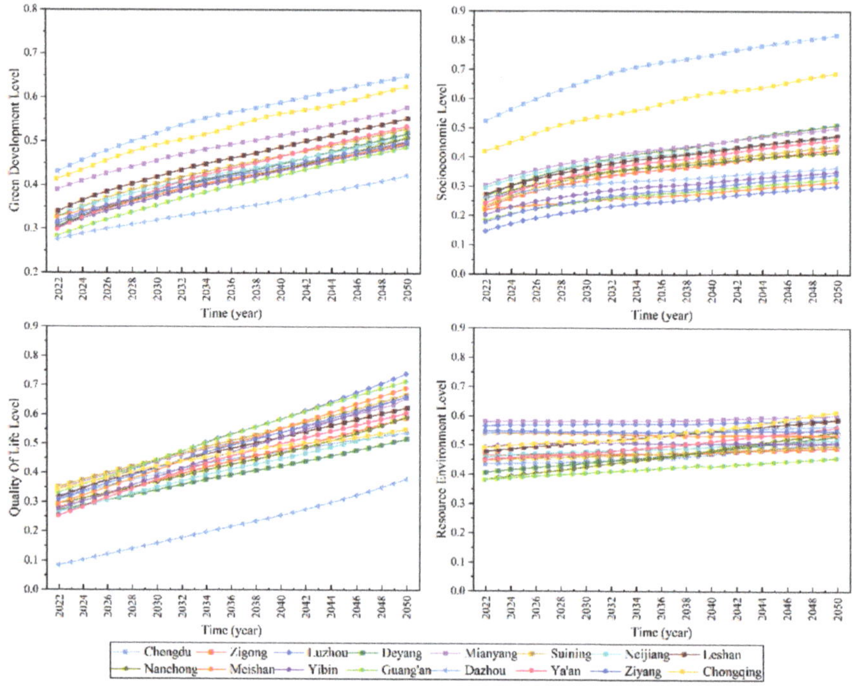

(e)

Figure 7. The green development level and the guideline level of each city in the future under 5 SSPs paths: (**a**) SSP1, (**b**) SSP2, (**c**) SSP3, (**d**) SSP4, (**e**) SSP5.

3.4.2. Analysis of the Overall Future Green Development Level of the Chengdu-Chongqing City Group

Based on the analysis of the future green development level of each city in the Chengdu-Chongqing City Group and the level of the criterion layer under the SSPs, the future green development level of the Chengdu-Chongqing City Group as a whole was calculated; it is shown in Figure 8.

From Figure 8, it can be seen that the future green development level and the guideline level of the Chengdu-Chongqing City Group as a whole will have a rising trend in all SSPs, and the growth rate will gradually decrease. In terms of green development, its level will be the highest under SSP1 and the lowest under SSP3, in which its growth rate will also be the slowest among all the SSPs considered. Moreover, the gap among the green development levels of SSP2, SSP4, and SSP5 will be narrow until 2035, with the highest level in SSP4 and the lowest in SSP2; after 2035, the growth rate under SSP4 will slow down and will be surpassed by that of SSP5, such that the level of green development under the latter will exceed that of the former and will be second only to SSP1. In terms of the socio-economic level of the Chengdu-Chongqing City Group, this will be the highest under SSP5, followed by SSP1, and it will be the lowest under SSP3. In terms of the resource and environment level, this will be the highest under SSP1 and the lowest under SSP3 and SSP5, with a small difference between the latter two. In terms of the quality-of-life level, this will be the highest under SSP1 and the lowest under SSP3. Moreover, under SSP4 it will be second only to SSP1 until 2039, after which that under SSP5 will grow relatively fast and will surpass SSP4.

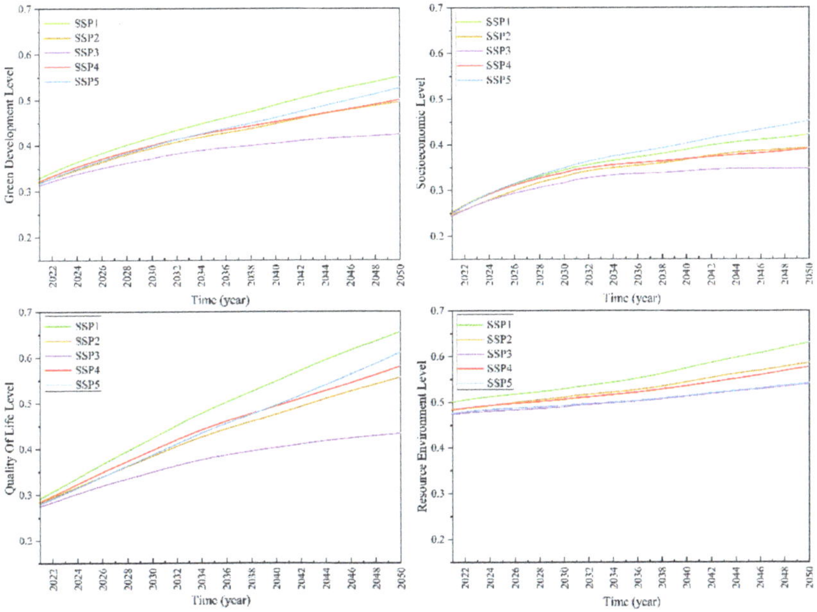

Figure 8. The overall green development level and criteria level of the Chengdu-Chongqing City Group based on SSP paths.

From the perspective of individual paths, the green development level, the resource and environment level, and the quality-of-life level of the Chengdu-Chongqing City Group under SSP1 will be the highest, and their growth rate will also be the highest. The green development level will rise relatively steadily, increasing by 1.62 times by 2050, with an average annual growth rate of about 1.7%. The level of resources and environment will increase by 1.1 times until 2035, with an average annual growth rate of 0.63%; from 2035 to 2050, the growth rate will slightly accelerate, increasing by 1.14 times, with an average annual growth rate of about 0.93%. The development advantage of the level of resources and environment will become increasingly evident under SSP1, and the gap with the other paths will gradually increase. The socio-economic level of the Chengdu-Chongqing City Group under the SSP1 path will be second only to that of SSP5 and will be characterized by an increase of 1.56 times by 2050 and an average annual growth rate of about 1.6%.

Under SSP2, the green development level of the Chengdu-Chongqing City Group ranks fourth, with an increase of 1.5 times by 2050 and an average annual growth rate of about 1.5%. Under this pathway, the socio-economic level ranks fourth with an average annual growth rate of about 2.0% until 2041, after which SSP2 will overtake SSP4 and rank third, with an average annual growth rate of about 0.6% from 2041 to 2050.Under SSP2, the resource and environment level will be basically the same as that of SSP4 until 2026; after 2026, it will rank second after SSP1 and will increase by 1.2 times until 2050, with an average annual growth rate of 0.66%. In this pathway, the quality-of-life level will be similar to that under SSP5 until 2030, after which a gap will open up, and it will gradually lag behind SSP5, ranking fourth, with an increase of 1.9 times by 2050 and an average annual growth rate of about 2.3%.

The green development level and the guideline level of the Chengdu-Chongqing City Group under SSP3 will be the lowest, and the growth rate will be relatively slow and will tend to stabilize. Under this pathway, the green development level will increase by 1.34 times until 2050, with an average annual growth rate of about 0.99%. The socio-economic level will grow by 1.35 times by 2050, with an average annual growth rate of about 1.1%. Moreover, the resource and environment level will be only slightly lower than in SSP5; it will increase by about 1.1 times by 2050, with an average annual growth rate of about 0.44%. The quality-of-life level will increase by about 1.5 times by 2050, with an average annual growth rate of about 1.5%.

The green development level and the guideline level of the Chengdu-Chongqing City Group under SSP4 will be at an average level. The former will rank second after SSP1 until 2030; after 2030, it will be surpassed by that of SSP5 and will, thus, rank third; after 2042, it will be very close to that of SSP2 and will be equal to 0.46 in 2042. Overall, from 2022 to 2050, the green development level under SSP4 will increase by 1.5 times, with an average annual growth rate of about 1.46%. Under this pathway, the socio-economic level will be very close to that of SSP1 and SSP5 until 2030, after which it will gradually diverge, ranking third until 2041. After that, it will be slightly surpassed by that of SSP2, ranking fourth, with an average annual growth rate of about 1.39% until 2050. The quality-of-life level under SSP4 will rank second after SSP1 until 2040, after which it will be overtaken by that of SSP5 and will rank third, with an increase of 1.9 times by 2050 and an average annual growth rate of 2.4%.

The green development level, the socio-economic level, and the quality-of-life level of the Chengdu-Chongqing City Group under SSP5 will be the highest among all the pathways considered, while the resource and environment level will be the lowest. Under this pathway, the green development level will rank third until 2030, after which it will overtake SSP4 to rank second, increasing from 0.33 in 2022 to 0.53 in 2050, with an increase of 1.61 times and an average annual growth rate of about 1.69%. The socio-economic level will always be the highest, and its gap with that of the other pathways will widen, growing by 1.67 times by 2050, with an average annual growth rate of about 1.92%, i.e., the fastest growth rate among the five SSPs. Moreover, the resource and environment level under SSP5 will be almost the same as that in SSP3 and will be the lowest among the five SSPs considered, although it will slowly increase by 1.12 times by 2050. The quality-of-life level under SSP5 will rank third until 2039, after which it will overtake SSP4 to rank second, right after the SSP1 path, and will increase by 2.1 times by 2050, with an average annual growth rate of 2.68%.

3.5. *Suggestions for the Future Development Path of the Chengdu-Chongqing City Group*

The comparative analysis of the green development level of the Chengdu-Chongqing City Group as a whole and for each city under the shared SSPs considered, showed that under SSP1 both the green development level and its growth rate will be the highest among all the SSPs. Moreover, the green development level of the Chengdu-Chongqing City Group under SSP2 will be very similar to that under SSP4 and SSP5 before 2025, and will be lower only than that under SSP1. The year 2025 is the closing year of China's 14th Five-Year Plan, which calls for the promotion of green development and the acceleration of the green transformation of the development methods. The period of 2025–2035 will be key for China to achieve a full modernization; moreover, the peak of carbon emissions is set to be reached before 2030, after which the steady decrease in carbon emissions will contribute to the achievement of the goal of building a beautiful China. Therefore, the transition from SSP2 to SSP1 should be completed by 2035 to achieve real sustainable development, and the latter pathway should be steadily followed from 2035 to 2050 so as to achieve carbon neutrality by 2060.

From the socio-economic point of view, the economic development of the Chengdu-Chongqing City Group will remain very uncoordinated and "polarized" for a long time in the future. The economic development level of Chengdu and Chongqing is far ahead of that of the other cities; the fundamental reason is that the regional GDP and the degree of openness of the other cities are relatively low, and the total import and export amount of Dazhou and Ya'an is only about 1% of that of Chengdu and Chongqing. Hence, it is important to narrow the economic differences among the cities in the Chengdu and Chongqing City Group to improve the overall green development level.

From the point of view of resources and environment, the resource and environment condition of the Chengdu-Chongqing City Group will not greatly improve in the future, which is in line with its development status; moreover, resource consumption and environmental pollution will increase together with the economic development. At present, the proportion of secondary industries in the Chengdu-Chongqing urban agglomeration is still very high, with more than 50% of the cities accounting for more than 45% of the secondary industries. The high energy consumption and pollution emissions of traditional heavy industries are the key factors hindering the green growth of the entire Chengdu-Chongqing City Group.

In terms of quality of life, the quality-of-life level of all the cities will steadily increase, although that of Dazhou will be considerably lower compared to the other cities. Although the population of Dazhou is gradually declining, its total population is higher than that of Mianyang, the second largest city in Sichuan, and its population pressure is also high. Accordingly, the resources at the disposal of each individual, such as medical care, roads, and green areas, are also relatively small.

4. Conclusions

This study constructed a green development SD model for the Chengdu-Chongqing City Group and set five simulation scenarios under the SSPs. The future green development level and the guideline level of each city and of the overall Chengdu-Chongqing City Group were simulated and calculated, with the comparison and analysis of the changes, obtaining the following main conclusions:

1. The future green development level of each city in the Chengdu-Chongqing City Group under the SSPs considered will have an upward trend and will tend to be stable under SSP3. In SSP2, the green development level of Chengdu and of Chongqing will be very close, while in the other pathways the former will be the highest, followed by the latter. The socio-economic level of each city will tend to be stable under SSP3, while it will slowly increase in the other pathways. In terms of resource and environment level, the cities investigated will show little change in their resource and environment level under all the SSPs and only a weak upward trend overall. In terms of quality-of-life level, all the cities will have a stable upward trend under all the SSPs, which will gradually stabilize under SSP3.
2. The results of the analysis of the changes in the green development level and the criterion level of the Chengdu-Chongqing City Group as a whole under the SSPs considered show that the green development level and the criterion level of the Chengdu-Chongqing City Group will increase under all of the five SSPs. The green development level, the resource and environment level, and the quality-of-life level will be the highest under SSP1 and the lowest under SSP3, and this gap will gradually increase. The socio-economic level will be the highest under SSP5, followed by SSP1, and it will be the lowest under SSP3.

Based on these findings, the Chengdu-Chongqing City Group should follow suitable pathways for future development. Specifically, SSP2 should be used as the basis for development until 2025, which entails the continuance of the current development trend, after which a transition should be gradually performed toward SSP1, which should be completed before 2035 to achieve real sustainable development. Afterwards, the Chengdu-Chongqing City Group should continue to develop according to SSP1 from 2035 to 2050.

Author Contributions: Conceptualization, Y.L., L.Z., M.L. and Y.X.; writing—original draft preparation, Y.L. and M.L.; writing—review and editing, L.Z. and Y.X.; supervision, L.Z. and Y.X.; funding acquisition, Y.X. and J.X. All authors have read and agreed to the published version of the manuscript.

Funding: This research is supported by the National Natural Science Foundation of China (Grant No.41890824, No.52279023).

Institutional Review Board Statement: Not applicable.

Informed Consent Statement: Not applicable.

Data Availability Statement: Not applicable.

Conflicts of Interest: The authors declare no conflict of interest.

Appendix A

Appendix A shows the historical statistical descriptions of the parameters and the initial values of the state variables of the green development level system dynamics model of the Chengdu-Chongqing City Group constructed in this study, together with the historical test results of the green development level system dynamics model of each city in the Chengdu-Chongqing City Group.

Table A1. Statistical description of model parameters in Chengdu and Chongqing.

City	Model Parameters	Maximum Value	Minimum Value	Average Value
Chengdu	Population growth rate	0.1222	0.0080	0.0230
	GDP growth rate	0.2013	0.0689	0.1337
	Rate of increase in total energy consumption	0.1878	−0.1519	0.0158
	Rate of increase in arable land area	0.2015	−0.0162	0.0238
	Increase rate of public green space	0.2089	−0.0153	0.0477
	Rate of increase in total water supply	0.1725	−0.0885	0.0473
Chongqing	Population growth rate	0.0110	−0.0012	0.0059
	GDP growth rate	0.2088	0.0412	0.1250
	Rate of increase in total energy consumption	0.1369	0.0209	0.0648
	Rate of increase in arable land area	0.0595	−0.1096	−0.0032
	Increase rate of public green space	0.2655	−0.0578	0.0969
	Rate of increase in total water supply	0.1128	−0.0863	0.0500

Table A2. Initial values of state variables of the SD model of the green development level of each city in the Chengdu-Chongqing City Group.

City	Total Population	Total GDP	Total Energy Consumption	Arable Land Area	Area of Public Green Space	Total Water Supply
Chengdu	1044.31	1705.2732	1155.32	362,600	7289.28	48,810
Chongqing	2606.26	2400.2470	2693.21	1,353,200	8053.34	71,100
Dazhou	637.81	239.1671	1259.78	270,540	685.08	2933
Deyang	380.59	328.6206	358.60	194,162	2620.89	14,993
Guang'an	448.54	170.3600	372.66	173,600	1957.75	2400
Leshan	347.63	215.5705	668.07	149,900	2443.84	9823
Luzhou	468.10	204.3567	384.19	211,000	2515.80	8794
Meishan	340.66	171.7821	470.97	176,000	665.53	3200
Mianyang	527.50	396.6000	548.47	283,100	4239.74	7191
Nanchong	717.73	239.3039	315.12	300,519	5155.67	12,017
Neijiang	421.25	182.3293	499.88	164,507	504.60	6000
Suining	376.60	148.4155	166.10	154,891	2564.46	6247
Ya'an	153.15	91.9400	88.22	60,000	1131.37	3100
Yibin	515.01	257.6936	422.67	242,406	792.27	4900
Ziyang	344.77	124.5088	183.24	193,000	340.97	4200
Zigong	315.30	192.4271	735.49	123,100	3244.46	7622

Table A3. Pass rate of historical test of SD model of the green development level of cities in the Chengdu-Chongqing City Group.

City	Total Population		Total GDP		Total Energy Consumption	
	10% Pass Rate (%)	20% Pass Rate (%)	10% Pass Rate (%)	20% Pass Rate (%)	10% Pass Rate (%)	20% Pass Rate (%)
Chengdu	100	100	100	100	93.3	100
Chongqing	100	100	93.3	100	100	100
Dazhou	100	100	100	100	66.7	80
Deyang	100	100	100	100	86.7	100
Guang'an	100	100	100	100	80	100
Leshan	100	100	100	100	80	100
Luzhou	100	100	93.3	100	60	86.7
Meishan	100	100	100	100	93.3	100
Mianyang	100	100	100	100	86.7	100
Nanchong	100	100	100	100	80	100
Neijiang	100	100	93.3	100	93.3	100
Suining	100	100	100	100	86.7	93.3
Ya'an	100	100	100	100	93.3	100
Yibin	100	100	100	100	80	86.7
Ziyang	100	100	93.3	100	86.7	100
Zigong	100	100	100	100	86.7	100
City	Arable Land Area		Area of Public Green Space		Total Water Supply	
	10% Pass Rate (%)	20% Pass Rate (%)	10% Pass Rate (%)	20% Pass Rate (%)	10% Pass Rate (%)	20% Pass Rate (%)
Chengdu	86.7	100	100	100	100	100
Chongqing	100	100	80	93.3	100	100
Dazhou	100	100	73.3	93.3	93.3	100
Deyang	100	100	100	100	86.7	93.3
Guang'an	100	100	93.3	100	80	100
Leshan	100	100	100	100	86.7	100
Luzhou	100	100	93.3	100	80	100
Meishan	100	100	80	93.3	100	100
Mianyang	100	100	86.7	100	86.7	100
Nanchong	100	100	100	100	100	100
Neijiang	100	100	80	100	80	93.3
Suining	100	100	93.3	93.3	93.3	93.3
Ya'an	100	100	73.3	73.3	86.7	100
Yibin	100	100	86.7	100	93.3	100
Ziyang	100	100	66.7	80	86.7	100
Zigong	100	100	86.7	100	86.7	100

Ten percent pass rate: the ratio of the simulated year to the simulated year for which the relative error between the simulated model value and the historical statistical value is within ±10%; twenty percent pass rate: the ratio of the simulated year to the simulated year for which the relative error between the simulated model value and the historical statistical value is within ±20%.

References

1. Wang, Z. On William Petty's Thought of Market Economy. *J. Suzhou Univ. Sci. Technol.* **1996**, *2*, 18–22.
2. Hu, A. Leading the Road to a Green Power with the "Red Line" of the Scientific Outlook on Development. *Environ. Prot.* **2009**, *1*, 47–49.
3. Han, M.; Yuan, Q.; Fahad, S.; Ma, T. Dynamic Evaluation of Green Development Level of ASEAN Region and Its Spatio-Temporal Patterns. *J. Clean Prod.* **2022**, *362*, 132402. [CrossRef]
4. Zhang, R.; Ma, Y.; Ren, J. Green Development Performance Evaluation Based on Dual Perspectives of Level and Efficiency: A Case Study of the Yangtze River Economic Belt, China. *Int. J. Environ. Res. Public Health* **2022**, *19*, 9306. [CrossRef] [PubMed]
5. Jiang, S.; Yu, H.; Li, Z.; Geng, B.; Li, T. Study on the Evolution of the Spatial-Temporal Pattern and the Influencing Mechanism of the Green Development Level of the Shandong Peninsula Urban Agglomeration. *Sustainability* **2022**, *14*, 9549. [CrossRef]
6. Wang, Z.; Zhang, X.; Zhang, C.; Yang, Q. How Regional Integration Affects Urban Green Development Efficiency: Evidence from Urban Agglomeration in the Middle Reaches of the Yangtze River. *Int. J. Environ. Res. Public Health* **2022**, *19*, 7937. [CrossRef]

7. Wu, W.; Luo, X.; Knopp, J.; Jones, L.; Banzhaf, E. A European-Chinese Exploration: Part 2-Urban Ecosystem Service Patterns, Processes, and Contributions to Environmental Equity under Different Scenarios. *Remote Sens.* **2022**, *14*, 3488. [CrossRef]
8. Gu, M.; Wang, F.; Wang, S. Simulation of China's Green Development Policy Based on System Dynamics. *Chin. J. Environ. Manag.* **2021**, *13*, 126–135. [CrossRef]
9. Hall, B.; Kerr, M.L. *Green Index 1991–1992: A State-by-State Guide to the Nation's Environmental Health*; Island Press: Washington, DC, USA, 1991.
10. Feng, C.; Wang, M.; Liu, G.; Huang, J. Green Development Performance and Its Influencing Factors: A Global Perspective. *J. Clean. Prod.* **2017**, *144*, 323–333. [CrossRef]
11. Li, X.; Liu, Y.; Song, T. Calculation of the Green Development Index. *Soc. Sci. China* **2014**, *6*, 69–95.
12. Wang, M.; Zhao, H.; Cui, J.; Fan, D.; Lv, B.; Wang, G.; Li, Z.; Zhou, G. Evaluating Green Development Level of Nine Cities within the Pearl River Delta, China. *J. Clean. Prod.* **2018**, *174*, 315–323. [CrossRef]
13. Guo, M.; Nowakowska-Grunt, J.; Gorbanyov, V.; Egorova, M. Green Technology and Sustainable Development: Assessment and Green Growth Frameworks. *Sustainability* **2020**, *12*, 6571. [CrossRef]
14. Shields, K.; Langer, H.; Watson, J.; Stelzner, K. *European Green City Index. Assessing the Environmental Impact of Europe's Major Cities*; Siemens AG: Muenchen, Germany, 2009.
15. Carli, R.; Dotoli, M.; Pellegrino, R. Multi-Criteria Decision-Making for Sustainable Metropolitan Cities Assessment. *J. Environ. Manag.* **2018**, *226*, 46–61. [CrossRef] [PubMed]
16. Tian, S.; Ding, Y. Greenization Measurement of Urban Agglomeration in Yangtze River Delta and Its Influence Factors: An Empirical Study Based on the Entropy-Tobit Model of 26 Cities' Panel Data. *Econ. Geogr.* **2019**, *39*, 94–103. [CrossRef]
17. Hu, S.; Zeng, G.; Cao, X.; Yuan, H.; Chen, B. Does Technological Innovation Promote Green Development? A Case Study of the Yangtze River Economic Belt in China. *IJERPH* **2021**, *18*, 6111. [CrossRef] [PubMed]
18. Forrester, J. Industrial Dynamics—A Major Breakthrough for Decision Makers. *Harv. Bus. Rev.* **1958**, *36*, 37.
19. Mass, N.J. *Readings in Urban Dynamics*, 2nd ed.; The MIT Press: Cambridge, MA, USA, 1977; ISBN 978-0-262-06067-7.
20. Alfeld, L.E.; Graham, A.K. *Introduction to Urban Dynamics*; Wright-Allen Press: MA, USA, 1976.
21. Rudneva, L.; Pchelintseva, I.; Gureva, M. Scenario Modelling of the "Green" Economy in an Economic Space. *Resour.-Basel* **2018**, *7*, 29. [CrossRef]
22. Yang, G.; Shi, Y.; Yang, F.; Zhang, F. Assessment of sustainable development of water resources carrying capacity based on system dynamics model: Case of Chongqing City. *Yangtze River* **2019**, *50*, 6–13+51. [CrossRef]
23. Zhou, J. Research on Assessment and Improving Path of Green Development in the Beijing-Tianjin-Hebei Urban Agglomeration. Ph.D. Thesis, China University of Mining and Technology, Beijing, China, 2019.
24. Wang, H.; Bao, C. Scenario Modeling of Ecological Security Index Using System Dynamics in Beijing-Tianjin-Hebei Urban Agglomeration. *Ecol. Indic.* **2021**, *125*, 107613. [CrossRef]
25. Li, J.; Sun, K. Research on the Industrial Green Development of Beijing-Tianjin-Hebei Based on System Dynamics. *Soft Sci.* **2018**, *32*, 113–119. [CrossRef]
26. Zhang, J.; Yang, D.; Zhu, J.; Sun, Z.; Yu, X. Research on the Green and Low-carbon Path of Urban Agglomeration on the Perspective of System Dynamics. *Sci. Technol. Ind.* **2021**, *21*, 307–314.
27. O'Keeffe, J.; Pluchinotta, I.; De Stercke, S.; Hinson, C.; Puchol-Salort, P.; Mijic, A.; Zimmermann, N.; Collins, A.M. Evaluating Natural Capital Performance of Urban Development through System Dynamics: A Case Study from London. *Sci. Total Environ.* **2022**, *824*, 153673. [CrossRef] [PubMed]
28. Gudlaugsson, B.; Ghanem, D.A.; Dawood, H.; Pillai, G.; Short, M. A Qualitative Based Causal-Loop Diagram for Understanding Policy Design Challenges for a Sustainable Transition Pathway: The Case of Tees Valley Region, UK. *Sustainability* **2022**, *14*, 4462. [CrossRef]
29. Wu, C. *Golden Waterway—Yangtze River Economic Belt*, 1st ed.; Chongqing University Press: Chongqing, China, 2020.
30. Wu, C.; Huang, L. Research on the Performance and the Synergistic Effect of Green Development of Yangtze River Economic Belt's Industry. *J. China Univ. Geosci. (Soc. Sci. Ed.)* **2018**, *18*, 46–55. [CrossRef]
31. Zhou, P.; Chen, S. Spatial and Economic Ties between Chengdu-Chongqing City Groups and Its Evolution. *J. China West Norm. Univ. (Nat. Sci.)* **2021**, 1–11.
32. Liu, H.; Liu, Y.; Meng, L.; Jiao, K.; Zhu, M.; Chen, Y.; Zhang, P. Research progress of entropy weight method in water resources and water environment. *J. Glaciol. Geocryol.* **2022**, *44*, 299–306.
33. Wei, X.; Zhao, Y.; Li, X.; Xue, Z.; Xia, S. Characteristics and Optimization of Geographical Space in Urban Agglomeration in the Upper Reaches of the Yangtze River Based on the Function of "Production-Living-Ecological". *Resour. Environ. Yangtze Basin* **2019**, *28*, 1070–1079.
34. You, Z. Evaluation and Analysis of Ecological Carrying Capacity of Huainan City Based on System Dynamics. Master's Thesis, Anhui Jianzhu University, Hefei, China, 2018.
35. Wang, Q. *System Dynamics*, revised ed.; Tsinghua University Press: Beijing, China, 1994; ISBN 978-7-302-01523-9.
36. Tian, L.; Zhang, H.; Zhang, X. A system dynamics approach for economic developing zone water demand forecasting:a case study of Tianjin Linkong area. *J. Tiangong Univ.* **2009**, *28*, 68–72.
37. Vensim User Manual—Baidu Library. Available online: https://wenku.baidu.com/view/54635a0390c69ec3d5bb75e0.html (accessed on 6 September 2022).

38. Lv, D. Research and Environment Carrying Capacity of Ordos:Based on System Dynamics Model. Master's Thesis, China University of Geosciences, China, Beijing, 2016.
39. Zhong, Y.; Jia, X.; Qian, Y. *System Dynamics*, 2nd ed.; Science Press: Beijing, China, 2009; ISBN 978-7-03-023535-0.
40. Moss, R.H.; Edmonds, J.A.; Hibbard, K.A.; Manning, M.R.; Rose, S.K.; van Vuuren, D.P.; Carter, T.R.; Emori, S.; Kainuma, M.; Kram, T.; et al. The next Generation of Scenarios for Climate Change Research and Assessment. *Nature* **2010**, *463*, 747–756. [CrossRef]
41. van Vuuren, D.P.; Edmonds, J.; Kainuma, M.; Riahi, K.; Thomson, A.; Hibbard, K.; Hurtt, G.C.; Kram, T.; Krey, V.; Lamarque, J.-F.; et al. The Representative Concentration Pathways: An Overview. *Clim. Chang.* **2011**, *109*, 5–31. [CrossRef]
42. van Vuuren, D.P.; Riahi, K.; Moss, R.; Edmonds, J.; Thomson, A.; Nakicenovic, N.; Kram, T.; Berkhout, F.; Swart, R.; Janetos, A.; et al. A Proposal for a New Scenario Framework to Support Research and Assessment in Different Climate Research Communities. *Glob. Environ. Change-Hum. Policy Dimens.* **2012**, *22*, 21–35. [CrossRef]
43. Zhang, L.; Chen, X.; Xin, X. Short commentary on CMIP6 Scenario Model Intercomparison Project (ScenarioMIP). *Clim. Chang. Res.* **2019**, *15*, 519–525.
44. Elsawah, S.; Pierce, S.A.; Hamilton, S.H.; van Delden, H.; Haase, D.; Elmandi, A.; Jakeman, A.J. An Overview of the System Dynamics Process for Integrated Modelling of Socio-Ecological Systems: Lessons on Good Modelling Practice from Five Case Studies. *Environ. Modell. Softw.* **2017**, *93*, 127–145. [CrossRef]
45. Jiang, T.; Zhao, J.; Jing, C.; Cao, L.G.; Sun, H.M.; Wang, A.Q.; Huang, J.L.; Su, B.D.; Wang, R. National and Provincial Population Projected to 2100 Under the Shared Socioeconomic Pathways in China. *Clim. Chang. Res.* **2017**, *13*, 128–137.
46. Jing, C.; Su, B.; Cao, Q.; Zhai, J.; Wang, Y.; Tao, H.; Jiang, T. Projection of urbanization and economy in the 'Belt and Road' countries based on the shared socioeconomic pathways. *China Popul. Resour. Environ.* **2019**, *29*, 21–31.
47. Wang, Y.; Jing, C.; Jiang, T.; Zhai, J.; Feng, Z.; Yang, C.; Zhang, R.; Su, B. Projection of provincial urban and rural population and its influencing factors in China's mainland (2015–2050). *J. Nanjing Univ. Inf. Sci. Technol. (Nat. Sci. Ed.)* **2020**, *12*, 395–405. [CrossRef]
48. Zhao, J. Study on the Changes of National Population and Economy Based on the Grid under the Shared Socioeconomic Pathways by IPCC. Master's Thesis, Nanjing University of Information Technology, Nanjing, China, 2018.
49. Pan, J.; Su, B.; Wang, Y.; Jing, C.; Zhai, J.; Jiang, T. Spatio-temporal changes of output value from the primary, secondary and tertiary industries for 2020-2050 under the Shared Socioeconomic Pathways. *Clim. Chang. Res.* **2020**, *16*, 725–737.
50. Jiang, T.; Zhao, J.; Cao, L.; Wang, Y.; Su, B.; Jing, C.; Wang, R.; Gao, C. Projection of national and provincial economy under the shared socioeconomic pathways in China. *Clim. Chang. Res.* **2018**, *14*, 50–58.
51. Xu, G.; Zou, J. The Method of System Dynamics: Principle, Characteristics and New Development. *J. Harbin Inst. Technol. (Soc. Sci. Ed.)* **2006**, *4*, 72–77. [CrossRef]
52. Mashaly, A.F.; Fernald, A.G. Analyzing and Assessing Dynamic Behavior of a Physical Supply and Demand System for Sustainable Water Management under a Semi-Arid Environment. *Water* **2022**, *14*, 1939. [CrossRef]
53. Li, F.; Sun, Y.; Li, X.; Hao, X.; Li, W.; Qian, Y.; Liu, H.; Sun, H. Research on the Sustainable Development of Green-Space in Beijing Using the Dynamic Systems Model. *Sustainability* **2016**, *8*, 965. [CrossRef]

water

MDPI

Article

Evaluation of Adaptive Utilization Capacity of Water Resources and Analysis of Driving Element: A Case Study of Tarim River Basin

Xing Li [1], Qiting Zuo [2], Jiaqi Zhai [1,*], Yong Zhao [1], Yong Wang [1] and Shuying Han [1]

[1] State Key Laboratory of Simulation and Regulation of Water Cycle in River Basin, China Institute of Water Resources and Hydropower Research, Beijing 100038, China
[2] School of Water Conservancy Engineering, Zhengzhou University, Zhengzhou 450001, China
* Correspondence: jiaqizhai@163.com; Tel.: +86-135-8161-8705

Abstract: The research on the adaptive utilization of water resources (AUWR) is of great significance to improve the coordinated development among water resources, economic society, and ecological environment in complex environments, and to promote the development of adaptive utilization of regional water resources. Based on the calculation method of harmony theory and the calculation method of the comprehensive co-evolution model, this paper obtains the harmony degree and adaptive utilization capacity of water resources (AUCWR) of each subsystem in the Tarim River Basin (TRB), analyzes the main factors affecting the AUCWR, and finally compares the two methods. The results show that: (1) From 2004 to 2018, the AUCWR in the TRB has gradually improved (harmony theory method: from 0.43 in 2004 to 0.56 in 2018, with a growth rate of 30.23%; comprehensive co-evolution model method: from 0.37 in 2004 to 0.62 in 2018, with a significant increase of 67.57%) and (2) From the perspective of indicators, indicators such as per capita GDP, the proportion of non-agricultural output value in GDP, and per capita net income of rural residents have a greater impact on the AUCWR in the TRB. Using different calculation methods to analyze the temporal and spatial distribution characteristics of the AUCWR in the TRB has important guiding significance for the future development and utilization of water resources, economic and social development, and ecological environment protection.

Keywords: Tarim River Basin; adaptive utilization capacity of water resources; harmony theory; comprehensive co-evolution model

Citation: Li, X.; Zuo, Q.; Zhai, J.; Zhao, Y.; Wang, Y.; Han, S. Evaluation of Adaptive Utilization Capacity of Water Resources and Analysis of Driving Element: A Case Study of Tarim River Basin. *Water* **2022**, *14*, 3820. https://doi.org/10.3390/w14233820

Academic Editor: Athanasios Loukas

Received: 7 November 2022
Accepted: 21 November 2022
Published: 23 November 2022

1. Introduction

Water resources are basic natural resources, which can provide human beings with clean drinking water, irrigation water, and ecological water [1]. Water is crucial to the sustainable development of societies [2]. Water resources are the major medium of climate change impacts on the environment, ecosystems, and humans, and are increasingly affecting the global economic, social and environmental development [3,4], and the accelerated economic development, population growth, and urban expansion have increased the water shortage, thus highlighting the global systemic risk of water shortage [5,6]. At the same time, the changes in the development and utilization of water resources will also affect the decision-makers' adjustment of water environment policies [7,8]. However, with climate change and economic and social development, the properties and functions of water resources are becoming more diverse, while the linkages with external systems, such as social, economic and ecological systems, are becoming more complex [9,10]. The adaptive development of water resources is a manifestation of their sustainability by adapting to environmental changes, with the increased demand for water resources brought about by increasing population, which leads to water scarcity, excessive groundwater extraction, water pollution, and other problems ensuing [11,12]. The International Association of

Hydrological Sciences (IASH) launched the P.R. (2013–2022) program in 2013, which emphasizes the intersection of nature and society to study human-water relationships, explore the synergistic evolution of human-water systems, and actively promote adaptive research on human-water relationships [13].

Since the 21st century, water resources adaptation research has become an important demand and a hot issue for global and national responses to environmental changes. Many scholars have conducted studies to investigate how water resources systems adapt to the complex and changing environment and the interactions with the economic and social environment [14,15]. The study of water resources adaptation involves many aspects, such as water resources, as well as the economic, social, and ecological environment. Water resources adaptation can be improved by research on optimal water resources allocation, developing water resources management strategies, establishing adaptation models, improving water resources carrying capacity, and reducing water resources-related risks [16–19]. Zhou proposed an integrated optimal allocation model that provides research ideas for complex adaptive systems for water resources management, and applied it in the Dongjiang River Basin in Guangdong Province, China [20,21]. Guided by the idea of adaptive utilization of water resources(AUWR), H.P. discussed how integrated water resources management can achieve adaptive water resources in response to environmental changes, and discuss the specific requirements on how to improve adaptive water resources management and governance [22]; the environmental adaptation of vulnerable water resource systems can be improved by assessing the status of regional water resources in the context of climate change using appropriate models [23].

However, they are all water resources adaptation responses and strategies proposed in response to environmental changes, without proposing water resources development and utilization strategies from the general height of the reciprocal feedback between water resources systems and environmental changes, and have not yet risen to a water resources adaptation and utilization model. Based on this, Zuo elaborated on the AUWR model, the theoretical system framework, and its application issues, and defined the concept of AUWR, the process of water resources development and utilization, following the laws of nature and social development, adapting to the impact of environmental changes such as human activities, climate change, and land surface changes, and ensuring the virtuous cycle of water systems, the chosen water resources utilization [24–26]. On this basis, the concept of adaptive utilization capacity of water resources (AUCWR) is proposed—under the guidance of the theory of adaptive utilization of water resources, based on the evaluation system of AUWR, the effect and overall level of AUWR obtained through quantitative evaluation method.

On the basis of the gradual improvement of the theoretical system of adaptive use of water resources, the quantitative study of the adaptive use of water resources has gradually become a hot issue. Zhang constructed a three-dimensional framework consisting of several risk factor indicators based on water resources resilience theory and established a set of water resources resilience assessment methods to evaluate the resilience of Beijing's water resources system [27]. Yao proposed a comprehensive co-evolution model, based on the conditions of the elements and on the mechanism of their interaction, to study the adaptive development of WRS, it was eventually applied to three rivers in Heilongjiang Province and Shandong Province [28,29]. Adaptive use of water resources is an efficient way to solve complex and uncertain ecosystems and compensate for the limitations of the human-water harmony theory.

In the TRB, artificial oases and desertification processes are increasing [30,31]. As a result, the area of desert-oasis ecological zones is rapidly decreasing and ecological problems are becoming more prominent. At the same time, due to the rapid urbanization of the TRB and the continuous socio-economic development, water demand is also increasing, leading to an increasing conflict between water resources, economic and social development, and ecological environmental protection. Therefore, it is necessary to evaluate the current level of AUCWR in the TRB and to find a reasonable model of water resources development and utilization.

At present, most studies focus on the allocation and regulation of reservoir water resources [32], the adaptive management of water resources for reservoir water resources management [20], and some policies-based water resources management measures are proposed [14]. However, there is insufficient research on the quantitative evaluation of the adaptive use of water resources, especially a set of systematic, perfect, and popularized quantitative evaluation methods. Based on this, this paper uses the team's harmony theory method to systematically evaluate the adaptive use of water resources in the TRB by constructing a system of indicators for evaluating the adaptive use of water resources, and at the same time conducts a comparative analysis with the comprehensive co-evolutionary model method to verify the reasonableness of its results with each other.

In this paper, four main parts of work are done: (a) Systematically proposed a theoretical system of AUWR; (b) Constructing a systematic and complete index system for assessing the adaptive use of water resources; (c) Proposing a method for evaluating the AUCWR in the TRB (harmony theory method), and compared the results with those of the well-established comprehensive co-evolution model method to verify each other; (d) Analyzing the main factors affecting the AUCWR.

2. Theoretical System of AUWR

2.1. Theoretical of AUWR

Adaptive utilization of water resources, sustainable use of water resources, and comprehensive use of water resources are all water resources development and utilization modes, the purpose of which is to ensure the virtuous cycle of water systems, in order to achieve the goal of human-water harmony, but the focus of the three is different. Adaptive use of water resources is a means to address the impact of environmental change, through human regulation measures to mitigate the adverse impact of climate change, human activities, and other water resources, economic, social, and ecological environment.

The theory of adaptive use of water resources takes the human-water system as the research object, through adaptive use of water resources, to achieve sustainable use of water resources and achieve the goal of human-water harmony. Human activities, climate change, and land surface change are the driving factors, which are the source driving force to promote the adaptive use of water resources and the main factors for scientific regulation. The dialectical relationship is that water resources development and protection coexist, the positive and negative impacts of water resources utilization coexist, and the supply side and demand side of the water system coexist and comply with the two laws, four principles, three tasks, and four functions. Adaptive use of water resources needs to consider the balance of human-water relationship transfer, and needs, through a series of regulatory means, to achieve a harmonious balance of adaptation to environmental change transfer, towards the direction of human-water harmony. Its theoretical approach includes a guiding theoretical approach and a basic theoretical approach [25]. As shown in Figure 1.

2.2. Mechanism of AUWR

The core of the mechanism of adaptive use of water resources is the interaction between the three subsystems of water resources, economy, society, and ecology under the influence of climate change and human activities.

The impact of climate change on the water resources-economic society-ecological environment system mainly comes from changes in precipitation, temperature, wind speed, humidity, radiation, and other basic meteorological factors caused by changes in atmospheric circulation: on the one hand, it leads to changes in the water cycle process, which in turn produces changes in the supply side and demand side of water resources, on the other hand, it changes the total amount of water resources and spatial and temporal distribution characteristics, thus increasing the risk of extremes. On the other hand, the change in the total water resources and the spatial and temporal distribution characteristics increase the risk of extreme weather events such as floods and droughts, which cause natural disasters and further affect the stability of the economic, social, and ecological environments.

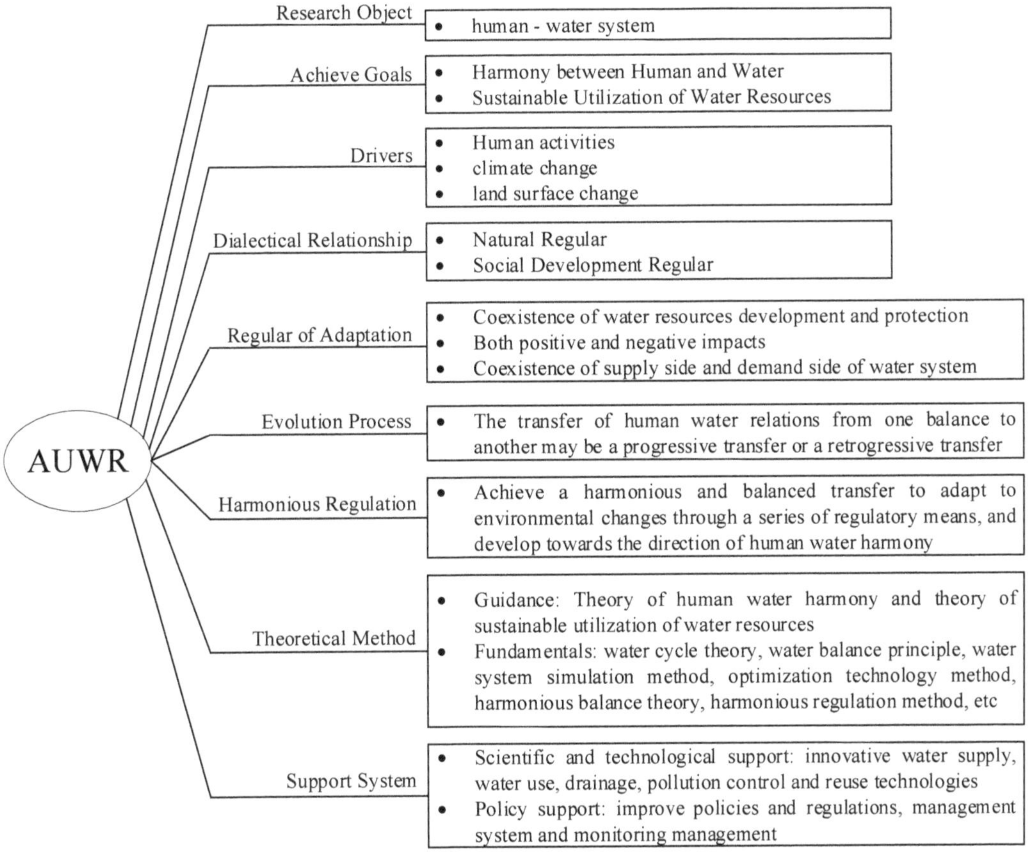

Figure 1. Framework of the theoretical system of AUWR.

The impact of human activities on the water resources-economic society-ecological environment system is: on the one hand, through the transformation of the natural environment to cause changes in water supply potential and natural ecology and environment, on the other hand, through the change of economic and social patterns to cause changes in production and lifestyle, which in turn affects the change of artificial consumption and drainage, resulting in the constant change of water resources and ecological environment state, leading to the imbalance of the original state of the whole system. As shown in Figure 2 [33].

2.3. Framework of Application Rules for AUWR

Adaptive use of water resources involves complex systems and rich contents, so it is necessary to follow certain rules to solve the problems faced by the adaptive use of water resources. In the literature [24], Zuo first proposed a framework of application rules for water resources adaptive use theory, which requires that when applying water resources adaptive use theoretical methods to solve practical problems, it should follow two major laws, conform to four major principles, shoulder three major tasks and have four major functions, as shown in Figure 3.

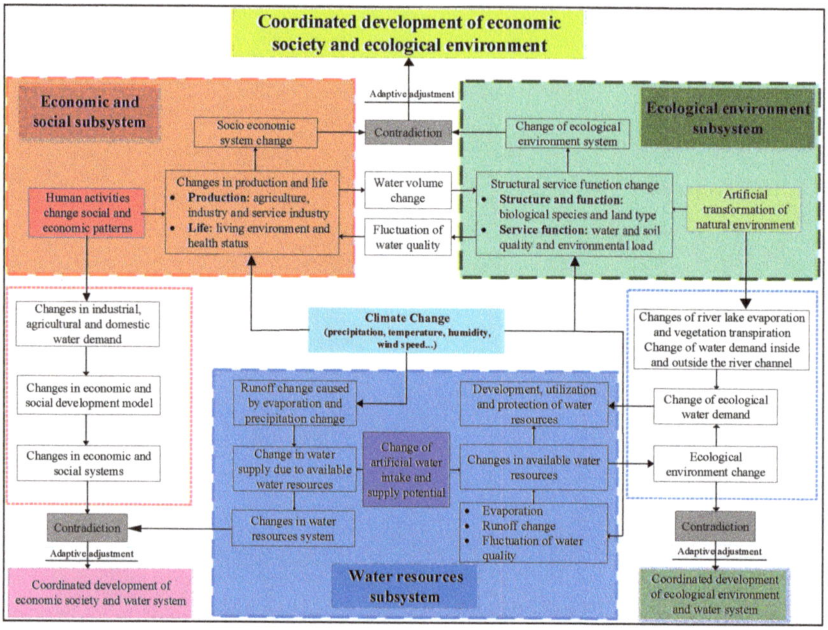

Figure 2. Interaction mechanism of water resources, economic and social systems, and ecological environment affected by climate change and human activities.

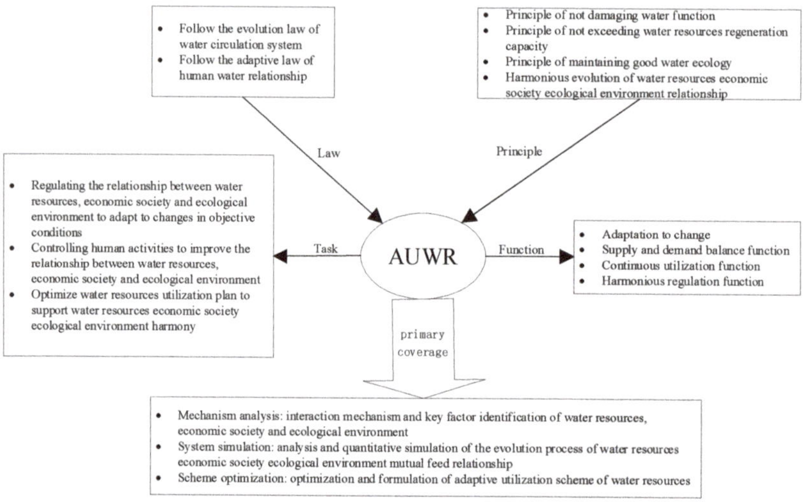

Figure 3. Framework of application rules for AUWR.

3. Materials and Methods

3.1. Study Area

The TRB is located in the northern Tarim Basin of Xinjiang Uyghur Autonomous Region. It originates from the Tianshan Mountains and the Karakorum Mountains, with a total length of 2179 km, making it the longest inland river in China and the fifth-largest inland river in the world [34]. The TRB is composed of three major headwaters, the Hotan River, the Yarkant River, and the Aksu River [35], with a basin area of 1.02 million square kilometers, including 42 counties in five prefectures and 45 regiments in four corps divisions, with a population of more than 12 million people living in the basin. The average annual natural runoff of the TRB is 39.83 billion cubic meters, and the total water resources of the basin are 42.9 billion cubic meters, the main source of runoff in the TRB is glacier melt, accounting for nearly 50% of the runoff, while the remaining runoff sources include precipitation from rain and snow and river base flow [36,37]. The irrational exploitation of water resources has caused a certain impact on the ecological environment and the sustainability of economic development in the TRB. To meet the demand for water for economic and social development and agricultural irrigation (the demand for water for agricultural irrigation is very high, accounting for about 96% of the total water consumption in the TRB) [38], the water resources in the main-stream of the TRB are over-exploited, which has affected the tributaries and the lower streams of the ecosystem, further compressing water for the ecological environment, leading to ecological degradation. The population of the TRB accounts for 46.85% of Xinjiang, the total GDP accounts for 27.68% of Xinjiang, the GDP per capita is far below the average level of Xinjiang, the urbanization level is low, and the economic and social development is generally backward. The study area is mainly composed of five prefectures in the basin, namely Aksu, Bayingol Mongolian Autonomous Prefecture (BMAP), Kizilsu Kirgiz Autonomous Prefecture (KKAP), Kashgar Prefecture (KP), and Hotan Prefecture (HP) (Figure 4).

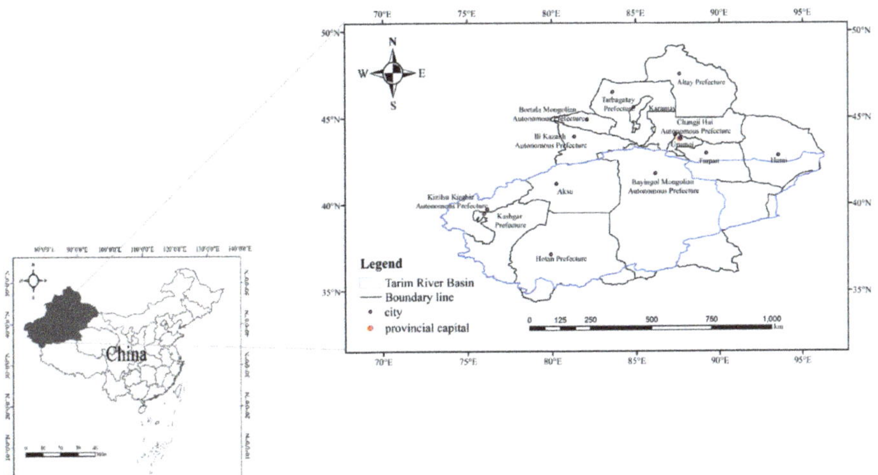

Figure 4. Schematic diagram of Xinjiang and TRB.

3.2. *Constructing the Framework of the Element System*

The assessment of AUCWR is the basis for rational development and the utilization of regional water resources, sustainable economic and social development, and ecological environmental protection. To assess the AUCWR, it is necessary to build a set of assessment index systems from the two laws, four principles, three tasks, and four functions of the adaptive utilization theory of water resources [24]. Taking into account the water resources endowment conditions, economic and social factors, and the ecological environment of the TRB, 25 evaluation indexes are finally selected, and the AUCWR is used as the target layer to build a system covering the assessment element system of AUCWR covering three guideline layers of water resources, economic society, and ecological environment is constructed, as shown in Table 1. In the table, (+) represents positive indicators and (−) represents negative indicators.

Table 1. Evaluation element system of AUCWR.

Target	Sub-Problem Domain	Element	Unit
Adaptive utilization capacity of water resources (AUCWR)	Water Resource	Precipitation depth (I_1) (+)	mm
		Water yielding modulus (I_2) (+)	10^4 m^3/km^2
		Average per capita water resources (I_3) (+)	m^3/person
		Exploitation rate of water resources (I_4) (−)	/
		Per capita water consumption (I_5) (−)	m^3/person
		Water consumption per 10,000 yuan of GDP (I_6) (−)	$m^3/10^4$ CNY
		Water consumption per 10,000 yuan of industrial added value (I_7) (−)	$m^3/10^4$ CNY
		Average irrigation water consumption per unit area of farmland (I_8) (−)	m^3/hm^2
		Per capita domestic water consumption (I_9) (−)	L/person
	Economic Society	Per capita GDP (I_{10}) (+)	10^4 CNY/person
		Proportion of non−agricultural output value in GDP (I_{11}) (+)	/
		Grain production per cubic meter of water (I_{12}) (+)	kg/m^3
		Per capita disposable income of urban residents (I_{13}) (+)	CNY/person
		Per capita net income of rural residents (I_{14}) (+)	CNY/person
		Urbanization rate (I_{15}) (+)	/
		Population density (I_{16}) (+)	person/km^2
		Natural population growth rate (I_{17}) (+)	/
		Water popularization rate of urban population (I_{18}) (+)	/
	Ecological Environment	Forest coverage rate (I_{19}) (+)	/
		Green coverage rate of built-up area (I_{20}) (+)	/
		Ecological environment water consumption rate (I_{21}) (+)	/
		COD emission per capita (I_{22}) (−)	t/10^4 person
		Ammonia nitrogen emissions per capita (I_{23}) (−)	t/10^4 person
		Per capita discharge of sewage and wastewater (I_{24}) (−)	m^3/person
		Fertilizer application intensity (I_{25}) (−)	kg/hm^2

3.3. Methods

3.3.1. Calculate Element Weights

The methods of determining the weights of the index system can be generally divided into two categories: subjective assignment method and objective assignment method. The objective assignment method includes such methods as the mean square difference method, principal component analysis method, entropy method, representative calculation method, etc. The subjective assignment method includes the subjective weighting method, expert survey method, hierarchical analysis method, comparative weighting method, multivariate analysis method, fuzzy statistics method, etc. In this paper, the entropy weighting method is used to determine the weights in the evaluation study of the effect of demonstration [39].

The entropy weighting method is used to calculate the objective weights [40]. Generally speaking, if the information entropy of an index is smaller, it indicates that the greater the degree of variation of the index value, the more information it provides, the greater the role it can play in the comprehensive evaluation, and the greater its weight. The steps to determine the weights by the entropy method are as follows.

1. The data are standardized and normalized.

$$Y_{ij} = \frac{X_{ij} - min(X_{ij})}{max(X_{ij}) - min(X_{ij})} \quad (1)$$

$$Z_{ij} = \frac{Y_{ij}}{\sum_1^n Y_{ij}} \quad (2)$$

2. Seek the information entropy of indicators

$$E_j = -\ln(n)^{-1}\sum_{1}^{n} Z_{ij} ln Z_{ij} \tag{3}$$

3. Determine the weight:

$$W_j = \frac{1 - E_j}{\sum 1 - E_j}, \ j = 1, 2, \ldots, m \tag{4}$$

where, W_j is the weight.

3.3.2. Harmony Methods

Through the evaluation of using the harmony methods [41], we can reflect the harmony degree on the whole, the state, and level, as well as the spatial and temporal changes, and provide the basis for the evaluation of harmony problems and the search for harmony strategies. It mainly adopts the evaluation method of "single indicator quantification—multiple indicators synthesis—multiple criteria integration", as follows:

1. Single-indicator quantification: It includes quantitative and qualitative indicators, and each indicator has a harmonious degree (called *SHD*) with the value range of [0, 1]. In order to facilitate calculation and comparative analysis, the quantitative description of single indicator harmony can be quantified by using segmented linear affiliation function quantification method for positive indicators, negative indicators, and bidirectional indicators respectively, and mapping each indicator to [0, 1] uniformly. Among them, the harmony degree of positive and negative indicators is calculated as follows [42].

$$SHD_k = \begin{cases} 0 & x_k \le a_k \\ 0.3(\frac{x_k - a_k}{b_k - a_k}) & a_k < x_k \le b_k \\ 0.3 + 0.3(\frac{x_k - b_k}{c_k - b_k}) & b_k < x_k \le c_k \\ 0.6 + 0.2(\frac{x_k - c_k}{d_k - c_k}) & c_k < x_k \le d_k \\ 0.8 + 0.2(\frac{x_k - d_k}{e_k - d_k}) & d_k < x_k \le e_k \\ 1 & e_k < x_k \end{cases} \quad SHD_k = \begin{cases} 1 & x_k \le e_k \\ 0.8 + 0.2(\frac{d_k - x_k}{d_k - e_k}) & e_k < x_k \le d_k \\ 0.6 + 0.2(\frac{x_k - x_k}{c_k - d_k}) & d_k < x_k \le c_k \\ 0.3 + 0.3(\frac{b_k - x_k}{b_k - c_k}) & c_k < x_k \le b_k \\ 0.3(\frac{a_k - x_k}{a_k - b_k}) & b_k < x_k \le a_k \\ 0 & a_k < x_k \end{cases} \tag{5}$$

where, SHD_k is the harmony degree of the k-th index, $k = 1, 2, \ldots, n$, n is the number of indicators; a_k, b_k, c_k, d_k, e_k is the worst value, poor value, pass value, better value and best value of the k-th index.

2. Multi-indicator synthesis: it can be calculated by multi-indicator weighting method, according to the single indicator affiliation weighted by the weight.

$$HD = \sum_{j=1}^{n} w_j \mu_j \in [0, 1] \tag{6}$$

where, μ_j is the harmony of the kth indicator SHD_j, w_j is the weight. It can also be calculated according to the single indicator affiliation weighted by exponential weights.

$$HD_t = \prod_{j=1}^{n} (\mu_j)^{w_j} \in [0, 1] \tag{7}$$

where, w_j is the weight.

3. Multi-criteria integration: It can be calculated using a weighted average or index weighting method.

$$\text{AUCWR} = \sum_{t=1}^{T} w_t HD_t \text{ or AUCWR} = \sum_{t=1}^{T} (HD_t)^{\beta_t} \tag{8}$$

where ω_t, β_t are the weights of the t-criteria, $\sum_{t=1}^{T} \omega_t = 1$, $\sum_{t=1}^{T} \beta_t = 1$, and the other symbols are the same as before.

On the basis of the constructed index system for assessing the adaptive capacity of water resources, the problem of adaptive use of water resources is understood as a dynamic and harmonious balance of water resources-economy-society-ecology-environment system. The goal is to maximize the harmony of the water resources-economic-social-ecological environment system. The overall harmony degree is calculated by using the comprehensive evaluation method of "Single Indicator Quantification—Multi-Indicator Integration—Multi-Criteria Integration" (SMI-P method) of the harmony theory. Firstly, we quantify each indicator and calculate the individual indicator harmony degree, then we assign and weight each indicator to calculate the harmony degree of each criterion layer, and finally, we weight each criterion layer to calculate the harmony degree.

3.3.3. Comprehensive Co-Evolution Model Methods

According to the comprehensive co-evolutionary model proposed in each reference [28,29], the adaptive capacity of the influencing factors to environmental changes is measured by calculating the absolute adaptability, and the relative adaptability is used to describe the adaptability of the interaction between the influencing factors, based on the characteristics of mutual adaptation between different influencing factors or indicators in the theory of adaptive use of water resources. The combination of absolute and relative adaptability is used to evaluate the AUCWR. The method is divided into the following steps.

1. Division of criterion layers and determination of weights

According to the index system established above, the criterion layer is divided into three aspects: water resources, economic and social factors, and ecological environment. The weights are determined using the entropy weighting method above to ensure that the weights of the influencing factors are consistent between the harmony theory and the comprehensive co-evolutionary model approach.

2. Calculation of absolute adaptability of factors

In order to effectively reduce the influence brought by the uncertainty of the relationship between factors, the gray correlation analysis method is first used to determine the correlation degree between individual factors; the gray correlation degree method is as follows.

$$\alpha_{ij} = \frac{\min_i \min_j |X_{oj} - X_{ij}| + \rho \max_i \max_j |X_{oj} - X_{ij}|}{|X_{oj} - X_{ij}| + \rho \max_i \max_j |X_{oj} - X_{ij}|} \tag{9}$$

where ρ denotes the resolution factor, usually taken as 0.5 [43] where X_{oj} represents the optimal value of the jth factor.

$\overline{\alpha_{ij}}$ as the average value of each point between X_{ij} and X_{oj}, $\overline{\alpha} = \frac{1}{n}\sum_{1}^{n} \alpha_{ij}$; $\epsilon_{ij} = \alpha_{ij} - \overline{\alpha_{ij}}$; where ϵ_{ij} is used to represent the fluctuation value between the factors α_{ij}, the system adaptation, and finally the absolute adaptation of the factors is derived as follows:

$$f_j^C = 1 - \sqrt{\left(\overline{\alpha_{ij}} - 1\right)^2 + \sum_{I=1}^{n} {\epsilon_{ij}}^2 * W_j} \tag{10}$$

where: f_j^C represents the absolute factor fitness; W_j represents the factor weights.

3. Calculation of the relative fitness of factors

$$f_j^R = \frac{0.5 + HD_j - AHD}{0.5 + HD_j} * f_j^C \tag{11}$$

where: f_j^R represents the absolute suitability of the factors; HD_j represents the Hemming distance between the actual and ideal values in the evaluation matrix; AHD represents the average of the hemming distance of each element; 0.5 represents the smoothing factor. HD and AHD are calculated by the formula between the original literature.

4. Factor adaptation calculation

The article combines the absolute and relative fitness of the factors with the weights to calculate the fitness of the factors with the following formula.

$$f_j^S = W_j * f_j^C + (1 - W_j) * f_j^R \tag{12}$$

where: f_j^S represents the adaptation of the factors.

5. Calculation of AUCWR

Based on the results of the factor adaptability, the calculated data are standardized to obtain the standard value X_{Ij}^* for each indicator and consequently the survival adaptability of the target layer. In order to maintain consistency with the Harmony Theory approach, the target layer is here designated as the AUCWR, and thus the formula for calculating the AUCWR is obtained as:

$$D_i = \sum_{j=1}^{m} \frac{f_j^S}{\sum_{j=1}^{m} f_j^S} * X_{Ij}^* \tag{13}$$

where D_i represents the AUCWR of the ith evaluation object, where i represents the calculation year (2004–2018); m represents the number of factors.

3.3.4. Obstacle Degree Model Methods

The obstacle degree model can assess the degree of influence of each factor on the final goal by analyzing the magnitude of the obstacle effect of different indicators in the assessment index system [44]. Obstacle degree models are widely used in assessing land use impact factor assessment, ecological security assessment, and other fields. In this paper, the obstacle degree model is introduced to analyze the degree of contribution of impact factors in order to better regulate the AUCWR. The specific steps are as follows.

The obstacle degree Q_i (the degree of influence of each subsystem or each indicator on the AUCWR is calculated by introducing the factor contribution degree w_j (the weight of a single indicator on the total target) and the indicator deviation degree I_i (the distance between the actual value of each indicator and the optimal value, expressed as the difference between 1 and the standardized value X_{ij} of each indicator), which is calculated as follows:

$$Q_i = \frac{I_i \times w_i}{(\sum_{i=1}^{m} I \times w_i)} \tag{14}$$

where $I_i = 1 - X_{ij}$, X_{ij} is the normalized value of the indicator.

3.4. Data Sources

The data used in this paper are all from Xinjiang and the Aksu, BMAP, KKAP, KP, and HP regions yearbooks from 2005–2019, and the statistics are from 2004–2018.

4. Results

4.1. Element Thresholds and Weights

4.1.1. Element Thresholds

According to the single indicator quantification in the harmony theory method, the thresholds of 25 indicators in the evaluation index system are divided, and the thresholds are divided into five nodes according to the single indicator quantification calculation formula, which are optimal, better, medium, worse and worst in order, and the final 25 indicator thresholds are divided in Table 2.

Table 2. Element threshold division table.

Element	Threshold Division					Element	Threshold Division				
	Worst	Poor	Moderate	Better	Best		Worst	Poor	Moderate	Better	Best
I_1	39	150	400	600	850	I_{14}	1000	4000	7000	10,000	13,000
I_2	3	6	9	12	15	I_{15}	0.2	0.4	0.6	0.8	1
I_3	1000	6000	12,500	19,000	25,000	I_{16}	5	11	17	24	30
I_4	1	0.8	0.6	0.4	0.2	I_{17}	3	6.5	10	20	30
I_5	7000	5500	3500	2000	500	I_{18}	0.2	0.4	0.6	0.8	1
I_6	10,000	7500	5000	3000	1000	I_{19}	0.005	0.02	0.04	0.1	0.16
I_7	1000	700	400	200	50	I_{20}	0.2	0.4	0.6	0.8	1
I_8	1200	950	700	450	200	I_{21}	0.005	0.01	0.02	0.035	0.05
I_9	150	125	100	75	50	I_{22}	200	150	100	75	50
I_{10}	0.5	1.75	3	5	7	I_{23}	20	15	10	6	2
I_{11}	0.2	0.4	0.6	0.8	1	I_{24}	100	65	30	17.5	5
I_{12}	0.1	0.2	0.3	0.4	0.5	I_{25}	100,000	75,000	50,000	30,000	10,000
I_{13}	5000	12,500	20,000	30,000	40,000						

4.1.2. Element Weights

According to the entropy weighting method, a total of 25 indicators in three subsystems of TRB, namely, water resources, economic and social factors, and ecological environment, are weighted as shown in Table 3.

Table 3. Water resources subsystem element weights.

System	Element Weight										
Water Resources subsystem	Element	I_1	I_2	I_3	I_4	I_5	I_6	I_7	I_8	I_9	Total
	Weight	0.043	0.038	0.049	0.036	0.032	0.035	0.027	0.037	0.028	0.325
Economic and Social subsystem	Element	I_{10}	I_{11}	I_{12}	I_{13}	I_{14}	I_{15}	I_{16}	I_{17}	I_{18}	Total
	Weight	0.075	0.055	0.042	0.028	0.050	0.043	0.045	0.041	0.045	0.424
Ecological Environment subsystem	Element	I_{19}	I_{20}	I_{21}	I_{22}	I_{23}	I_{24}	I_{25}			Total
	Weight	0.039	0.044	0.056	0.027	0.026	0.027	0.032			0.251

Among them, the economic and social subsystem weight is larger, accounting for 0.424, the water resources subsystem has the second largest weight, accounting for 0.325, and the ecological environment subsystem has the smallest weight of 0.251. Among the indicators, the per capita water resources in the water resources subsystem has the largest weight of 0.049, in the economic and social subsystem, the per capita GDP has the largest weight of 0.075, and in the ecological environment subsystem, ecological environmental water use rate, the largest weight is 0.056.

4.2. Evaluation of AUCWR

4.2.1. Temporal and Spatial Variation Characteristics of AUCWR in TRB

The results obtained based on the harmony theory method are shown in Figure 5a. In general, the AUCWR in the TRB demonstrates a fluctuating upward trend, the results show that this trend is in line with the current development situation of the Tarim River Basin [45]. The AUCWR in the TRB increased from 0.43 in 2004 to 0.56 in 2018, with a growth rate of 30.23%. The AUCWR is mainly concentrated in the range of 0.40–0.60, with an annual average value of 0.497, which indicates that the adaptability among water resources, economic and social, and ecological environment subsystems is in the near-adaptation stage, and the level of adaptive development in the basin is moderate. The adaptive use capacity levels of water resources from 2004 to 2018 are all in the near-adaptation stage. According to the growth rate of the AUCWR, the development of the AUCWR in the basin demonstrates an increasing trend from 2004 to 2006 (average annual growth rate of 2.57%); during the period of 2006–2010, the AUCWR in the basin reveals a fluctuating, increasing trend (average annual growth rate of 1.38%); from 2010–2012, the AUCWR in the basin indicates a fluctuating downward trend (average annual decrease rate of 2.25%); from 2012 to 2018, the AUCWR in the basin verifies an upward trend (average annual growth rate of 1.17%). The adaptive development level of AUCWR in the basin has increased during the period 2004–2018 (average annual growth rate of 0.08%), but the rising level is low.

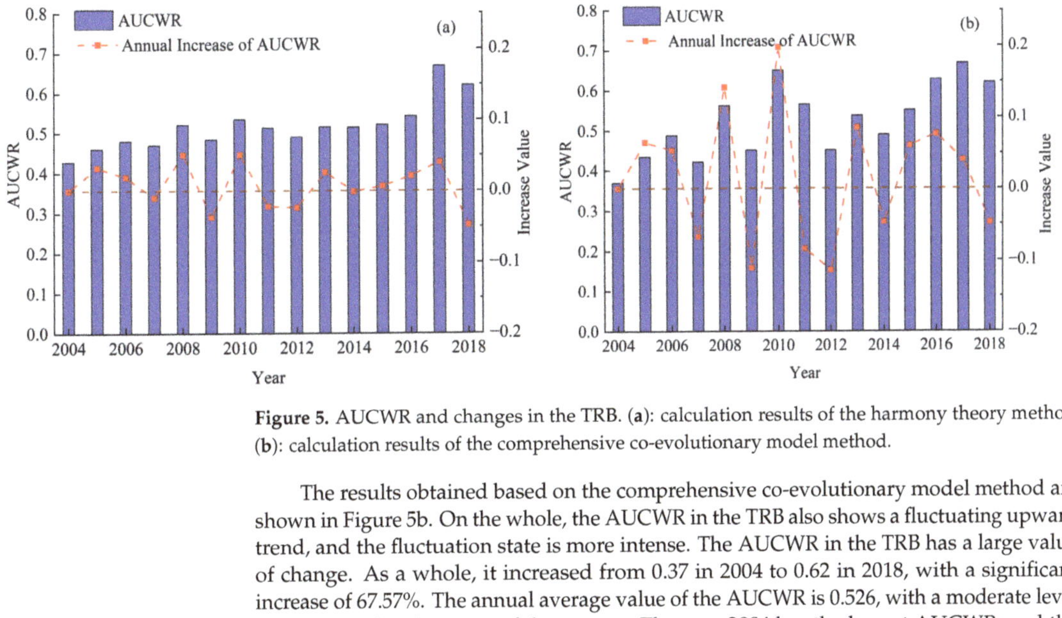

Figure 5. AUCWR and changes in the TRB. (**a**): calculation results of the harmony theory method. (**b**): calculation results of the comprehensive co-evolutionary model method.

The results obtained based on the comprehensive co-evolutionary model method are shown in Figure 5b. On the whole, the AUCWR in the TRB also shows a fluctuating upward trend, and the fluctuation state is more intense. The AUCWR in the TRB has a large value of change. As a whole, it increased from 0.37 in 2004 to 0.62 in 2018, with a significant increase of 67.57%. The annual average value of the AUCWR is 0.526, with a moderate level of adaptive development of the system. The year 2004 has the lowest AUCWR, and the adaptive level is at the basic non-adaptive stage; 2015–2009 and 2011–2015 are at the near adaptive stage, while 2010 and 2016–2018 are at the AUCWR. The AUCWR in 2010 and 2016–2018 are all at the basic adaptive stage. Based on the magnitude of changes in adaptive capacity, the harmony theory method calculations show similar trends: a gradual increase during 2004–2006 (with an average annual increase of 5.93%), a fluctuating increase from 2006–2010 (with an average annual increase of 4.04%), a gradual decrease from 2010–2012 (with an average annual decrease of 9.89%), and a fluctuating up (with an average annual increase of 1.91%), and 2014–2017 gradually up (with an average annual increase of 5.91%). By and large, the level of adaptive development of the AUCWR in the basin increased during 2004–2018 (average annual growth rate of 0.2%), but the level of increase is limited.

The AUCWR in the TRB is assessed by the harmony theory method and the comprehensive co-evolutionary model method, and the results of both calculation methods show that the AUCWR in the TRB is not high during 2004–2018 (mean value of the harmony theory method: 0.497; the mean value of the comprehensive co-evolutionary model method: 0.526), but the development trend is good and the capacity gradually improved. The adaptive use of water resources in the TRB is limited, and the AUCWR is around 0.6 after improvement (calculated by the harmony theory method: 0.56; calculated by the comprehensive co-evolutionary model method: 0.62), which is near the passing level. The current problems should be addressed, and solutions should be proposed to improve the overall AUCWR in the TRB.

The results of the AUCWR assessment of the TRB calculated by the two methods are shown in Figure 6a,b. The analysis reveals that the calculated overall change trends of the TRB and each prefecture are consistent and demonstrate an increasing trend; secondly, the average value of the AUCWR in the TRB from 2004 to 2018 calculated by the harmony theory method is 0.497, and the result calculated by the integrated coevolutionary model method is 0.526, which is basically similar to the water resources of each prefecture. The results are similar to the AUCWR in each state. In general, the results of the two calculation methods are consistent, and the results of the two methods can be combined to make a comprehensive assessment of the AUCWR in the TRB and each state. Therefore, in the following assessment of the AUCWR in each state, in order to focus on the analysis of the changes between the states, the calculation results are averaged using the calculation results of the two methods (Figure 6c).

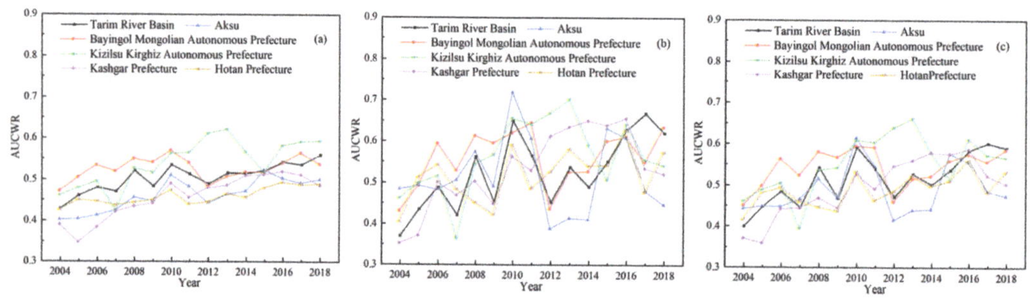

Figure 6. AUCWR and changes of various prefectures in the TRB. (**a**,**b**): calculation results of the harmony theory and the comprehensive co-evolutionary model method. (**c**): average of the results calculated by the two methods.

Using the two methods, by analyzing the AUCWR in the five prefectures (Aksu, BMAP, KKAP, KP, and HP regions) belonging to the TRB from 2004 to 2018, we obtained the trend graph of the AUCWR in each prefecture, as shown in Figure 6a,b, and combined the results of the two calculations to obtain the trend graph of the AUCWR in each prefecture, as shown in Figure 6c.

The results verify that, in terms of temporal trends, the AUCWR in all states of the TRB has similar trends, with all five regions showing fluctuating upward trends. The growth rates of Aksu, BMAP, KKAP, KP, and HP regions are 14.01%, 24.45%, 28.87%, 24.76%, and 14.81%, respectively, with the largest increase in the KKAP region and the smallest increase in the Aksu region. The fluctuation of the KKAP region is more dramatic, and its standard deviation of AUCWR from 2004 to 2018 reaches 0.056, which is larger than the remaining four prefectures. At the same time, there is little difference in the mean value of AUCWR in each prefecture. The average AUCWR of the TRB from 2004 to 2018 is 0.52. The average AUCWR of the BMAP and KKAP regions is larger than that of the TRB, 0.54 and 0.55, respectively, while the average AUCWR of the Aksu, KP, and HP regions is smaller than that of the TRB, 0.49, 0.50, and 0.49, respectively. By analyzing the trends and average values of the AUCWR in the TRB as a whole and in each state, we found that the AUCWR in each state is not high at present and still has great potential for development. The trend of fluctuating growth is the same as that of the TRB, but the growth rate is not large.

4.2.2. Temporal and Spatial Variation Characteristics of System Adaptability in TRB

The system adaptability of the three subsystems of water resources, economic and social factors, and ecological environment in the TRB and the five prefectures is obtained according to Equation (5) to Equation (8), as shown in Figure 7.

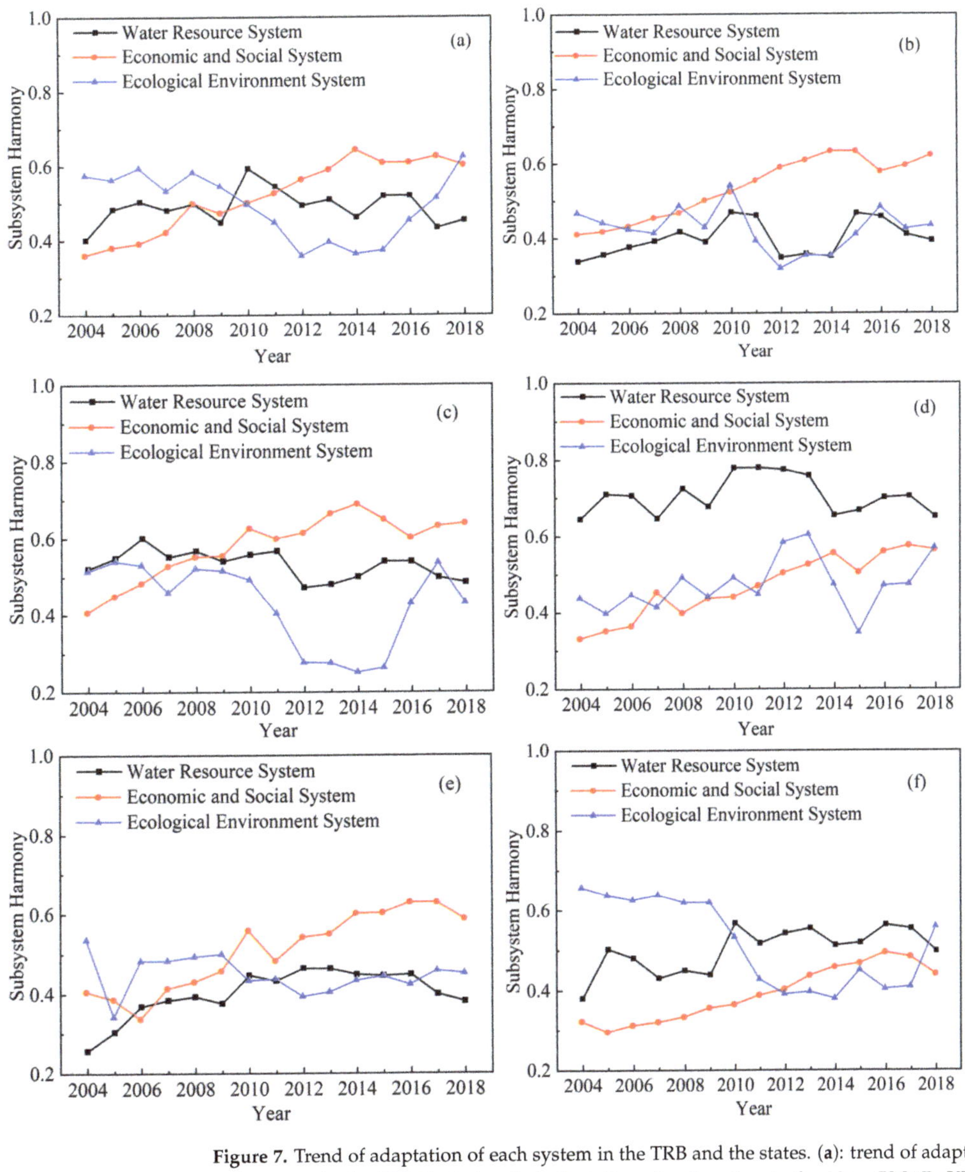

Figure 7. Trend of adaptation of each system in the TRB and the states. (**a**): trend of adaptation of each system in the TRB. (**b**–**f**): trend of adaptation of each system in the Aksu, BMAP , KKAP, KP, and HP regions.

Figure 7a confirms the adaptability of each subsystem in the TRB, which ranges from 0.4 to 0.6, indicating that the adaptability of each subsystem is at a medium level in the study area. From 2004 to 2018, the change in the adaptability of the water resources system is small (average annual growth rate of 5.2%), and although the water resources subsystem reveals an upward trend, the trend is not obvious, and by 2018, the adaptability of the water resources subsystem is significantly lower than that of the economic and social and ecosystem subsystems. It can no longer meet the needs of economic and social development and ecological protection. During 2004–2018, the economic and social subsystem adaptation degree indicates a rapid upward trend (average annual growth rate of 24%), and the level of economic and social development steadily increased during this period, but after 2014, it decreased, which may be due to the fact that with the increase of ecological environmental protection, the development of certain environmentally crude enterprises is restricted to a certain extent, which caused the growth of economic and social development certain impact, but generally speaking, the momentum of economic and social development is good. During the period of 2004–2014, the adaptability of the ecological and environmental subsystem indicates a decreasing trend (the average annual decrease rate is 21%), but during the period of 2014–2018, the adaptability of the ecological and environmental subsystem indicates an upward trend and an obvious upward trend (the average annual growth rate is 26%). It indicates that before 2014, the economic and social development of the TRB might be to a certain extent at the expense of the ecological environment. Strongly affected by human activities, the ecological environment is damaged to some extent, the ecological environment is becoming worse and worse, the ecological carrying capacity is gradually increasing [46], and the research shows that the changes in human activities and climate have a significant impact on the ecological environment and oasis changes in the TRB [47]. With the introduction of the policy of ecological protection, the ecological environment is obviously improved and implies a good development trend after increasing ecological protection and management.

Figure 7b–f show the changes in the adaptability of each subsystem in the five prefectures, among which, the adaptability trends of each subsystem in BMAP, Aksu, KP, and HP regions are consistent with those of the TRB, all showing a decreasing trend of the adaptability of the ecological environment subsystem, and an increasing trend of the adaptability of the water resources and economic and social subsystems, while the adaptability of the water resources, economic and social, and ecological environment subsystems in KP. The adaptation of water resources and economic and social subsystems in the KKAP regions shows an increasing trend, which indicates that the KKAP region is better than the other four states in environmental protection. Relevant research results also show that this trend is in line with the actual situation [48,49].

4.3. *Element Analysis*

4.3.1. Analysis of Element Change Characteristics

The adaptability of the subsystems is influenced by the changes in their internal elements. Figure 8 shows the average growth rate of each subsystem index, and the results show that the main factors affecting the adaptive development of the water resources, economic and social, and ecological environment subsystems are water consumption per 10,000 yuan of industrial added value (I_7), per capita domestic water consumption (I_9), water consumption of 10,000 yuan of GDP (I_6), per capita net income of rural residents (I_{14}), per capita GDP (I_{10}), per capita disposable income of urban residents (I_{13}), ecological environment water consumption rate (I_{21}), and fertilizer application intensity (I_{25}).

Since the increase in per capita domestic water consumption (I_9, growth rate 7.57%) is significantly higher than the annual precipitation depth of the water resources subsystem (I_1, growth rate 1.04%), it may lead to the crowding out of a large amount of ecological and environmental water and a significant decrease in the ecological environmental water consumption rate (I_{21}, decrease rate 5.97%), together with the inadequate environmental protection measures, all these combined effects may lead to the adaptation of the ecological and environmental subsystem declining. Additionally, to solve these problems, while developing and utilizing water resources and promoting economic development, we should strengthen ecological environmental protection and promote the integrated development of water resources, society, and ecology.

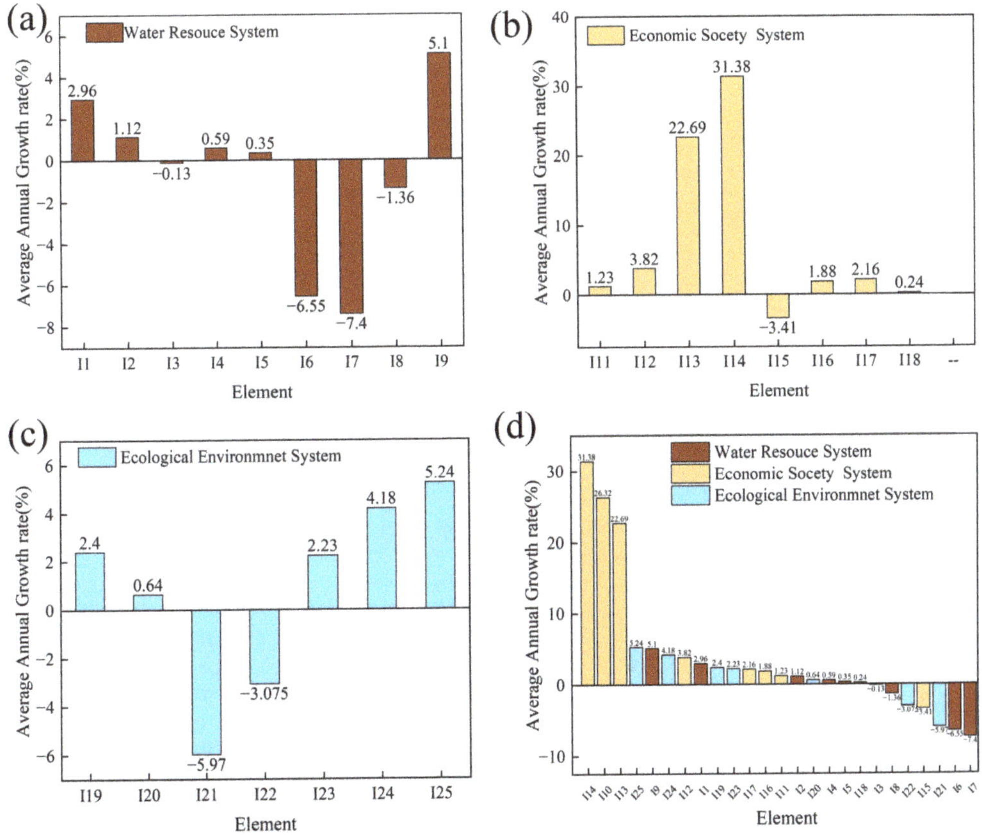

Figure 8. Average annual growth rate of each subsystem element. (**a**–**c**): the average annual growth rate of three subsystems. (**d**): ranking of average annual growth rate of three subsystems.

With industrial upgrading and progress in water conservation technology, rapid economic development can also be driven without affecting basic domestic water consumption, which is mainly reflected in the water consumption per 10,000 yuan of GDP (I_6, decline rate 7.01%), water consumption per 10,000 yuan of industrial added value (I_7, decline rate 7.78%) and average irrigation water consumption per unit area of farmland (I_8, decline rate 0.43%), per capita GDP (I_{10}, growth rate of 26.32%), per capita disposable income of urban residents (I_{13}, growth rate of 22.69%), and per capita net income of rural residents (I_{14}, growth rate of 31.38%) on the indicators, which are also the main reasons for the improvement of the economic and social subsystem adaptation.

Due to the western development strategy, the arable land area has been expanding while the economy is developing rapidly, with an increased rate of 83.8% in 2018 compared to 2004, which has increased the water demand to some extent. Although the amount of water resources in the TRB has increased, the water resources development and utilization rate remain high (72.64% on average), which is already higher than the internationally accepted limit of 40% and has reached a bottleneck. Therefore, it is necessary to improve water use efficiency, strengthen water conservation measures and improve water saving efficiency. The changes in water consumption per 10,000 yuan of GDP (I_6, decline rate 7.78%), water consumption per 10,000 yuan of industrial added value (I_7, decline rate 7.78%), and average irrigation water consumption per unit area of farmland (I_8, decline rate 0.43%) are precisely the expression of water use efficiency improvement and the increase in the adaptation of water resources subsystem. However, the increasing water demand and the unreasonable allocation of water resources are also the main reasons for limiting the further improvement of the water resources subsystem.

In order to promote the coordinated development of water resources, economy, society, and ecological environment, it is necessary to actively carry out economic restructuring while developing and utilizing water resources, taking into account the endowment conditions of water resources, and driving the sustainable and stable development of the economy. At the same time, it is also necessary to focus on protecting the ecological environment, limiting unreasonable development of arable land, accelerating the construction of grassland and other ecological projects, reasonably allocating and dispatching water resources, realizing the healthy development of rivers, and coordinating the coordination between economic development and ecological protection. From the root cause, the water resources management system should be strengthened to coordinate the harmonious relationship between economic development and ecological environment, source and tributaries, upstream and downstream, based on water resources development and utilization, and improve the management system and system to realize the rational use of water resources.

4.3.2. Element Sensitivity Analysis

Using the barrier degree model to calculate the barrier degree of impact factors, the barrier degree of each subsystem and each indicator in the TRB from 2004 to 2018 is obtained, and the results are shown in Table 4. From Table 4, it can be concluded that there are differences in the barrier degrees of water resources, economic and social, and ecological environment subsystems on the AUCWR. In terms of temporal changes, the barrier degree of the water resources system increases year by year, but the growth rate is small, with an average annual growth rate of only 0.105%; the barrier degree of the economic and social subsystem gradually decreases, with an average annual reduction rate of 0.081%; the barrier degree of the ecological environment subsystem fluctuates more, first decreasing and then increasing, with an overall upward trend, but there is a large decrease in 2018. From the analysis of the three major subsystem barrier degree values, the economic and social subsystem has the largest barrier degree with an average value of 51.45%, followed by the water resources subsystem with an average barrier degree of 30.37%, while the ecological environment subsystem has the lowest barrier degree with an average value of only 18.18%. This shows that the economic and social subsystem is the main constraint subsystem affecting the improvement of the AUCWR in the TRB. Therefore, in order to further improve the AUCWR in the TRB, we should focus on the economic and social subsystem, further consider the development and utilization of water resources, ecological and environmental protection and economic and social development, effectively improve the level of coupled and coordinated development among water resources, economic and social factors, and the ecological environment, promote the healthy and sustainable development of the basin, and continuously improve the AUCWR.

Taking the 2018 data as an example, the barrier degree of each indicator to the overall system of the basin is analyzed, and the results are obtained as shown in Table 5. In terms of the barrier degree values of each indicator, the top indicators are mainly the economic and social subsystem indicators, and the top five indicators in the barrier degree of this system are per capita GDP (I_{10}), the proportion of non-agricultural output value in GDP (I_{11}), per capita net income of rural residents (I_{14}), population density (I_{16}), and water penetration rate of the urban population (I_{18}); the top three indicators in the barrier degree of the water resources system. The top three obstacles in the water resources system are average per capita water resources (I_3), precipitation depth (I_1), and water yielding modulus (I_2); the top three obstacles in the ecological environment subsystem are ecological environment water consumption rate (I_{21}), green coverage rate of built-up area (I_{20}), and forest coverage rate (I_{19}). On the whole, the indicators with a higher barrier degree have a greater impact on the AUCWR. Therefore, when analyzing and regulating the AUCWR in the future, the indicators with a higher barrier degree can be regulated.

Table 4. Subsystem level barriers to AUCWR in 2004–2018 (%).

Year	Water Resources	Economic and Social	Ecological Environment	Year	Water Resources	Economic and Social	Ecological Environment
2004	29.51	52.30	18.19	2012	30.53	51.24	18.23
2005	29.74	52.32	17.94	2013	30.60	51.14	18.25
2006	29.95	52.18	17.88	2014	30.86	50.79	18.35
2007	30.08	51.84	18.08	2015	30.70	51.00	18.30
2008	30.32	51.51	18.17	2016	30.57	50.98	18.45
2009	30.36	51.51	18.13	2017	30.60	50.94	18.46
2010	30.31	51.46	18.23	2018	30.98	51.17	17.86
2011	30.51	51.31	18.18				

Table 5. Subsystem level barriers to AUCWR in 2018 (%).

Element	Obstacle	Element	Obstacle	Element	Obstacle	Element	Obstacle	Element	Obstacle
I_1	4.04	I_6	3.42	I_{11}	6.72	I_{16}	5.53	I_{21}	3.69
I_2	3.65	I_7	2.68	I_{12}	5.06	I_{17}	5.19	I_{22}	1.97
I_3	4.69	I_8	3.53	I_{13}	3.28	I_{18}	5.46	I_{23}	1.91
I_4	3.40	I_9	2.51	I_{14}	5.80	I_{19}	2.80	I_{24}	1.98
I_5	3.04	I_{10}	8.76	I_{15}	5.36	I_{20}	3.24	I_{25}	2.26

5. Discussion

The AUCWR in the TRB is calculated by the harmony theory method and the comprehensive co-evolutionary model method, and the results are compared and analyzed as shown in Figure 9. According to Figure 9, the range of AUCWR in the TRB calculated by the two methods is not very different, with the range of 0.4–0.6 for the harmony theory calculation and 0.33–0.67 for the coevolutionary model method. The results calculated by both methods show a fluctuating upward trend from the overall time period of 2004–2016, followed by a consistent trend every two years, such as a gradual increase from 2004–2006, a fluctuating trend from 2006–2011, and then a gradual increase from 2012–2016. Taken together, the results of AUCWR calculated by the two methods can corroborate each other and increase the reliability of the results. Further, from the viewpoint of the magnitude of change, the harmony theory method has a small change and shows a steady upward trend overall, with the largest change in the two time periods of 2007–2008 and 2009–2010, with a change of 10.9% and 9.7% respectively. The comprehensive co-evolutionary model method has a larger change, and overall, the change from 0.34 in 2004 to 0.58 in 2016 is 0.22, which is much larger than the change value of 0.11 for the harmony theory method. Among them, the comprehensive co-evolutionary model method has the largest change before and after 2010, and the change before and after is 48.8% and 22.6%, respectively. Continuing to analyze the calculation results of the two methods in each region of the TRB, the mean values calculated by the comprehensive co-evolution model method are both higher than those calculated by the harmony theory, but the calculations are closer and the differences are not significant. While the fluctuation ranges are both larger than those calculated by the harmony theory method.

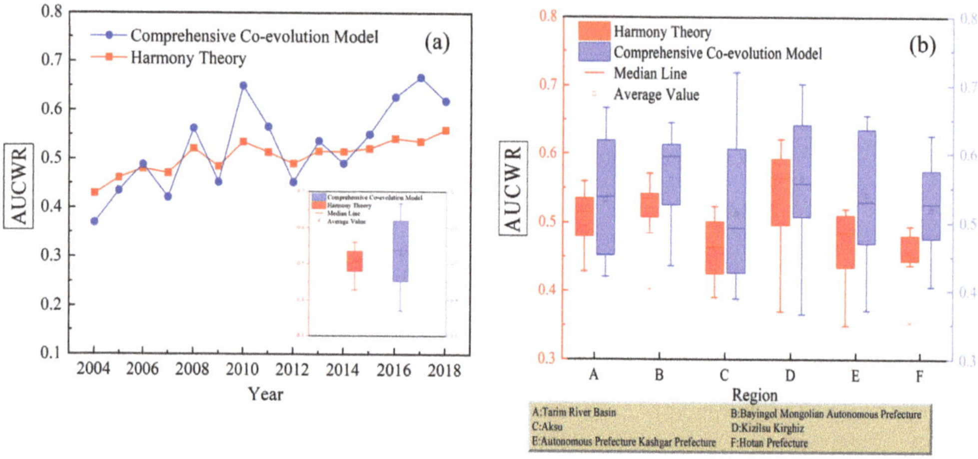

Figure 9. Comparative analysis of AUCWR in the TRB. (**a**): trend comparison of calculation results. (**b**): box-plot of calculation results in different regions.

A comparative analysis of the calculation results for the two algorithms can verify the reliability of the calculation results. From the calculation process of the two methods, in which the establishment of an evaluation index system, the determination of index weights, and the determination of evaluation index thresholds are the basic contents of the two methods. The indicator system proposed in the paper takes into account the indicators of the three dimensions of water resources, economy and society, and ecological environment. There are many types of indicators selected, which are more representative [46]. Among them, the harmony theory method uses the index quantification and criterion integration method to calculate the AUCWR, and the algorithm is relatively simple and easy to calculate; the coevolutionary model method has clear ideas, but the calculation formula is more complicated compared with the harmony theory, in which the whole calculation process involves the gray correlation method and the calculation of Hemming distance, which increases its calculation volume [50]. The evaluation index system and index weights are consistent in the two methods, and the evaluation index threshold method is used to a different extent in the two methods. In the harmony theory method, as long as the quantification of each index relies on the index threshold, the final weighting is integrated to obtain the final results, so the division of the index threshold has a greater impact on the calculation results of the harmony theory to a certain extent; the comprehensive co-evolutionary model method in which only the optimal value of the index threshold is used. Therefore, the division of indicator thresholds has relatively less influence on the calculation results of the comprehensive co-evolutionary model. Through comprehensive comparison and analysis, the two calculation results are basically reliable, and each calculation method has its own advantages. This paper evaluates the AUCWR in the TRB based on the two methods, and the evaluation results are also basically in line with the reality.

6. Conclusions

The AUCWR in the TRB and its five prefectures is assessed using the harmony theory method and the comprehensive co-evolutionary model method, and the key factors affecting the AUCWR are analyzed, finally, the applicability of the two assessment methods are discussed at the end. The following conclusions were drawn:

(1) The AUCWR in the TRB demonstrates a fluctuating upward trend from 2004 to 2018 (the harmony theory method assessment results: from 0.43 in 2004 to 0.56 in 2018, with a growth rate of 30.23%; the comprehensive co-evolutionary model method assessment results: from 0.37 in 2004 to 0.62 in 2018, with a significant increase of

67.57%). The development trend is good, but the current level of AUCWR in the TRB is still not high, and there is a lot of room for improvement.

(2) There are differences in the adaptability of subsystems in the TRB, mainly in that the adaptability of the water resources subsystem changes less, the economic and social subsystem increases significantly, and the ecological environment subsystem indicates a decreasing and then increasing trend. The trend of subsystem adaptations in BMAP, Aksu, KP, and HP is consistent with that of TRB, the adaptations of water resources and economic and social subsystems are increasing, while the adaptations of ecological environment subsystems are decreasing. While the adaptations of water resources, economic and social subsystems, and ecological environment subsystems in KP are increasing.

(3) By analyzing the factors, the change characteristics of each factor and the degree of influence on the AUCWR are obtained. Among them, the indicators with large changes from 2004 to 2018 are mainly: water consumption per 10,000 yuan of industrial added value (I_7), per capita domestic water consumption (I_9), water consumption of 10,000 yuan of GDP (I_6), per capita net income of rural residents (I_{14}), per capita GDP (I_{10}), per capita disposable income of urban residents (I_{13}), ecological environment water consumption rate (I_{21}), and fertilizer application intensity (I_{25}). While the analysis of the barrier degree model obtained that the economic and social subsystem had the largest barrier degree with a mean value of 51.45% at subsystem level. From the perspective of indicators, indicators such as per capita GDP (I_{10}), the proportion of non-agricultural output value in GDP (I_{11}), per capita net income of rural residents (I_{14}), population density (I_{16}), and water popularization rate of urban population (I_{18}).

Author Contributions: Conceptualization, Q.Z. and X.L.; methodology, Q.Z. and X.L.; formal analysis, X.L. and Q.Z.; investigation, Q.Z., Y.Z. and X.L.; resources, X.L., Y.W. and S.H.; data curation, X.L.; writing—original draft preparation, Q.Z. and X.L.; writing—review and editing, Q.Z., Y.Z. and J.Z.; visualization, X.L. and S.H.; supervision, Q.Z. and J.Z.; project administration, J.Z.; funding acquisition, Q.Z., Y.Z., J.Z. and Y.W. All authors have read and agreed to the published version of the manuscript.

Funding: The research was supported by the National Key Research and Development Program of China (Grant No. 2021YFC3200205) and The Natural Sciences Foundation of China (No. U1803241, 52025093, 51979284, and 52109044).

Institutional Review Board Statement: Not applicable.

Informed Consent Statement: Not applicable.

Data Availability Statement: The data that support the finding of this study are available from the corresponding author upon reasonable request.

Acknowledgments: We are very grateful to the editor's comments and suggestions. The provided comments have contributed substantially to improve the paper.

Conflicts of Interest: The authors declare no conflict of interest.

References

1. Wang, W.; Tang, D.; Pilgrim, M.; Liu, J. Water Resources Compound Systems: A Macro Approach to Analysing Water Resource Issues under Changing Situations. *Water* **2016**, *8*, 2. [CrossRef]
2. Rezaee, A.; Bozorg-Haddad, O.; Chu, X. Reallocation of water resources according to social, economic, and environmental parameters. *Sci. Rep.* **2021**, *11*, 17514. [CrossRef]
3. Honkonen, T. Water Security and Climate Change: The Need for Adaptive Governance. *Potchefstroom Electron. Law J.* **2017**, *20*, 1–26. [CrossRef]
4. Liu, L. Assessment of water resource security in karst area of Guizhou Province, China. *Sci. Rep.* **2021**, *11*, 7641. [CrossRef]
5. Li, D.; Zuo, Q.; Wu, Q.; Li, Q.; Ma, J. Achieving the tradeoffs between pollutant discharge and economic benefit of the Henan section of the South-to-North Water Diversion Project through water resources-environment system management under uncertainty. *J. Clean. Prod.* **2021**, *321*, 128857. [CrossRef]

6. Li, D.; Zuo, Q.; Zhang, Z. A new assessment method of sustainable water resources utilization considering fairness-efficiency-security: A case study of 31 provinces and cities in China. *Sustain. Cities Soc.* **2022**, *81*, 103839. [CrossRef]
7. Zeng, X.T.; Huang, G.H.; Yang, X.L.; Wang, X.; Fu, H.; Li, Y.P.; Li, Z. A developed fuzzy-stochastic optimization for coordinating human activity and eco-environmental protection in a regional wetland ecosystem under uncertainties. *Ecol. Eng.* **2016**, *97*, 207–230. [CrossRef]
8. Zeng, X.; Xiang, H.; Liu, J.; Xue, Y.; Zhu, J.; Xu, Y. Identification of Policies Based on Assessment-Optimization Model to Confront Vulnerable Resources System with Large Population Scale in a Big City. *Int. J. Environ. Res. Public Health* **2021**, *18*, 13097. [CrossRef]
9. Kundzewicz, Z.W. Water resources for sustainable development. *Hydrol. Sci. J.* **1997**, *42*, 467–480. [CrossRef]
10. Wang, J.; Xiao, W.; Wang, H.; Chai, Z.; Niu, C.; Li, W. Integrated simulation and assessment of water quantity and quality for a river under changing environmental conditions. *Chin. Sci. Bull.* **2013**, *58*, 3340–3347. [CrossRef]
11. He, Y.; Yang, J.; Chen, X.; Lin, K.; Zheng, Y.; Wang, Z. A Two-stage Approach to Basin-scale Water Demand Prediction. *Water Resour. Manag.* **2018**, *32*, 401–416. [CrossRef]
12. Wheater, H.S.; Gober, P. Water security and the science agenda. *Water Resour. Res.* **2015**, *51*, 5406–5424. [CrossRef]
13. Montanari, A.; Young, G.; Savenije, H.H.G.; Hughes, D.; Wagener, T.; Ren, L.L.; Koutsoyiannis, D.; Cudennec, C.; Toth, E.; Grimaldi, S.; et al. "Panta Rhei—Everything Flows": Change in hydrology and society—The IAHS Scientific Decade 2013–2022. *Hydrol. Sci. J.* **2013**, *58*, 1256–1275. [CrossRef]
14. Ha, T.P.; Dieperink, C.; Dang Tri, V.P.; Otter, H.S.; Hoekstra, P. Governance conditions for adaptive freshwater management in the Vietnamese Mekong Delta. *J. Hydrol.* **2018**, *557*, 116–127. [CrossRef]
15. Kummu, M.; Guillaume, J.H.A.; de Moel, H.; Eisner, S.; Flörke, M.; Porkka, M.; Siebert, S.; Veldkamp, T.I.E.; Ward, P.J. The world's road to water scarcity: Shortage and stress in the 20th century and pathways towards sustainability. *Sci. Rep.* **2016**, *6*, 38495. [CrossRef]
16. Cheng, K.; Fu, Q.; Chen, X.; Li, T.; Jiang, Q.; Ma, X.; Zhao, K. Adaptive Allocation Modeling for a Complex System of Regional Water and Land Resources Based on Information Entropy and its Application. *Water Resour. Manag.* **2015**, *29*, 4977–4993. [CrossRef]
17. Herrera-Pantoja, M.; Hiscock, K.M. Projected impacts of climate change on water availability indicators in a semi-arid region of central Mexico. *Environ. Sci. Policy* **2015**, *54*, 81–89. [CrossRef]
18. Ren, C.; Guo, P.; Li, M.; Li, R. An innovative method for water resources carrying capacity research—Metabolic theory of regional water resources. *J. Environ. Manag.* **2016**, *167*, 139–146. [CrossRef]
19. Sauchyn, D.J.; St-Jacques, J.-M.; Barrow, E.; Nemeth, M.W.; MacDonald, R.J.; Sheer, A.M.S.; Sheer, D.P. Adaptive Water Resource Planning in the South Saskatchewan River Basin: Use of Scenarios of Hydroclimatic Variability and Extremes. *JAWRA J. Am. Water Resour. Assoc.* **2016**, *52*, 222–240. [CrossRef]
20. Zhou, Y.; Guo, S.; Xu, C.-Y.; Liu, D.; Chen, L.; Wang, D. Integrated optimal allocation model for complex adaptive system of water resources management (II): Case study. *J. Hydrol.* **2015**, *531*, 977–991. [CrossRef]
21. Zhou, Y.; Guo, S.; Xu, C.-Y.; Liu, D.; Chen, L.; Ye, Y. Integrated optimal allocation model for complex adaptive system of water resources management (I): Methodologies. *J. Hydrol.* **2015**, *531*, 964–976. [CrossRef]
22. Herrfahrdt-Pähle, E. Integrated and adaptive governance of water resources: The case of South Africa. *Reg. Environ. Chang.* **2013**, *13*, 551–561. [CrossRef]
23. Xia, J.; Qiu, B.; Li, Y. Water resources vulnerability and adaptive management in the Huang, Huai and Hai river basins of China. *Water Int.* **2012**, *37*, 523–536. [CrossRef]
24. Zuo, Q. Application rules and key issues in theory of adaptive utilization of water resources. *Arid. Land Geogr.* **2017**, *40*, 925–932.
25. Zuo, Q. Theory of adaptive utilization of water resources and its application prospect in water management practices. *South-North Water Transf. Water Sci. Technol.* **2017**, *15*, 18–24.
26. Zuo, Q.; Han, S.; Han, C.; Luo, Z. Research frame of adaptive utilization allocation-regulation model of water resources in Xinjiang region based on RS. *Water Resour. Hydropower Eng.* **2019**, *50*, 52–57.
27. Zhang, J.; Li, L.W.; Zhang, Y.N.; Liu, Y.F.; Ma, W.L.; Zhang, Z.M. Using a fuzzy approach to assess adaptive capacity for urban water resources. *Int. J. Environ. Sci. Technol.* **2019**, *16*, 1571–1580. [CrossRef]
28. Yao, J.; Ren, Y.; Wei, S.; Pei, W. Assessing the complex adaptability of regional water security systems based on a unified co-evolutionary model. *J. Hydroinform.* **2017**, *20*, 34–48. [CrossRef]
29. Yao, J.; Wang, G.; Xue, W.; Yao, Z.; Xue, B. Assessing the Adaptability of Water Resources System in Shandong Province, China, Using a Novel Comprehensive Co-evolution Model. *Water Resour. Manag.* **2019**, *33*, 657–675. [CrossRef]
30. Fu, A.; Li, W.; Chen, Y.; Wang, Y.; Hao, H.; Li, Y.; Sun, F.; Zhou, H.; Zhu, C.; Hao, X. The effects of ecological rehabilitation projects on the resilience of an extremely drought-prone desert riparian forest ecosystem in the Tarim River Basin, Xinjiang, China. *Sci. Rep.* **2021**, *11*, 18485. [CrossRef]
31. Zhao, R.; Chen, Y.; Shi, P.; Zhang, L.; Pan, J.; Zhao, H. Land use and land cover change and driving mechanism in the arid inland river basin: A case study of Tarim River, Xinjiang, China. *Environ. Earth Sci.* **2013**, *68*, 591–604. [CrossRef]
32. van Vliet, M.T.H.; van Beek, L.P.H.; Eisner, S.; Flörke, M.; Wada, Y.; Bierkens, M.F.P. Multi-model assessment of global hydropower and cooling water discharge potential under climate change. *Glob. Environ. Chang.* **2016**, *40*, 156–170. [CrossRef]

33. Han, S. Research on Action Mechanism and Quantitative Method of Adaptive Utilization of Water Resources. Master's Thesis, Zhengzhou University, Zhengzhou, China, 2020.
34. Zuo, Q.; Zhao, H.; Mao, C.; Ma, J.; Cui, G. Quantitative Analysis of Human-Water Relationships and Harmony-Based Regulation in the Tarim River Basin. *J. Hydrol. Eng.* **2015**, *20*, 1–11. [CrossRef]
35. Fang, G.; Yang, J.; Chen, Y.; Li, Z.; Ji, H.; De Maeyer, P. How Hydrologic Processes Differ Spatially in a Large Basin: Multisite and Multiobjective Modeling in the Tarim River Basin. *J. Geophys. Res. Atmos.* **2018**, *123*, 7098–7113. [CrossRef]
36. Xu, Z.; Chen, Y.; Li, J. Impact of Climate Change on Water Resources in the Tarim River Basin. *Water Resour. Manag.* **2004**, *18*, 439–458. [CrossRef]
37. Yaning, C.; Changchun, X.; Xingming, H.; Weihong, L.; Yapeng, C.; Chenggang, Z.; Zhaoxia, Y. Fifty-year climate change and its effect on annual runoff in the Tarim River Basin, China. *Quat. Int.* **2009**, *208*, 53–61. [CrossRef]
38. Wang, F.; Chen, Y.; Li, Z.; Fang, G.; Li, Y.; Xia, Z. Assessment of the Irrigation Water Requirement and Water Supply Risk in the Tarim River Basin, Northwest China. *Sustainability* **2019**, *11*, 4941. [CrossRef]
39. Gao, P.; Wang, X.; Wang, H.; Cheng, C. Viewpoint: A correction to the entropy weight coefficient method by Shen et al. for accessing urban sustainability [Cities 42 (2015) 186–194]. *Cities* **2020**, *103*, 102742. [CrossRef]
40. Cheng, W.; Xi, H.; Sindikubwabo, C.; Si, J.; Zhao, C.; Yu, T.; Li, A.; Wu, T. Ecosystem health assessment of desert nature reserve with entropy weight and fuzzy mathematics methods: A case study of Badain Jaran Desert. *Ecol. Indic.* **2020**, *119*, 106843. [CrossRef]
41. Luo, Z.; Zuo, Q.; Shao, Q. A new framework for assessing river ecosystem health with consideration of human service demand. *Sci. Total Environ.* **2018**, *640–641*, 442–453. [CrossRef]
42. Zuo, Q.; Zhang, Y.; Lin, P. Index system and quantification method for human-water harmony. *J. Hydraul. Eng.* **2008**, *39*, 440–447.
43. Sun, G.; Guan, X.; Yi, X.; Zhou, Z. Grey relational analysis between hesitant fuzzy sets with applications to pattern recognition. *Expert Syst. Appl.* **2018**, *92*, 521–532. [CrossRef]
44. Wang, D.; Li, Y.; Yang, X.; Zhang, Z.; Gao, S.; Zhou, Q.; Zhuo, Y.; Wen, X.; Guo, Z. Evaluating urban ecological civilization and its obstacle factors based on integrated model of PSR-EVW-TOPSIS: A case study of 13 cities in Jiangsu Province, China. *Ecol. Indic.* **2021**, *133*, 108431. [CrossRef]
45. Yang, Y.; Hu, N. The spatial and temporal evolution of coordinated ecological and socioeconomic development in the provinces along the Silk Road Economic Belt in China. *Sustain. Cities Soc.* **2019**, *47*, 101466. [CrossRef]
46. Xue, L.; Wang, J.; Zhang, L.; Wei, G.; Zhu, B. Spatiotemporal analysis of ecological vulnerability and management in the Tarim River Basin, China. *Sci. Total Environ.* **2019**, *649*, 876–888. [CrossRef]
47. Ling, H.; Xu, H.; Fu, J. Changes in intra-annual runoff and its response to climate change and human activities in the headstream areas of the Tarim River Basin, China. *Quat. Int.* **2014**, *336*, 158–170. [CrossRef]
48. Wu, G.; Li, L.; Ahmad, S.; Chen, X.; Pan, X. A Dynamic Model for Vulnerability Assessment of Regional Water Resources in Arid Areas: A Case Study of Bayingolin, China. *Water Resour. Manag.* **2013**, *27*, 3085–3101. [CrossRef]
49. Huang, S.; Wortmann, M.; Duethmann, D.; Menz, C.; Shi, F.; Zhao, C.; Su, B.; Krysanova, V. Adaptation strategies of agriculture and water management to climate change in the Upper Tarim River basin, NW China. *Agric. Water Manag.* **2018**, *203*, 207–224. [CrossRef]
50. Ehrlich, P.R.; Raven, P.H. Butterflies and plants: A study in coevolution. *Evolution* **1964**, *18*, 586–608. [CrossRef]

water

MDPI

Article

Evaluation of Regional Water-Saving Level Based on Support Vector Machine Optimized by Genetic Algorithm

Wenge Zhang [1], Shengling Hou [2,*], Huijuan Yin [1], Lingqi Li [1] and Kai Wu [1]

1 Water Resources Institute of Yellow River Hydraulic Research Academy, Zhengzhou 450003, China
2 School of Water Conservancy Engineering, Zhengzhou University, Zhengzhou 450001, China
* Correspondence: hh13369770689@163.com

Abstract: The evaluation of regional water-saving level can provide scientific theoretical support for steadily promoting the implementation of a national water-saving priority strategy. Referring to the water consumption statistics of 31 provinces (except Hong Kong, Macao and Taiwan) in China in 2018, 14 easily accessible and comprehensive indexes were selected to establish an index system of regional water-saving level and a water-saving level evaluation model based on support vector machine optimized by genetic algorithm (GA-SVM) was constructed to analyze the national regional water-saving level from different perspectives. The results showed that the water-saving level in China presented a spatial distribution characteristic with Beijing City, Henan Province and Zhejiang Province as the center and gradually decreased outward. From the perspective of regionalization, the water-saving level in North China, Central China and Southeast China was higher, while the water-saving level in Northwest China, Southwest China and Northeast China need to be improved. Therefore, the national water-saving level is generally at a medium level and effective water-saving work and water-saving schemes should be carried out according to different regions and industries.

Keywords: genetic algorithm; support vector machine; index system; water-saving level

Citation: Zhang, W.; Hou, S.; Yin, H.; Li, L.; Wu, K. Evaluation of Regional Water-Saving Level Based on Support Vector Machine Optimized by Genetic Algorithm. *Water* **2022**, *14*, 2615. https://doi.org/10.3390/w14172615

Academic Editors: Qiting Zuo, Fuqiang Wang, Jiaqi Zhai, Xiuyu Zhang, Dunxian She, Lei Zou, Rong Gan and Zengliang Luo

Received: 13 June 2022
Accepted: 22 August 2022
Published: 25 August 2022

1. Introduction

Water, as a resource, is an indispensable and irreplaceable natural resource for individuals, society and even the whole earth. However, with the development and progress of human society and science and technology economy, available water resources are decreasing. Since the 1980s, China's urbanization construction and social economic development process have been restricted by the shortage of water resources. Water saving and emission reduction are the only ways to reduce the consumption of limited water resources and maximize the benefits of available water resources [1]. In 2014, General Secretary Xi Jinping proposed "water saving priority" as the primary theoretical content in the 16-character new period water control work idea [2]. The evaluation of water-saving level is a basic link in the national priority water-saving strategy. Carrying out research on regional water-saving level evaluation can provide a theoretical basis and technical support for the scientific guidance of water-saving priorities and steady promotion of water-saving measures [3].

In foreign countries, initial studies on water-saving evaluation did not carry out comprehensive evaluations, but evaluated water-saving level from different fields, including efficient farmland irrigation, total water consumption control for industrial production, urban domestic water and water-saving management. American scholars Ben and Sammis [4] evaluated agricultural water-saving potential and regional evapotranspiration under a drip irrigation system through comparative experiments in 1975. In 1988, Thompson et al. [5] studied and analyzed the level of industrial water saving by using the industrial water consumption index and concluded that increasing the reuse rate of industrial wastewater could improve the reuse rate of industrial water. In 2016, French scholar Rinaudo [6] proposed the tradable water-saving certificate system to improve the efficiency of water resource allocation among drinking-water enterprises in the basin. Although this model still has some improvements, it has a certain promotion effect on the development of urban water saving.

The evaluation of water-saving level in China began at the end of the 20th century. Some scholars carried out research on urban water saving in terms of water consumption index, water-saving efficiency and water shortage [7]. Meanwhile, some achievements have also been made in efficient farmland irrigation, total water control for industrial production and water-saving management [8–10]. There are not only regional water-saving level evaluations for provincial, municipal and county administrative regions [11], but industry (product) water-saving level evaluation for enterprises, communities, schools, products and other specific water users [12]. The evaluation method evolved from the original single index evaluation method, including mean multi-variable index evaluation, single variable index evaluation and subjective index evaluation to multi-index comprehensive evaluation method, including "pressure state response" (PSR) model, analytic hierarchy process, order relation analysis, fuzzy comprehensive evaluation and technique for order preference by similarity to ideal solution (TOPSIS) method [13,14]. Traditional evaluation methods have certain subjectivity and cumbersome calculations and the evaluation factors are relatively limited. When the data are too complicated, it is difficult to reflect the comprehensive characteristics of high-dimensional data, while machine learning theory, such as support vector machine, has good learning ability and generalization in dealing with nonlinear, nonlocal and nonconvex high-dimensional problems by constructing scientific and objective general function [15]. At present, the evaluation research on water-saving level mainly focuses on the analysis of water-saving level of a single industry or a single region. On the contrary, there are few studies on the water-saving level of a national region as well as water-saving level research models.

On the basis of the existing research, a representative, independent, comparable and operable evaluation index system was constructed in this paper. Taking the machine learning model and intelligent algorithm as a theoretical basis, a national regional water-saving level evaluation method based on a genetic algorithm optimized by support vector machine is proposed, and the water-saving level is analyzed from different regions, different indexes and factors affecting the water-saving level. The research results are expected to provide new theoretical ideas for the scientific research and development of regional water-saving level evaluation.

2. Methods

2.1. Establishment of Evaluation Index System

China's water-saving problems can be mainly divided into five categories, including comprehensive problems, industrial water-saving problems, agricultural water-saving problems, domestic water-saving problems and water-saving management problems [16]. In accordance with the requirements of total amount control, efficiency control and emission reduction control, the library of alternative indexes for water-saving evaluation was established based on scientific and systematic principles.

To screen the water-saving level evaluation indexes suitable for this paper from the library of alternative indexes, we first consulted the relevant national, industry, and collective standards and regulations, including "The 13th Five-Year Plan for Building a Water-saving Society" [17], "National Festival Water Action Plan" [18] and "Water-saving Society Evaluation Index System and Evaluation Method" [19] to ensure the authority and reliability of the indexes. Secondly, quantitative indexes are adopted as far as possible to make qualitative evaluation on water-saving problems that cannot be easily quantified. Finally, combined with the requirements of representativeness, independence, comparability and operability, the evaluation indexes of water-saving level were sorted out and simplified from the five perspectives of comprehensive water saving, industrial water saving, agricultural water saving, urban water saving and water-saving management. Fourteen evaluation indexes that have great influence on regional water-saving level are obtained and the constructed water-saving level evaluation index system is shown in Table 1.

Table 1. Comprehensive evaluation index system of regional water-saving level.

Index Type		Index	Unit	Evaluation Direction
Comprehensive evaluation index of regional water-saving level	Comprehensive water-saving indicators	Total water consumption control degree (A1)	%	The smaller the better
		Water consumption per ten thousand yuan GDP (A2)	m^3/ten thousand yuan	The smaller the better
		Decline rate of water consumption per ten thousand yuan of GDP (A3)	%	The bigger the better
		Ratio of unconventional water resource consumption (A4)	%	The bigger the better
	Industrial water-saving indicators	Water consumption per ten thousand yuan of industrial added value (B1)	m^3/ten thousand yuan	The smaller the better
		Utilization rate of water for irrigation (B2)	%	The bigger the better
	Agricultural water-saving indicators	Efficient utilization coefficient of irrigation water (C1)	Dimensionless	The bigger the better
		Proportion of high-efficiency water-saving irrigation area (C2)	%	The bigger the better
	Water-saving indicators for urban life	Leakage rate of urban public water supply pipe network (D1)	%	The smaller the better
		Penetration rate of water-saving appliances (D2)	%	The bigger the better
		Centralized rate of urban sewage treatment (D3)	%	The bigger the better
	Water-saving management indicators	Plan water rate (E1)	%	The bigger the better
		Installation rate of metering facilities (E2)	%	The bigger the better
		Standard quota timeliness (E3)	Dimensionless	The bigger the better

2.2. Evaluation Method of Water-Saving Level Based on Genetic-Algorithm-Optimized Support Vector Machine

In 1995, support vector machine (SVM) was proposed by Vapnik and other scholars to analyze nonlinear regression and classification problems [20] and was later widely used in industrial production, health care, urban development and other fields. The core of support vector machine is support vector and its basic principle is to construct an optimal hyperplane in the high-dimensional space as the boundary of sample classification to ensure the distance from each sample to this boundary can be maximized and then the global optimal solution can be obtained [21].

In this paper, the 68 data samples collected were randomly divided into 37 training set samples and 31 test set samples. The input of the SVM model is the sample set x_i of each water-saving index and the output of the model is the water-saving classification level y_i. The two-dimensional sample set x_i is converted into the high-dimensional space $\varphi(x)$ for processing and the classification function is constructed as follows:

$$f(x) = sign(\omega^* \cdot \varphi(x) + b^*) \tag{1}$$

where ω^* is the weight vector and b^* is the classification threshold.

Considering the maximization of the interval between the sample set and the classification hyperplane and the fitting error, the problem of maximization for the interval is transformed into the problem of minimizing objective functions ω^* and b^* by introducing penalty factor C, relaxation factor ξ_i and relaxation factor ξ_i^*.

$$\begin{cases} min\left\{\frac{1}{2}\|\omega^*\|^2 + C\sum_{i=1}^{l}(\xi_i + \xi_i^*)\right\} \\ s.t.\begin{cases} f(x) - \omega^*\cdot\varphi(x) - b^* \le \xi_i \\ \omega^*\cdot\varphi(x) + b^* - f(x) \ge \xi_i^* \\ \xi_i, \xi_i^* \ge 0 \end{cases} \end{cases} \tag{2}$$

By adopting Lagrange multiplier method, the dual form of the objective function is obtained:

$$\begin{cases} min\left\{\sum_{i=1}^{l} y_i(\alpha_i - \alpha_i^*) - \frac{1}{2}\sum_{i,j=1}^{l}(\alpha_i - \alpha_i^*)\left(\alpha_j - \alpha_j^*\right)K(x_i, x_j)\right\} \\ s.t. \sum_{i=1}^{l}(\alpha_i - \alpha_i^*) = 0, \quad 0 \le \alpha_i, \alpha_i^* \le C,\ i = 1,2,3\ldots l \end{cases} \tag{3}$$

where α_i and α_i^* are Lagrange multiplier factors ad $K(x_i, x_j)$ is the RBF kernel functions.

$$K(x_i, x_j) = exp\left(-\frac{\|x_i - x_j\|^2}{2\sigma^2}\right) = exp\left(-\gamma x_i - x_j{}^2\right) \tag{4}$$

where γ is the kernel function parameter $\gamma = \frac{1}{2\sigma^2}$ and $\|x_i - x_j\|$ is the Euclidean norm.

The value of the model parameters will affect the results and accuracy of the model and the penalty factor γ and the kernel parameter γ are important parameters in SVM classifier. Since the learning function of SVM does not have a clear functional relationship with model parameters, traditional support vector machines adopt cross-validation or reliance on experience to determine relevant parameters, but the effect is not ideal and there is a certain degree of subjectivity and blindness. To make SVM achieve the highest mathematical analysis accuracy, genetic algorithm is applied to optimize the important parameters to improve the accuracy of SVM classifier. Genetic algorithm was first proposed by Professor Holland in 1975. It is an intelligent algorithm based on Darwinian evolution theory to obtain the optimal solution of the problem by simulating the biological evolution process [22]. The main process is to transform the mathematical solution process into an artificial evolution model, including individual selection, crossover, mutation and other operations [23]. Through continuous iterative calculation, the goal is comprehensively searched and then the optimal solution for the model is obtained.

The steps of water-saving level classification based on genetic-algorithm-optimized support vector machine are as follows and a flow chart is shown in Figure 1:

(1) Extraction of sample data.

The data for 14 indexes from 68 groups of the samples were extracted to obtain the distribution of the same index from different samples and the samples of training set and test set were selected.

(2) Data preprocessing.

Different indexes have different dimensions. To achieve the standardization and practicability of the sample data, normalized calculations were made on the original data.

$$x^* = \frac{x - x_{min}}{x_{max} - x_{min}} \tag{5}$$

where x is the measured value and x^* is the processed value.

(3) Genetic algorithm optimization.

In the training process of water-saving level classification based on support vector machine, genetic algorithm was introduced to obtain the optimal parameters. The new population group is obtained by selecting, crossover, mutation and other operations for each individual population. When the fitness of the new population group or the classification accuracy meets the stop condition, the iteration process is stopped and retreated and the optimal penalty factor and the kernel parameters with the highest fitness are obtained.

(4) Classification of the test set based on optimal parameters.

The optimal parameter $[C, \gamma]$ was substituted into RBF kernel function and GA-SVM water-saving level classification model. The output classification results included the actual classification of test set and the optimal classification based on GA-SVM and the classification accuracy was calculated.

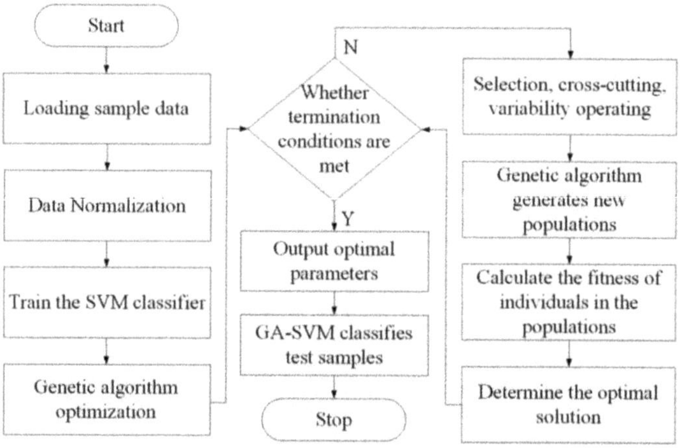

Figure 1. Genetic-algorithm-optimized support vector machine water-saving level classification flowchart.

3. Case Analysis

3.1. Data Sources

The data in the paper are extracted from the "China Water Resources Bulletin", "China Statistical Yearbook", "China Urban and Rural Construction Statistical Yearbook" and "China Water Statistics Yearbook" and the most stringent water resource management system assessment and publication statistics of each province (region, city) in 2018, from which the relevant data of 14 indexes in 31 provinces (autonomous regions and municipalities) in 2018 are obtained to carry out the example application of regional water-saving level evaluation.

3.2. Evaluation Index Analysis

To intuitively analyze the commonality and difference in indexes in different provinces, a scatter distribution diagram of each indicator data is drawn in Figure 2.

As can be seen from Figure 2, the distribution of the five indexes, including total water consumption control degree (A1), the decline rate of water consumption per 10,000 yuan of GDP (A3), water consumption per 10,000 yuan of industrial added value (B1), water consumption per 10,000 yuan of GDP (A2) and leakage rate of urban public water supply pipe network (D1), is relatively concentrated, with average values of 87.3%, 22.3%, 43.4 ($m^3/10{,}000$ CNY), 94.9 ($m^3/10{,}000$ CNY) and 14.9%, respectively. The five indexes of most provinces (autonomous regions, municipalities) are at a better level. The distribution of the proportion of high-efficiency water-saving irrigation area (C2), the reuse rate of industrial water (B2), the penetration rate of water-saving appliances (D2), the installation rate of metering facilities (E2), the planned water consumption rate (E1) and the centralized treatment rate of urban sewage (D3) are scattered. The differences in the indexes between the maximum value and the minimum value are large. For example, the maximum value of industrial water reuse rate is 95.8%, the minimum value is 18.3%, the maximum value of efficient water-saving irrigation area is 95.54%, the minimum value is 2.51%, the maximum value of planned water use rate is 97.26%, the minimum value is 1.2%, indicating that there are great differences in water-saving work in industry, agriculture and urban life among provinces (regions, cities). It is worth noting that the utilization ratio of unconventional water (A4) and the effective utilization coefficient of farmland irrigation water (C1) are relatively concentrated, with an average value of 2.7% and 0.54, respectively, but both of them are at a low level.

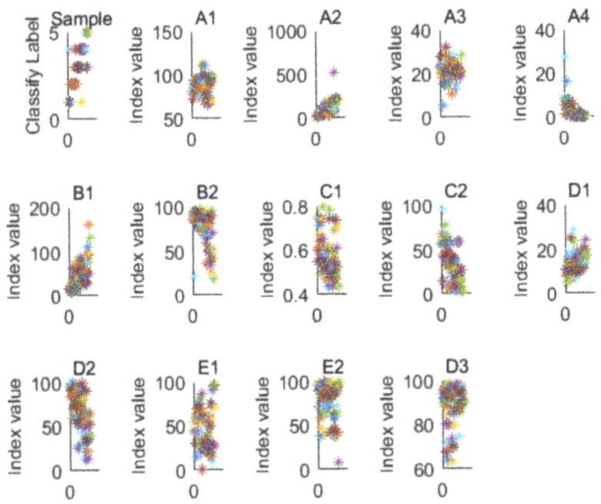

Figure 2. Scattered distribution chart of each indicator (The explanation of A1–D3 is shown in Table 1).

Referring to the standards for building water-saving societies in counties (Water Resources [2017] No. 184), "Technical Guidelines for Evaluation of Water-saving Communities" and "Urban Water-saving Evaluation Standards" [24], the threshold division of the indexes at each level is determined (Table 2) and then a quantitative distribution diagram of provinces (autonomous regions and municipalities) at different levels of each index is obtained (Figure 3).

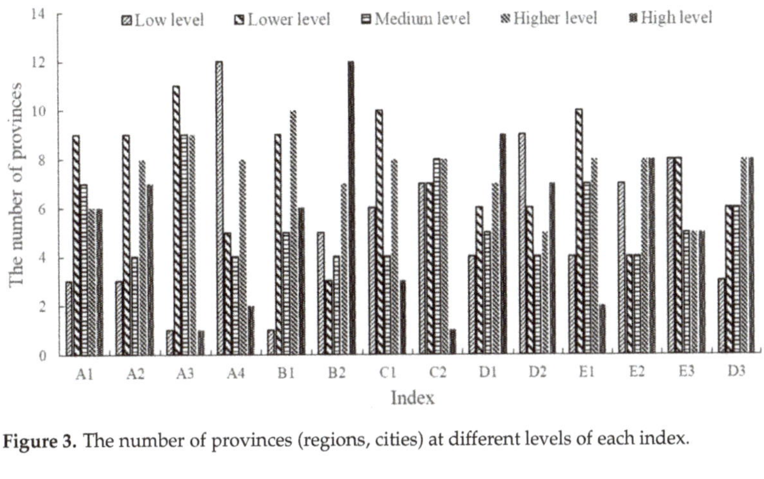

Figure 3. The number of provinces (regions, cities) at different levels of each index.

Table 2. Threshold division of each indicator at each level.

Index	Value Distribution of Each Level				
	High Level	**Higher Level**	**Medium Level**	**Lower Level**	**Low Level**
Total water consumption control degree (%)	60~80	80~85	85~90	90~100	100~120
Water consumption per ten thousand yuan GDP (m^3/ten thousand yuan)	10~40	40~70	70~90	90~200	200~520
Decline rate of water consumption per ten thousand yuan of GDP (%)	30~35	25~30	20~25	10~20	0~10
Ratio of unconventional water resource consumption (%)	10~30	2.0~10	1.0~2.0	0.6~1.0	0~0.6
Water consumption per ten thousand yuan of industrial added value (m^3/ten thousand yuan)	7~20	20~40	40~50	50~100	100~120
Utilization rate of water for irrigation (%)	90~100	80~90	60~80	40~60	15~40
Efficient utilization coefficient of irrigation water	0.7~0.8	0.6~0.7	0.55~0.6	0.5~0.55	0.4~0.5
Proportion of high-efficiency water-saving irrigation area (%)	70~100	40~70	20~40	10~20	0~10
Leakage rate of urban public water supply pipe network (%)	0~13	13~14	14~15	15~20	20~30
Penetration rate of water-saving appliances (%)	90~100	70~90	60~70	50~60	0~50
Plan water rate (%)	80~100	60~80	40~60	20~40	0~20
Installation rate of metering facilities (%)	90~100	80~90	70~80	50~70	0~50
Standard quota timeliness	2020	2019	2018	2016~2017	2014~2015
Centralized rate of urban sewage treatment (%)	97~100	95~97	93~95	90~93	80~90

The results show that most of the provinces (autonomous regions and cities) are at the medium or above level in terms of the eight indexes, including total water consumption control degree (A1), the decline rate of water consumption per CNY 10,000 of GDP (A3), water consumption per CNY 10,000 of industrial added value (B1), water consumption per CNY 10,000 of GDP (A2), centralized rate of urban sewage treatment (D3), leakage rate of urban public water supply pipe network (D1) and the installation rate of metering facilities (E2). More than half of the provinces (autonomous regions and municipalities directly under the central government) are at or below the primary level in terms of the proportion of high-efficiency water-saving irrigation area (C2), the penetration rate of water-saving appliances (D2) and the timeliness of standard quotas (E3). However, the effective utilization coefficient of farmland irrigation water (C1) and the utilization ratio of unconventional water sources (A4) are at a low level in most provinces (autonomous regions and municipalities).

3.3. Evaluation Model Parameter Optimization Results

By searching, collecting and sorting out the relevant database information, a total of 25 papers was queried and 22 groups of index data were sorted out. The relevant research results covered six typical water-saving cities (region, city), which can be used as an effective supplement to the water-saving-level evaluation data. Taking the collected 22 groups of index data and 15 groups of provincial (region, city) index data, a total of 37 sets of samples as the training set samples province (region, city) index data and 31 groups of provincial (autonomous regions, municipalities) index data as the test set samples, the relevant research in this paper was carried out.

The input of the model is water-saving index x_i and the output of the model is water-saving classification level y_i, so the training sample set of 37 groups is $D = \{(x_i, y_i)|i = 1, 2, \ldots, l, l = 37\}$. In the water-saving-level evaluation model of GA-SVM, the maximum genetic times of the population were set as 100, the population size was set as 20 and the variation range of parameters C and γ was between 0 and 100. In the iterative process of genetic algorithm, the fitness of individual population is the classification accuracy [25]. When the classification accuracy is the highest, 1 and 2 are the optimal parameters. As can be seen from Figure 4 for the genetic algorithm iteration, the optimal accuracy rate has been improved four-times and finally stabilized at 97.2973%. In this case, the optimal parameter value is obtained, C = 3.7053, γ = 0.61817.

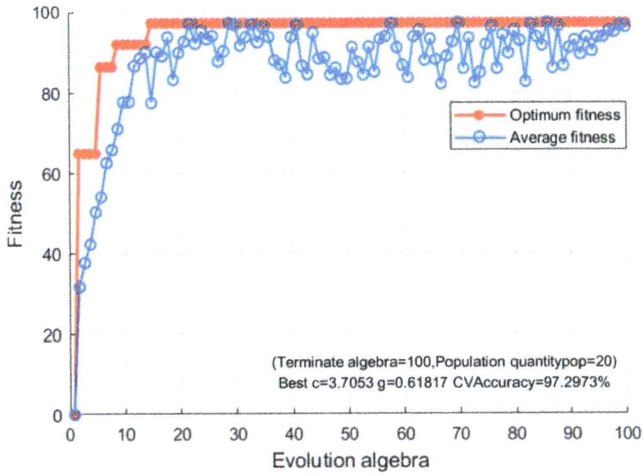

Figure 4. Genetic algorithm iteration graph.

3.4. Results Discussion

3.4.1. Analysis of Water-Saving Level in Various Provinces

The normalized data for 31 test sets were substituted into the algorithm program of GA-SVM to complete the classification and evaluation of water-saving level in provinces (regions, cities), as shown in Figure 5.

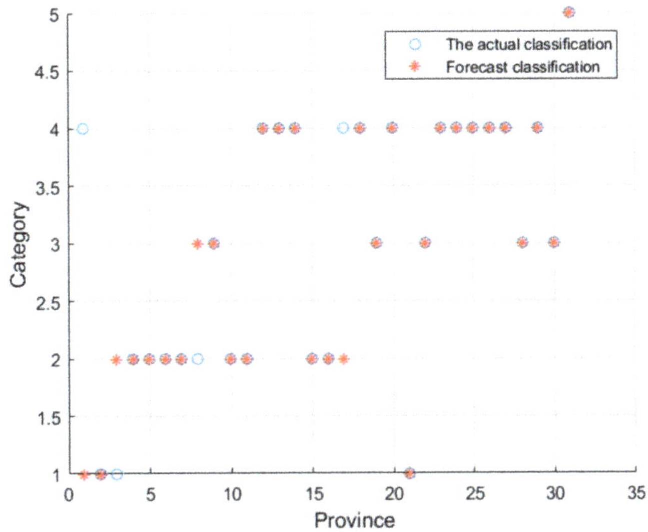

Figure 5. GA-SVM algorithm program result graph.

According to Figure 5, the classification results for water-saving levels in provinces (autonomous regions and municipalities) can be obtained. In China, there are 3 higher-level provinces (autonomous regions and municipalities), 10 high-level provinces, 6 medium-level provinces, 11 lower-level provinces and 1 low-level province. The spatial distribution characteristics of water-saving levels in all provinces (regions and cities) are shown in Figure 6.

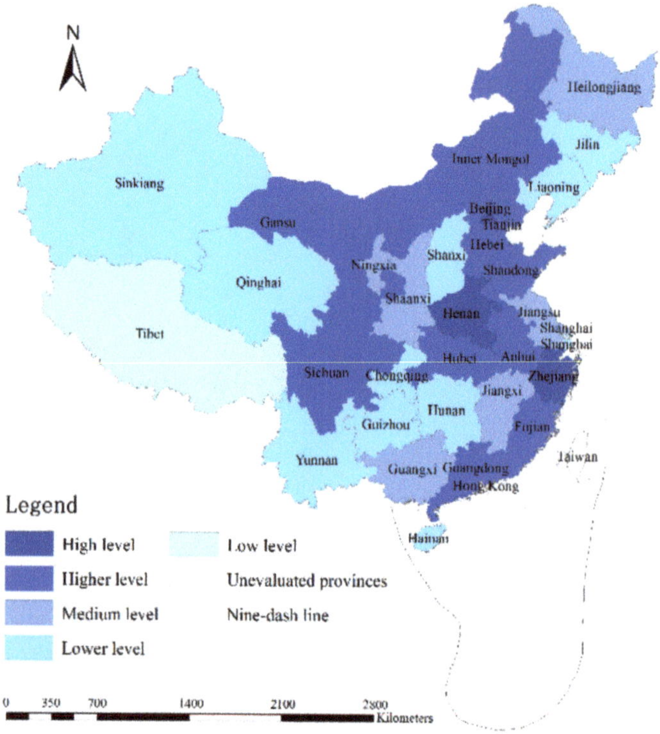

Figure 6. Spatial distribution characteristics of regional water-saving levels in China.

From Figure 6, more than half of China's provinces (autonomous regions and cities) are above the medium level, accounting for 61.3% of the total, and few provinces (autonomous regions and cities) reach the low level or high level. Generally speaking, the water-saving level in China presents the spatial distribution characteristics with Beijing, Henan and Zhejiang Province as the center and gradually decreases outward. Beijing, as a super-large city with a high economic level, and Henan Province, as a province with a large population in China [26], have made great progress in water saving by further strengthening water-saving management, implementing the national priority policy of water saving, promoting the development of water saving and vigorously achieving the sustainable utilization of water resources [27]. Zhejiang Province, located in the southeast coastal area, with abundant rainfall and sufficient water resources per capita, attaches great importance to water saving and pollution control and develops non-traditional water resources to improve water-saving level. There are many rivers in Tibet, but most of the rivers are intermittent [28]. The utilization rate of water resources in Tibet is extremely low and the waste phenomenon is extremely serious; the area is vast but the population is sparse and water-saving technologies and funds are scarce, making it the only low-level water-saving area.

3.4.2. Analysis of Water-Saving Level by District

According to the "13th Five-Year Plan" for the construction of water-saving society [29], the national water-saving work is divided into six regions, including Northeast, Southwest, North China, Central China, Northwest and Southeast. Based on this, the water-saving level in the six regions is determined (Figure 7).

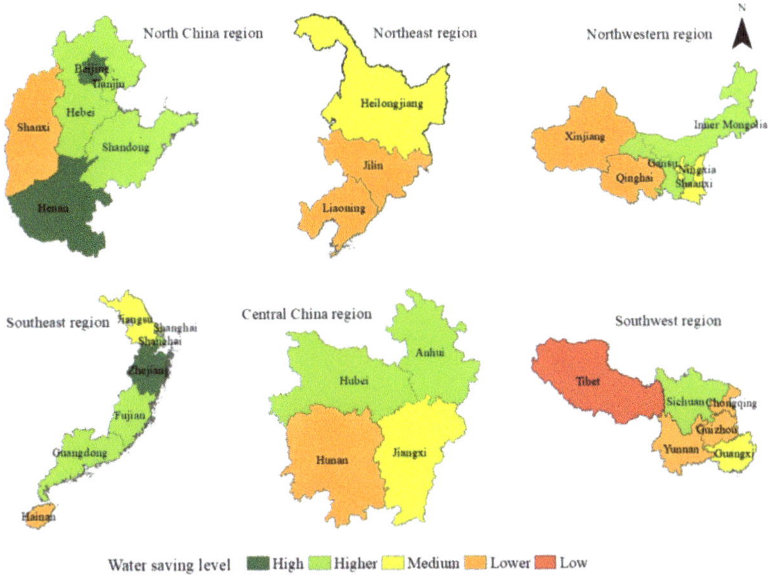

Figure 7. Spatial distribution of regional water-saving levels by district.

As a whole, North China, Central China and Southeast China have high economic levels, dense population and a large amount of water resources. Therefore, all provinces and autonomous regions should insist on giving priority to conservation [30], aiming at promoting water saving through structural adjustment to improve water-saving efficiency and keep the water-saving level at medium or above. In Southwest China, Guangxi, Chongqing and Yunnan provinces have complex terrain structure and abundant regional water resources [31]. However, the economic and social development is relatively backward and the water-saving level is low, so there is great potential for water saving. In view of the characteristics of regional resources and the disadvantages of water-saving work, it is necessary to strengthen the promotion of water-saving work to promote the sustainable development of water resources. Due to the impact of geographical environment, Northwest China is faced with the problems of water shortage and ecosystem imbalance [32] and the backward economy also restricts water-saving management. Compared with the Northwest regions, Jiangsu, Jiangxi and Hunan provinces in the Southern region have abundant water resources, but the regional and seasonal drought and water pollution problems are serious. Therefore, there is a lot of room for improvement in the construction of water-saving society in the south. Different regions have different priorities for promoting water-saving work, so each region should summarize the restrictive factors in the process of promoting water-saving work and formulate water-saving work implementation plans in combination with natural geographical conditions and socio-economic development characteristics.

3.4.3. Analysis of Factors Affecting Water-Saving Level

According to the analysis of evaluation indexes, it can be seen that the utilization ratio of unconventional water sources and the effective utilization coefficient of farmland irrigation water have a great impact on regional water-saving level, as illustrated in Figure 8.

From the perspective of the proportion of unconventional water sources, only North China has a high water-saving level, while the water-saving level in the rest of the regions is at a lower level. In case of a shortage of conventional water resources, most provinces (autonomous regions and municipalities) must pay attention to the theoretical research and engineering construction of unconventional water resources to increase the total utilization of unconventional water resources. In terms of effective utilization coefficient of farmland irrigation water, the vast majority of regions are at a low level, which indicates that most regions in China have high water-saving potential for agricultural water use. Due to the low cost of farmland irrigation at present, farmers' awareness of water conservation is weak and agricultural water-saving irrigation projects have not covered the whole country [33], so water-saving benefits are always at a low level. Therefore, it is necessary to develop a perfect legal system and implement high-efficiency water-saving irrigation technology to promote the development of agriculture.

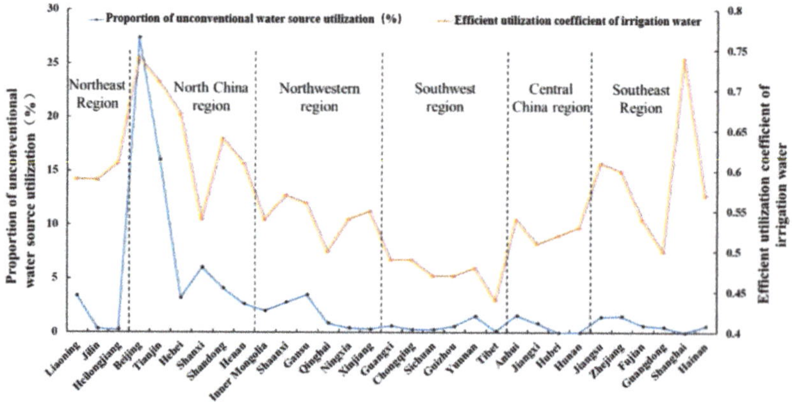

Figure 8. Low-level indicators district statistics chart.

4. Conclusions

Evaluation of the water-saving level is basic work in water resource planning, development, utilization, protection and management. Scientific and objective evaluation results can provide a decision-making basis for the sustainable utilization of resources and provide technical support to steadily promote the implementation of water conservation. This paper takes 31 provinces (region, city) in China (except Hong Kong, Macao and Taiwan) as the research object, establishes an evaluation index system of water saving, constructs an evaluation model of water-saving level based on the support vector machine optimized by genetic algorithm and discusses the water-saving level in provinces (region, city) and the factors affecting the water-saving level. The following preliminary conclusions are obtained: in 2018, the water-saving level in China presented spatial distribution characteristics with Beijing, Henan and Zhejiang Province as the center and gradually decreases outward; in terms of regions, the water-saving level in North China, Central China and Southeast China is relatively high, while the water-saving level in Northwest China, Southwest China and Northeast China needs to be improved. However, the effective utilization coefficient of farmland irrigation water and the proportion of unconventional water use have a relatively high impact on regional water-saving level, so each province (autonomous regions and municipalities) still needs to strengthen the water-saving level of unconventional water use and high-efficiency agricultural irrigation. Considering natural resource conditions and economic development, different regions should formulate water-saving construction plans suitable for their own regions and strengthen the construction of water-saving supervision systems and mechanisms to achieve the improvement in the level of water-saving in the whole region.

In terms of methods, the evaluation method of genetic algorithm optimization support vector machine has good performance in water-saving level evaluation and it is applied in China, a country with large regional differences in water-saving level, with high precision, which can be used as a reference for water-saving level evaluation in other countries. In this paper, only 14 evaluation indexes for 31 provinces (regions and cities) in 2018 are selected for analysis. They are greatly influenced by sample time series and spatial sequence. In the follow-up study, index data of different levels of multiple time series can be selected and the regions refined to make the evaluation results more specific and effective.

Author Contributions: All authors contributed to the study conception and design. Material preparation was performed by L.L. The data collection was performed by and H.Y., S.H. was responsible for data analysis. K.W. reviewed and modified the manuscript. The first draft of the manuscript was written by W.Z. and all authors commented on previous versions of the manuscript. All authors have read and agreed to the published version of the manuscript.

Funding: This study was funded by the National Natural Science Foundation of China (No. 51979119) and Basic R&D Special Fund of Central Government for Non-profit Research Institutes (HKY-JBYW-2020-17). The funders are Zhang Wenge.

Institutional Review Board Statement: Not applicable.

Informed Consent Statement: Not applicable.

Data Availability Statement: The original data used during the study were provided by a third party. The data that support the findings of this study are available in "China Water Resources Bulletin", "China Statistical Yearbook", "China Urban and Rural Construction Statistical Yearbook" and "China Water Statistics Yearbook". These data were derived from the following resources available in the public domain: China Water Resources Bulletin: http://www.mwr.gov.cn/sj/tjgb/szygb/201907/t20190712_1349118.html (accessed on 1 May 2020), China Statistical Yearbook: http://www.stats.gov.cn/tjsj/ndsj/2018/indexch.htm(accessed on 1 May 2020), China Urban and Rural Construction Statistical Yearbook: http://www.mohurd.gov.cn/xytj/tjzljsxytjgb/jstjnj/ (accessed on 1 June 2020), China Water Statistics Yearbook: https://www.yearbookchina.com/navibooklist-n3020041501-1.html (accessed on 1 June 2020). The authors have made sure that all data and materials as well as software application or custom code comply with field standards.

Acknowledgments: The authors thank the National Natural Science Foundation of China (No. 51979119) and Basic R&D Special Fund of Central Government for Non-profit Research Institutes (HKY-JBYW-2020-17) for their partial financial support that made this project possible. The authors would also like to thank the support of Water Resources Institute of Yellow River Hydraulic Research Academy and School of Water Conservancy Engineering in Zhengzhou University.

Conflicts of Interest: The authors have no conflict of interest to declare that are relevant to the content of this article.

References

1. Hopwood, J.L. The contribution of roadside grassland restorations to native bee conservation. *Biol. Conserv.* **2008**, *141*, 2632–2640. [CrossRef]
2. Li, G. Thoughts on the implementation of the water saving priority water control policy-Do not forget the original heart, keep in mind the mission theme education research report. *Hebei Water Resour.* **2019**, *8*, 10–11. (In Chinese)
3. Fang, H.; Gan, S.; Yu, Y. Comprehensive analysis of drought duration characteristics of regional water supply system. *Adv. Water Sci.* **2007**, *1*, 95–101. (In Chinese)
4. Ben, A.J.; Sammis, T.W. Radiation and energy balance of a trickle-irrigated lemon grove. *Agron. J.* **1978**, *70*, 568.
5. Thompson, R.L.; Lee, S.; Geib, S.J.; Cooper, N.J. Intramolecular bridge/terminal oxo exchange within oxomolybdenum [MoV2O3] 4+ complexes containing linear oxo bridges. *Inorg. Chem.* **1993**, *32*, 6067–6075. [CrossRef]
6. Rinaudo, J.D.; Javier, C.; Marine, V.B. Tradable water saving certificates to improve urban water use efficiency: An ex-ante evaluation in a French case study. *Aust. J. Agric. Resour. Econ.* **2016**, *60*, 422–441. [CrossRef]
7. Fidar, A.M.; Memon, F.A.; Butler, D. Performance evaluation of conventional and water saving taps. *Sci. Total Environ.* **2016**, *541*, 815–824. [CrossRef]
8. Bryla, D.R.; Gartung, J.L.; Strik, B.C. Evaluation of Irrigation Methods for Highbush Blueberry—I. Growth and Water Requirements of Young Plants. *HortScience* **2011**, *46*, 95–101. [CrossRef]

9. Fontenot, B.E.; Hildenbrand, Z.L.; Carlton, D.D.; Walton, J.L.; Schug, K.A. Response to comment on An evaluation of water quality in private drinking water wells near natural gas extraction sites in the barnett shale formation. *Environ. Sci. Technol.* **2014**, *48*, 3597–3599. [CrossRef]
10. Mendelsohn, R.; Bennett, L.L. Global Warming and Water Management: Water Allocation and Project Evaluation. *Clim. Chang.* **1997**, *37*, 271–290. [CrossRef]
11. Dong, H.; Geng, Y.; Sarkis, J.; Fujita, T.; Okadera, T.; Xue, B. Regional water footprint evaluation in China: A case of Liaoning. *Sci. Total Environ.* **2013**, *442*, 215–224. [CrossRef]
12. Sušnik, J.; Vamvakeridou-Lyroudia, L.S.; Savić, D.A.; Kapelan, Z. Integrated System Dynamics Modelling for water scarcity assessment: Case study of the Kairouan region. *Sci. Total Environ.* **2012**, *440*, 290–306. [CrossRef]
13. Sener, E.; Davraz, A. Assessment of groundwater vulnerability based on a modified DRASTIC model, GIS and an analytic hierarchy process (AHP) method: The case of Egirdir Lake basin (Isparta, Turkey). *Hydrogeol. J.* **2013**, *21*, 701–714. [CrossRef]
14. Han, L.; Song, Y.; Duan, L.; Yuan, P. Risk assessment methodology for Shenyang Chemical Industrial Park based on fuzzy comprehensive evaluation. *Environ. Earth Sci.* **2015**, *73*, 5185–5192. [CrossRef]
15. Uwe, T.; Bülent, U.; Melssen, W.J.; Buydens, L.M. Multivariate calibration with least-squares support vector machines. *Anal. Chem.* **2004**, *76*, 3099–3105.
16. Zhang, Y.; Liu, W.; Cai, Y.; Khan, S.U.; Zhao, M. Decoupling analysis of water use and economic development in arid region of Chin—Based on quantity and quality of water use. *Sci. Total Environ.* **2020**, *761*, 143275. [CrossRef]
17. Mei, J.; Li, J.; Ding, Y. Play the leading role of planning to implement the water saving priority policy—Reflections on the preparation of the 13th Five-Year water-saving society construction plan. *China Water Resour.* **2015**, *7*, 11–13. (In Chinese)
18. Kang, S. Implementing the national water-saving action plan to promote the development of suitable agricultural water and green and efficient water saving. *China Water Resour.* **2019**, *13*, 1–6. (In Chinese)
19. Qin, F. Interpretation of Water-saving Society Evaluation Index System and Evaluation Method. *Public Stand.* **2012**, *06*, 10–12. (In Chinese)
20. Zeng, J.; Zou, B.; Qin, Y.; Chen, Q.; Xu, J.; Yin, L.; Jiang, H. Generalization ability of online pairwise support vector machine. *J. Math. Anal. Appl.* **2021**, *497*, 124914. [CrossRef]
21. Hu, L.; Lu, S.; Wang, X. A new and informative active learning approach for support vector machine. *Inf. Sci.* **2013**, *244*, 142–160. [CrossRef]
22. Sadeghi, M.; Roghanian, E. Reliability optimization for non-repairable series-parallel systems with a choice of redundancy strategies: Erlang time-to-failure distribution. *Proc. Inst. Mech. Eng.* **2017**, *231*, 587–604. [CrossRef]
23. Zou, D.; Li, S.; Kong, X.; Ouyang, H.; Li, Z. Solving the combined heat and power economic dispatch problems by an improved genetic algorithm and a new constraint handling strategy. *Appl. Energy* **2019**, *237*, 646–670. [CrossRef]
24. Yu, Q.; Sun, S.; Liu, J. Practice and exploration of the construction of water-saving society in counties in my country. *China Water Resour.* **2020**, *7*, 14–19. (In Chinese)
25. Pendharkar, P. Misclassification Cost Minimizing Fitness Functions for Genetic Algorithm-Based Artificial Neural Network Classifiers. *J. Oper. Res. Soc.* **2009**, *60*, 1123–1134. [CrossRef]
26. Li, Y.; Chen, B.; Chen, G.; Meng, J.; Hayat, T. An embodied energy perspective of urban economy: A three-scale analysis for Beijing 2002–2012 with headquarter effect. *Sci. Total Environ.* **2020**, *732*, 139097. [CrossRef]
27. Ndagijimana, M.; Kessler, A.; Asseldonk, M.V. Understanding farmers' investments in sustainable land management in Burundi: A case-study in the provinces of Gitega and Muyinga. *Land Degrad. Dev.* **2019**, *30*, 417–425. [CrossRef]
28. Feng, S.; Lu, H.; Yao, T.; Liu, Y.; Tian, P.; Lu, J. Microplastic footprints in the Qinghai-Tibet Plateau and their implications to the Yangtze River Basin. *J. Hazard. Mater.* **2021**, *407*, 124776. [CrossRef]
29. Li, W. Thirteenth Five-Year Plan for the Construction of a Water-saving Society is issued and requires the improvement of the water-saving standard system. *Eng. Constr. Stand.* **2017**, *2*, 7. (In Chinese)
30. Du, M.; Liao, L.; Wang, B.; Chen, Z. Evaluating the effectiveness of the water-saving society construction in China: A quasi-natural experiment. *J. Environ. Manag.* **2021**, *277*, 111394. [CrossRef]
31. Lu, H.; Wu, Y.; Li, Y.; Liu, Y. Effects of meteorological droughts on agricultural water resources in southern China. *J. Hydrol.* **2017**, *548*, 419–435. [CrossRef]
32. Alvarez, V.M.; Leyva, J.C.; Valero, J.F.M.; Górriz, B.M. Economic assessment of shade-cloth covers for agricultural irrigation reservoirs in a semi-arid climate. *Agric. Water Manag.* **2009**, *96*, 1351–1359. [CrossRef]
33. Belder, P.; Bouman, B.A.M.; Cabangon, R.; Guoan, L.; Quilang, E.J.P.; Yuanhua, L.; Spiertz, J.H.J.; Tuong, T.P. Effect of water-saving irrigation on rice yield and water use in typical lowland conditions in Asia. *Agric. Water Manag.* **2003**, *65*, 193–210. [CrossRef]

water

MDPI

Article

Integrated Evaluation of Rivers Based upon the River Happiness Index (RHI): Happy Rivers in China

Qianqian Ju, Changshun Liu * and Shan Jiang

State Key Laboratory of Simulation and Regulation of Water Cycle in River Basin, China Institute of Water Resources and Hydropower Research, Beijing 100038, China
* Correspondence: liucs@iwhr.com; Tel.: +86-13621283372

Abstract: How to better harmonize the relationship between humans and rivers is a global issue of widespread concern at home and abroad, and science-based and integrated evaluation of rivers themselves is crucial to river management. Based on Maslow's hierarchy of needs and according to the World Happiness Report and the 2030 Agenda for Sustainable Development, this paper argues that a happy river is a river that can maintain its own health, support high-quality economic and social development in the river basin and the region, reflect harmony between humans and water, and give people in the river basin a high sense of security and the ability to gain and satisfaction. This paper also analyzes happy rivers at five levels, including water security, water resources, water environment, water ecology, and water culture, and develops the River Happiness Index (RHI) and its indicator system, as well as assesses the overall river happiness in China's 10 first-grade water resource zones. The results show that China's RHI is at a medium level, with flood control capacity at a near-good level. On the grounds of the RHI evaluation results, the paper puts forward targeted measures for river basin governance, and provides a systematic solution to national river protection and governance.

Keywords: hierarchy of needs; connotation; RHI; indicator system; systematic solution

Citation: Ju, Q.; Liu, C.; Jiang, S. Integrated Evaluation of Rivers Based upon the River Happiness Index (RHI): Happy Rivers in China. *Water* **2022**, *14*, 2568. https://doi.org/10.3390/w14162568

Academic Editor: Roko Andricevic

Received: 30 June 2022
Accepted: 15 August 2022
Published: 20 August 2022

1. Introduction

Rivers nurtured material and spiritual civilizations of mankind, as well as played a decisive role in the origin, spread, and development of the world's civilizations [1]. As the ecological health of rivers is closely linked to society, economy and environment, river protection and restoration is a hot area of domestic and international research. What is the relationship between humans and rivers, how to evaluate it, and how to better harmonize the relationship between human and rivers is a global issue of widespread concern at home and abroad.

European and American countries were the first to realize the importance of river governance in the middle of the 20th century. The United States set up the River Restoration Centre in the 1990s [2], which proposed the evaluation criteria for successful ecological restoration of rivers [3]. European countries established the European Centre for River Restoration (ECRR) and carried out a lot of work for river improvement and restoration, such as in the Rhine, the Mississippi, and the Colorado [4–6]. The 2030 Agenda for Sustainable Development, adopted in 2015, sets the goal of "ensuring availability and sustainable management of water and sanitation for all", calling on countries to take action to protect and restore water-related ecosystems including rivers and lakes.

The Chinese government also attaches great importance to river basin governance and ecological protection and restoration. China fully implemented the river and lake chief system, and achieved initial results in six areas, including water resource conservation, water shoreline management, water pollution control, water environment management, water ecology restoration, and law enforcement [7–10]. On this basis, Beijing voiced the call to "make the Yellow River a happy river that benefits the people" in September 2019 [11]. The "Happy River" call, which builds itself upon a major national strategy and takes into account the bigger picture, has higher requirements and rich connotations. It is a new direction and requirement for river and lake governance in China, and also the outcome of China's long-standing experience, practices, theories, and techniques in water management. It is not only of special importance to the Yellow River, but also of significant reference value for other river basins [12]. The call is a Chinese approach to the sustainable development agenda and a new path for river governance under the guidance of sustainable development.

Most of the studies on rivers at home and abroad focus on the evaluation of river health [13,14], and the indicators characterizing the health of rivers mainly measure water quality, aquatic life, and water environment. At present, the main methods for river health evaluation are predictive models and multimetrics. Predictive models, such as the River Invertebrate Prediction and Classification System (RIVPACS) [15] and the Australian River Assessment System (AUSRIVAS) [16], are river status evaluation models that monitor the biodiversity and functionality of river invertebrates. As for multimetrics, the US Index of Biological Integrity (IBI) [17] evaluates river health based on 12 indicators of river fish species, such as richness, diversity, and nutrition types; the Index of Stream Condition (ISCO) [18], developed by the Australian Department of Natural Resources and Environment, evaluates river health and the long-term effectiveness of restoration using 19 specific indicators in five categories, namely hydrology, physical form, riparian zone, water quality, and aquatic life; the UK River Habitat Survey (RHS) [19] evaluates the characteristics and habitat quality of a river by investigating channel data, the type of vegetation structure, riparian zone characteristics, land use, and other indicators based on the physical structure of the river. However, few of the indicators for river health evaluation involve socioeconomic dimensions, and the evaluation criteria can hardly be determined. Based on the concept of river health, the idea of building harmony between humans and water was gradually formed in the research on Chinese river evaluation starting in 2004, which posits that the human-water harmony is a well-coordinated virtuous cycle, and these two systems influence and adapt to each other over time, and jointly promote the overall harmonious development of human-water system [20]. Furthermore, an indicator system following the Health-Development-Harmony (H-D-H) principle was developed to measure the Human–Water Harmony Degree (HWHD) [21] and evaluate the effectiveness of river governance by integrating water resources with economic and social indicators.

Therefore, the objective of this paper is to establish a more integrated river evaluation system and clarify the complex relationship between human happiness and rivers. To this end, based on the existing river evaluation methods, the paper analyzes the happy river from the five levels of security, resources, environment, ecology, and culture, clarifies the scientific connotation of the happy river, develops the River Happiness Index (RHI) and its evaluation system, evaluates the overall situation of river happiness in 10 first-class water resource zones in China, and puts forward suggestions for the popularization and application of the RHI.

2. What Is a Happy River?

According to Maslow's hierarchy of needs, humans have five levels of needs, namely physiological needs, security and safety needs, social belonging needs, esteem needs, and self-actualization needs, ranked from low to high [22]. A happy river is a river for the benefit of the people, which matches the hierarchy of human needs (Table 1).

Table 1. Hierarchical framework of a happy river.

Happy River Indicators	Happy River Objectives		Hierarchy of Needs	Maslow's Hierarchy of Needs
	Vision	Target		
river of security	A peaceful life with secure rivers	Low disaster losses, strong defense and high resilience	Basic level	Security and safety needs
river of prosperity	An affluent life with reliable water supply	Good water resource endowment, effectively guaranteed water use, development not restricted by water shortage, an affluent life		Physiological needs
river of livability	A livable environment with clear waters and green banks	Improving the water environment of natural rivers and lakes; increasing the environmental quality of urban and rural water	Higher level	Social belonging needs
river of ecology	A symbiotic harmony with fish swimming in shallows	Maintaining river health; achieving human–water harmony		
river of culture	A spiritual homeland with river civilizations	Respecting and protecting rivers; promoting the prosperity, flourishing and development of water culture, passing on historical water culture, and enriching the connotation of modern water culture	Highest level	Esteem needs Self-actualization needs

In the light of factors influencing and measuring human happiness in the first World Happiness Report [23] released by the United Nations in 2012, the definition of a happy river, which is "a river that can maintain its own health, support high-quality economic and social development in the river basin and the region, reflect harmony between human and water and give people in the river basin a high sense of security, gain and satisfaction", imposes stricter and more requirements on rivers [24]. A happy river is a river that meets the five needs of the people, including flood security, quality water resources, livable water environment, healthy water ecology, and advanced water culture, as well as achieves the unity of security, prosperity, livability, ecology, and culture [25,26].

In 2015, the United Nations Sustainable Development Summit officially adopted *Transforming our world: the 2030 Agenda for Sustainable Development* [27]. The agenda is composed of 17 Sustainable Development Goals (SGDs) and 169 targets, among which Goals 6, 9, 11, 13, and 15 are related to rivers (Table 2). As an integrated program for sustainable development in the area of river protection and restoration, the happy river indicator system is essentially consistent with the agenda.

Based on natural conditions and human needs, rivers that make humans happy should be people-centered and observe the concept of respecting, living in harmony with, and protecting nature; in other words, these rivers can maintain their own health, support high-quality economic and social development in corresponding river basins, reflect harmony between humans and water, and give people a high sense of security, gain, and satisfaction. The RHI evaluation system is more integrated than previous river evaluation indicators and theoretically innovative.

Table 2. A comparison between happy river indicators and the SDGs.

Happy River Indicators	SDGs	
	Goal	Target
river of security	Goal 9.	9.1 Develop quality, reliable, sustainable, and resilient infrastructure.
	Goal 11.	11.5 Significantly reduce the number of deaths and the number of people affected and substantially decrease the direct economic losses caused by disasters, including water-related disasters.
	Goal 13.	13.1 Strengthen resilience and the adaptive capacity to climate-related hazards and natural disasters.
river of prosperity	Goal 6.	6.1 Achieve universal and equitable access to safe and affordable drinking water for all.
		6.4 Substantially increase water use efficiency across all sectors and ensure sustainable withdrawals and supply of freshwater to address water scarcity and substantially reduce the number of people suffering from water scarcity.
		6.5 Substantially increase water use efficiency across all sectors and ensure sustainable withdrawals and supply of freshwater to address water scarcity and substantially reduce the number of people suffering from water scarcity.
river of livability	Goal 6.	6.3 Improve water quality by reducing pollution, eliminating dumping, and minimizing the release of hazardous chemicals and materials, halving the proportion of untreated wastewater and substantially increasing recycling and safe reuse globally.
river of ecology	Goal 6.	6.6 Protect and restore water-related ecosystems, including mountains, forests, wetlands, rivers, aquifers, and lakes.
	Goal 15.	15.1 Ensure the conservation, restoration and sustainable use of terrestrial and inland freshwater ecosystems and their services, in particular forests, wetlands, mountains, and drylands.
		15.5 Take urgent and significant action to reduce the degradation of natural habitats, halt the loss of biodiversity, and protect threatened species
A river of culture	Goal 6.	6.b Support and strengthen the participation of local communities in improving water and sanitation management.
	Goal 11.	11.4 Strengthen efforts to protect and safeguard the world's cultural and natural heritage.

3. Materials and Methods

Figure 1 shows the calculation process of the River Happiness Index evaluation system.

3.1. Indicator System

A happy river is a river of security, prosperity, livability, ecology, and culture. Hence, this paper develops evaluation indicators from these five perspectives to give a more complete and accurate evaluation of rivers, and proposes state indicators that characterize human experience of happiness and river health, as well as indicators of capacity for achieving or maintaining a good state (Figure 2).

1. Flood Control Capacity (FCC): FCC means the extent to which water disasters are prevented and controlled. In line with the vision of "a peaceful life with secure rivers", FCC is characterized by state indicators, such as Flood-induced Mortality Rate (FMR), Economic Loss Rate (ELR), and engineering and management capacity indicators, such as Rate of flood control Works with Accepted capacity (RWA) and post-Disaster Recovery Capability (DRC).
2. Water Resources Reliability (WRR): WRR means the capacity of water resources for supporting sustainable economic and social development. In line with the vision of "an affluent life with reliable water supply", indicators such as Available Water volume Per capita (AWP) and Water Supply Reliability (WSR) are selected to characterize the

condition of water resources and the availability of water resources; indicators such as Capacity for Supporting high-quality Development (CSD) and the Life Satisfaction Index (LSI) are selected to characterize the extent to which development and happiness are constrained by water resources.

3. Water Environment Livability (WEL): WEL refers to the degree of protection and improvement of the water environment of natural rivers and lakes in urban and rural areas. In line with the vision of "a livable environment with clear waters and green banks", WEL is characterized by state indicators, such as the Water Quality Index (WQI), Qualification rate of surface centralized Drinking water Source (QDS), and the Groundwater Protection Index (GPI), and experience indicators, such as the Water Entertainment Index (WEI).
4. Aquatic Ecosystem Health (AEH): AEH refers to the extent to which the health of river ecosystems is maintained and the quality and stability of river ecosystems is improved. Given the vision of "a symbiotic harmony with fish swimming in shallows", shrinking rivers and lakes, wetland degradation and biodiversity decline remain weak links. In this sense, AEH is characterized by indicators, such as the Rate of major rivers and lakes with accepted Ecological Flow (REF), Natural aquatic Habitat Retention rate (NHR), Index of Biological Integrity (IBI), and Soil and Water Conservation rate (SWC) in terms of flow, habitat, biology and land area, respectively.
5. Water Culture Prosperity (WCP): WCP means the extent to which water culture is promoted. In line with the vision of "a spiritual homeland with river civilizations", WCP is characterized by capacity indicators, such as the water Culture Protection and inheritance Index (CPI) and the Modern water culture Creation and Innovation Index (MCI), as well as human experience indicators, such as the Water Landscape impact Index (WLI) and Public Awareness and Engagement in water governance (PAE).

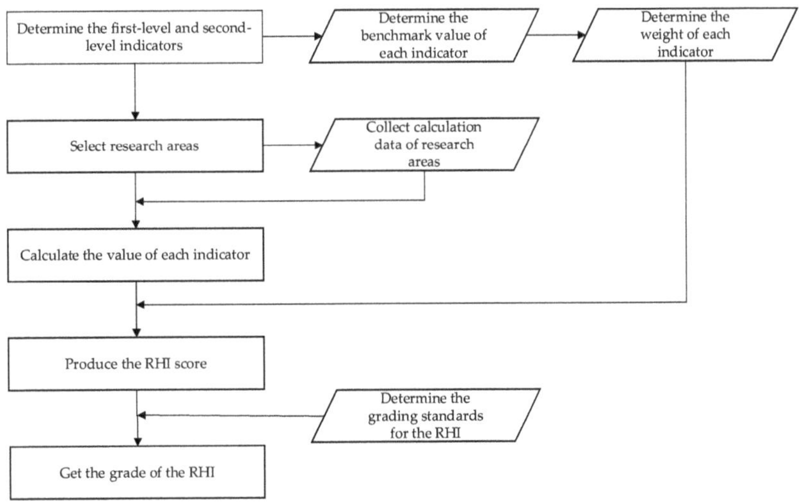

Figure 1. Flowchart of the RHI evaluation system.

In summary, indicators in the five dimensions, such as water security, water resources, water environment, water ecology, and water culture are broken down into 20 second-level indicators and 18 third-level indicators. See Table 3 for the framework of the RHI indicator system.

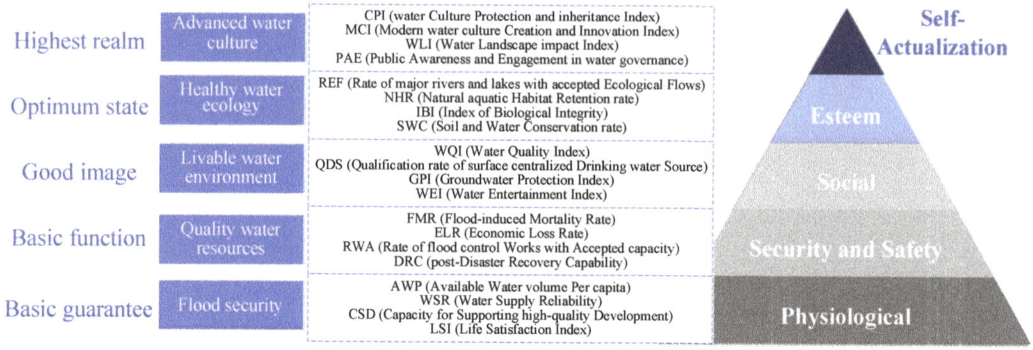

Figure 2. Framework of the RHI.

3.2. Assessment Criteria

The River Happiness Index (*RHI*) is an integrated index that reflects the capacity and level of rivers and lakes to stay in good condition, meet human needs, or provide services, and is specifically measured by five indicators, namely water security, water resources, water environment, water ecology, and water culture. The *RHI* is calculated through the following formulas:

$$RHI = \sum_{i=1}^{5} F_i w_i^f \quad (1)$$

$$F_i = \sum_{j=1}^{4} S_{i,j} w_{i,j}^s \quad (2)$$

$$S_{i,j} = \sum_{k=1}^{K} T_{i,j,k} w_{i,j,k}^t \quad (3)$$

where, *RHI* means River Happiness Index; F_i is the score of the first-level indicator i, i is the subscript of the first-level indicator, which ranges from 1 to 5, indicating FCC, WRR, WEL, AEH, and WCP, respectively; w_i^f is the weight of the first-level indicator i; $S_{i,j}$ is the score of the second-level indicator j of the first-level indicator i, j is the subscript of the second-level indicator, which ranges from 1 to 4; $w_{i,j}^S$ is the weight of the second-level indicator j of the first-level indicator i; $T_{i,j,k}$ is the score of the third-level indicator k of the second-level indicator j of the first-level indicator i, and k is the subscript of the third-level indicator, which ranges from 1 to K; and $w_{i,j,k}^t$ is the weight of the third-level indicator k of the second-level indicator j of the first-level indicator i.

The RHI is scored on a 100-point scale. A RHI score of 85 points or above indicates a "happy river/lake" (Table 4). If indicators at all levels score 85 points or above, then a river/lake is in a good state (Table 5).

Table 3. Indicator system of the RHI.

First-Level Indicator	Second-Level Indicator	Third-Level Indicator	Nature of Indicator		
			State Indicator		Capacity Indicator
			River Health	Human Experience	
Security of Flood Control (SFC)	1. Flood-induced Mortality Rate (FMR)	-	√		
	2. Economic Loss Rate (ELR)	-	√		
	3. Rate of flood control Works with Accepted capacity (RWA)	Rate of Accepted Dikes (RAD)			√
		Rate of Accepted Reservoirs (RAR)			
		Rate of Accepted flood detention Basins (RAB)			
	4. post-Disaster Recovery Capability (DRC)	-			√
Water Resources Reliability (WRR)	5. Available Water volume Per capita (AWP)	-		√	
	6. Water Supply Reliability (WSR)	Water Supply Coverage (WSC)			√
		Rate of actual Irrigated Areas (RIA)			
	7. Capacity for Supporting high-quality Development (CSD)	Water resources Utilization Rate (WUR)			√
		GDP Output per cubic meter of Water use (GOW			
	8. Life Satisfaction Index (LSI)	GDP Per Capita (GPC)		√	
		Engel's Coefficient (ENC)			
		Average Life Expectancy (ALE)			
Water Environment Livability (WEL)	9. Water Quality Index (WQI)	(River water Quality Index (RQI)	√		
		Rate of Eutrophic Lakes and reservoirs (REL)			
	10. Qualification rate of surface centralized Drinking water Source (QDS)	-	√		
	11. Groundwater Protection Index (GPI)	-	√		
	12. Water Entertainment Index (WEI)	-		√	
Aquatic Ecosystem Health (AEH)	13. Rate of major rivers and lakes with accepted Ecological Flows (REF)	-	√		
	14. Natural aquatic Habitat Retention rate (NHR)	Retention Rate of Waters (RRW)	√		
		River longitudinal Connectivity Index (RCI)			
	15. Index of Biological Integrity (IBI)	-	√		
	16. Soil and Water Conservation rate (SWC)	-	√		
Water Culture Prosperity (WCP)	17. water Culture Protection and inheritance Index (CPI)	water Heritage Protection capacity Index (HPI)			√
		Historical water culture Communication Capacity (HCC)			
	18. Modern water culture Creation and Innovation Index (MCI)	-			√
	19. Water Landscape impact Index (WLI)	-		√	
	20. Public Awareness and Engagement in water governance (PAE)	public Awareness Rate of Water (ARW)		√	
		public Engagement Rate in Water governance (ERW)			

Table 4. Grading standard for the *RHI*.

RHI	Grade		
$RHI \geq 95$	Very happy		
$95 > RHI \geq 85$	Happy		
$85 > RHI \geq 60$	Medium	$85 > RHI \geq 80$	Medium high
		$80 > RHI \geq 70$	Middle medium
		$70 > RHI \geq 60$	Medium low
$RHI < 60$	Unhappy		

Table 5. Grading standard for evaluation indicators of the RHI.

Score of Indicator *V* *	Grade		
$V \geq 95$	Excellent		
$95 > V \geq 85$	Good		
$85 > V \geq 60$	Medium	$85 > V \geq 80$	Medium high
		$80 > V \geq 70$	Middle medium
		$70 > V \geq 60$	Medium low
$V < 60$	Poor	$60 > V \geq 30$	Poor
		$V < 30$	Very poor

* *V* indicates F_i, $S_{i,j}$ or $T_{i,j,k}$.

3.3. Indicator Weight Calculation and Evaluation Method

The evaluation method involves five steps:

1. Determine the benchmark value of each indicator. The benchmark values are determined in line with national policies, economic and social development plans, spatial plans of national land, and technical standards; comprehensive and special plans for river basins; advanced levels at home and abroad; and research results of authoritative organizations and research institutes at home and abroad.
2. Calculate the value of each indicator. The values are calculated according to relevant statistical yearbooks, field survey data, and plans.
3. Determine the weight of each indicator. The weights are determined by taking into account the characteristics of river basins, socioeconomic conditions, and the opinions of the people. Weights of first-level and second-level indicators are listed in Table 6, which are mainly determined by the expert comprehensive evaluation method. The weights were initially determined by 24 experts and scholars present through a questionnaire, then validated by another 35 experts.
4. Produce the RHI score. The score of each indicator is calculated by multiplying the value of each indicator by its weight and the sum of the scores of all indicators is the RHI score.
5. Determine the grade of the RHI according to the grading standards for the RHI.

3.4. Data

With 2019 selected as the research year, this paper studies China's 10 first-grade water resource zones, including the Songhua River, the Liaohe River, the Haihe River, the Yellow River, the Huaihe River, the Yangtze River, the Taihu Lake, southeastern rivers, the Pearl River, southwestern rivers, and northwestern rivers, and calculates the values of the indicators by reference to relevant statistical yearbooks, field survey data, and plans (Table 7).

Table 6. Weights of first-level and second-level indicators of the RHI.

First-Level Indicator	Second-Level Indicator	Weight	
Security of Flood Control (SFC)	1. Flood-induced Mortality Rate (FMR)	0.25	0.30
	2. Economic Loss Rate (ELR)		0.30
	3. Rate of flood control Works with Accepted capacity (RWA)		0.30
	4. post-Disaster Recovery Capability (DRC)		0.10
Water Resources Reliability (WRR)	5. Available Water volume Per capita (AWP)	0.25	0.20
	6. Water Supply Reliability (WSR)		0.30
	7. Capacity for Supporting high-quality Development (CSD)		0.25
	8. Life Satisfaction Index (LSI)		0.25
Water Environment Livability (WEL)	9. Water Quality Index (WQI)	0.20	0.30
	10. Qualification rate of surface centralized Drinking water Source (QDS)		0.30
	11. Groundwater Protection Index (GPI)		0.20
	12. Water Entertainment Index (WEI)		0.20
Aquatic Ecosystem Health (AEH)	13. Rate of major rivers and lakes with accepted Ecological Flows (REF)	0.20	0.30
	14. Natural aquatic Habitat Retention rate (NHR)		0.25
	15. Index of Biological Integrity (IBI)		0.20
	16. Soil and Water Conservation rate (SWC)		0.25
Water Culture Prosperity (WCP)	17. water Culture Protection and inheritance Index (CPI)	0.10	0.25
	18. Modern water culture Creation and Innovation Index (MCI)		0.25
	19. Water Landscape impact Index (WLI)		0.25
	20. Public Awareness and Engagement in water governance (PAE)		0.25

Table 7. Indicator calculation methods and sources of data.

Criterion Layer	Indicator Layer		Calculation Method	Source
FDS	FMR		The number of flood-induced deaths and missings/total population, ppm	Bulletin of Flood and Drought Disasters in China
	ELR		Direct economic losses from flood disasters/regional GDP during the same period, %	
	RWA	RAL	The length of main stream dikes that meet flood control standards set forth in relevant plans/total length of planned main stream dikes, %	Reports on comprehensive planning for river basins
		PAR	The number of reservoirs working well in flood control/total number of planned reservoirs, %	
		PAD	The number of flood detention basins working well in flood discharging, storage and detention/total number of planned flood detention basins in the river basin, %	
	DRC		The capacity for restoring post-disaster production and life to an orderly state according to experts	River basin authorities
RWR	AWP		Water resources per capita, m^3/per	China Water Resources Bulletin
	WSR	WSC	The number of population having access to tap water/total population, %	China Urban-Rural Construction Statistical Yearbook
		PIA	Actual irrigation area/farmland irrigation area, %	China Water Statistical Yearbook
	CSD	WER	Water supply/total water resources, %	China Water Resources Bulletin
		GOW	GDP/water consumption, RMB/m^3	

Table 7. *Cont.*

Criterion Layer	Indicator Layer		Calculation Method	Source
	LSI	GDP	GDP in the river basin/population in the river basin, %	China Statistical Yearbook
		ENC	Total food expenditures/total personal consumption expenditures, %	
		ALE	The average age at death of the entire live-born population, year	China Health Statistical Yearbook
LWE	WQI	RQI	The length of rivers with Class III water quality and above/the length of rivers evaluated, %	China Water Resources Bulletin
		PEL	The number of eutrophic lakes and reservoirs/the number of lakes and reservoirs evaluated, %	
	QDS		The number of qualified surface centralized drinking water sources/total number of surface centralized drinking water sources, %	
	GPI		Total regional shallow groundwater withdrawal/total regional allowable groundwater withdrawal, %	Data from national water resources survey and evaluation
	WEI		The number of national water parks per 100,000 km^2 of area	
HAE	REF		The number of control sections (points) meeting the ecological flow target/the number of sections (points) evaluated, %	Comprehensive planning of water resources
	NHR	SAR	Water space area/historical reference area, %	Results of remote sensing interpretation of land use
		LCI	Barrier coefficient × position correction factor/the length of the river	Outcomes of national water censuses
	IBI		Index of Biological Integrity (IBI)/the number of cases	National river and lake health assessment
	SWC		Land area of mild soil erosion or below/land area evaluated, %	National dynamic monitoring of water and soil loss
PWC	CPI	HPC	(The number of provincial heritage sites + 2 × the number of national heritage sites + 5 × the number of world heritage sites)/drainage area/100,000 km^2	Relevant heritage lists
		CDC	(2 × the number of national museums or bases + the number of provincial museums or bases)/drainage area/100,000 km^2	
	MCI		(2 × the number of national (laws and regulations + standards + awards + patents for invention) + the number of provincial (laws and regulations + standards + awards + patents for invention))/drainage area/100,000 km^2	
	WLI		(5 × (the number of world-class) + 2 × (the number of national) natural heritage water parks + the number of provincial (wetland parks + national parks))/population in the river basin	Natural heritage lists
	PAE	ARW	public Awareness Rate of Water (ARW)/reference value	Public water awareness questionnaires
		PER	public Engagement Rate in Water governance (ERW)/reference value	Special survey reports on public engagement in water governance

4. Results

4.1. Flood Control Capacity (FCC)

FCC scores 84.9 points nationwide, which is at a medium-high level. The overall flood control capacity nationwide is: the overall RWA reached a medium-high level, and although DRC is still at a medium level, the impact of floods on the security of lives and properties of the people along the rivers was significantly reduced. Nationally, the sense of security of the people along the rivers was effectively guaranteed, laying a foundation for achieving the vision of "a peaceful life with secure rivers".

FCC evaluation results for the first-grade water resource zones are shown in Figure 3. Specifically, FCC scores for the Taihu Lake, the Yellow River, the Huaihe River, the Songhua River, and the Yangtze River all exceed 85 points, reaching a good level, and exceed 80 points for the Pearl River, the Haihe River, the Liaohe River, southeastern rivers, and northwestern rivers, staying at a medium-high level. Southwestern rivers score less than 80 points on FCC, representing a medium level.

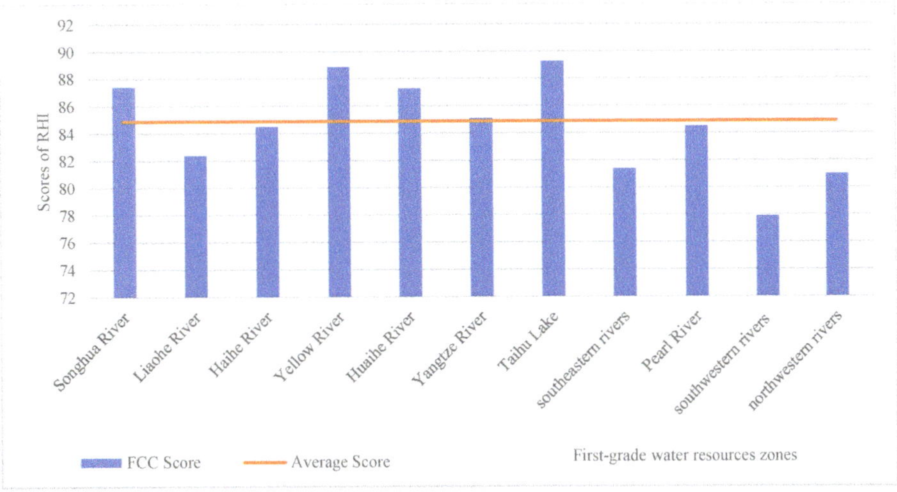

Figure 3. FCC evaluation results for China's first-grade water resource zones.

4.2. Water Resources Reliability (WRR)

WRR scores 77.1 points nationwide, which is generally at a medium level. The overall water resource reliability nationwide is: AWP stands at a medium-high level against the international warning line for water shortage, WSR is at a good level, and WUR is well below 40%, but water resources are unevenly distributed across regions, which is incompatible with population distribution and allocation of productive forces; the level of water conservancy is yet to be further improved, and there is a clear gap in GDP Output per cubic meter of Water use (GOW) compared with high-income countries; and CSD is generally at a medium-low level. Continuing to play the basic role of happy rivers and lakes to achieve the vision of "an affluent life with reliable water supply" is still on the way.

WRR evaluation results for the first-grade water resource zones are shown in Figure 4. In general, scores in the north are lower than in the south. Specifically, scores for the Yellow River, the Haihe River, the Huaihe River, northwestern rivers, and the Liaohe River are all below 70 points, at a medium-low level. Scores for southeastern rivers and the Taihu Lake are higher than 80 points, reaching a medium-high level.

4.3. Water Environment Livability (WEL)

WEL scores 70.4 points nationwide, which is generally at a medium level. The overall water environment livability nationwide is: the water quality of rivers generally reaches a good level and QDS is at a near-good level, but groundwater resources are poorly conserved, lakes and reservoirs are seriously eutrophic and there is a gap between the environment of urban and rural waters closely related to people's daily life and their water entertainment demand. The vision of "a livable environment with clear waters and green banks" is still far away.

WEL evaluation results for the first-grade water resource zones are shown in Figure 5. Specifically, southeastern rivers score 90.7 points, performing best and reaching a good level; both the Yangtze River and southwestern rivers score 80+ points, staying at a medium-high level; the Pearl River and northwestern rivers score 70–80 points, at a medium level; the Taihu Lake, the Yellow River and the Huaihe River score 60–70 points, at a medium-low level; while the Songhua River, the Haihe River, and the Liaohe River score less than 60 points, at a poor level.

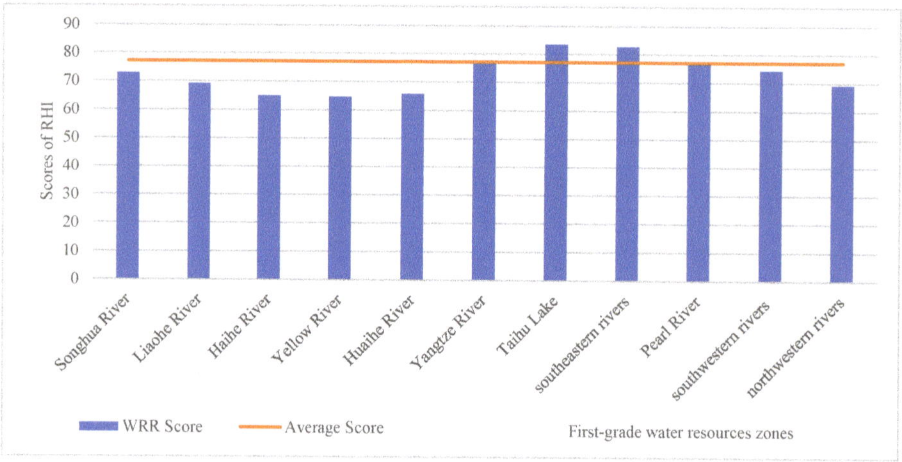

Figure 4. WRR evaluation results for China's first-grade water resource zone.

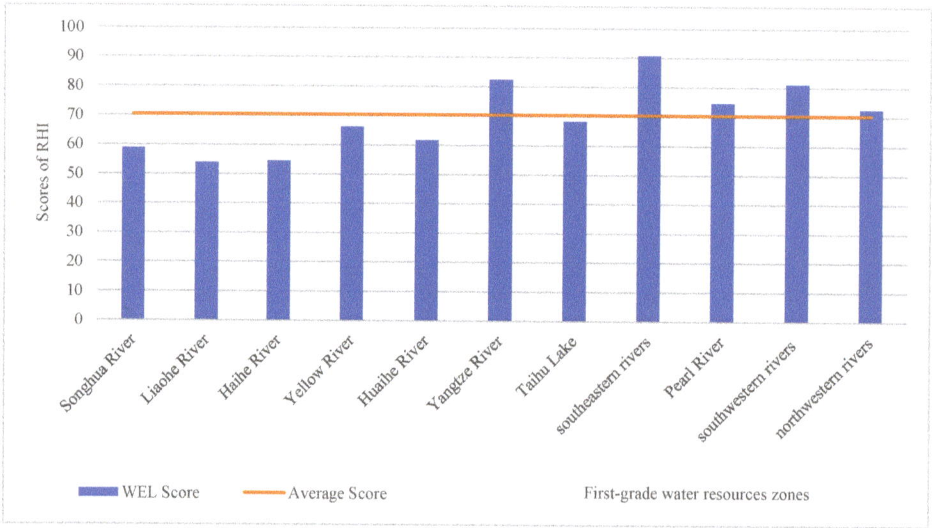

Figure 5. WEL evaluation results for China's first-grade water resource zones.

4.4. *Aquatic Ecosystem Health (AEH)*

AEH scores 74.1 points nationwide, which is generally at a medium level. The overall aquatic ecosystem health nationwide is: owing to fruitful water and soil conservation efforts, SWC reaches a good level, and under the support of steady guarantee for ecological flows and management through the river/lake chief system, REF generally improved to a medium level; however, both NHR and IBI are still at a medium-low level. The quality and stability of river and lake ecosystems should be systematically improved before the vision of "a symbiotic harmony with fish swimming in shallows" can be achieved.

AEH evaluation results for the first-grade water resource zones are shown in Figure 6. Southwestern rivers and the Songhua River score 85+ points, the highest among the first-class water resource zones, reaching a good level; scores for the Taihu Lake, the Pearl River, the Yangtze River, southeastern rivers, the Huaihe River, and northwestern rivers fall between 70 and 80 points, all at a medium level; the Liaohe River and the Haihe River score 60–70 points, both at a medium-low level; the Yellow River scores 56.8 points, the lowest, at a poor level.

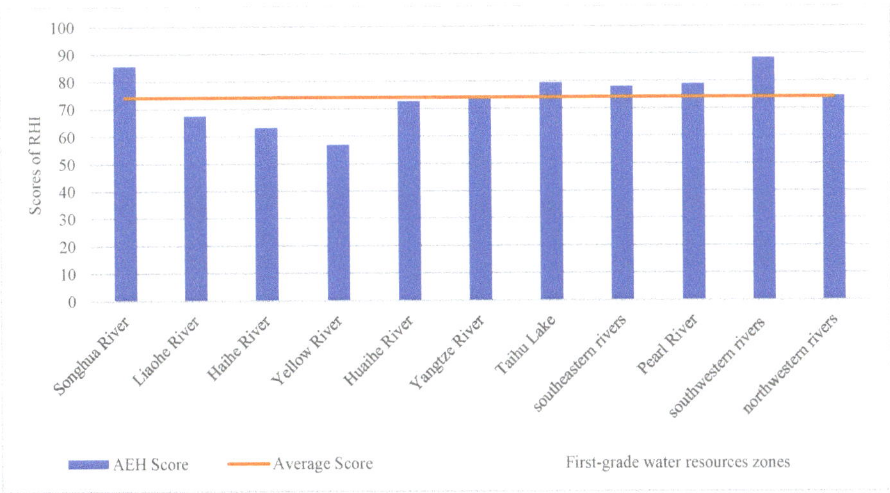

Figure 6. AEH evaluation results for China's first-grade water resource zones.

4.5. Water Culture Prosperity (WCP)

WCP scores 77.0 points nationwide, which is generally at a medium level. The overall water culture prosperity nationwide is: China boasts a profound historical background of water culture, which represents a precious treasure for achieving the vision of "a spiritual homeland with river civilizations"; however, both CPI and MCI are low, and PAE, which indicates the respect for and protection of rivers, is generally low and still at a medium level relative to people's increasing demand for cultural life, which is yet to improve.

WCP evaluation results are shown in Figure 7. Scores for the Yangtze River, the Yellow River, the Huaihe River, southeastern rivers, the Pearl River, and the Taihu Lake all exceed 80 points, reaching a medium-high level; the Haihe River, southwestern rivers, and northwestern rivers score 70–80 points on WCP at a medium level; scores for the Songhua River and the Liaohe River are relatively low, which are 68.8 and 67.5 points, respectively, at a medium-low level.

4.6. Overall Evaluation

In 2019, the FCC score was the highest nationwide, reaching a near-good level, while scores of WRR, WEL, AEH, and WCP fell between 70 and 80 points, all at a medium level (Figure 8).

In 2019, China's RHI scored 77.1 points at a medium level. On the whole, among the secondary indicators, the mortality rate, water supply rate, and soil and water conservation rate of flood disaster were relatively the highest, reaching a good grade. The economic loss rate of flood disasters, the standard rate of flood control projects, the per capita water resources availability, and the river and lake water quality index scored the second highest, which belonged to the medium-high grade. The degree of protection of groundwater resources and the degree of water entertainment score was low, and the evaluation grade was poor. The score for water resources supporting development ability, natural habitat retention rate, and aquatic biological integrity was the second lowest, which was medium-low level. The conditions of rivers and lakes in China calculated by RHI are consistent with the actual situation. The overall RHI evaluation results are shown in Figure 9. Specifically, RHI scores for southeastern rivers and the Taihu Lake were the highest, at a medium-high level, and RHI scores for the Yangtze River, southwestern rivers, and the Pearl River were

79.9, 79.4, and 79.3 points, respectively, all at a medium level. RHI scores for these southern first-grade water resource zones were higher than the national average. In the north, the Songhua River had the highest RHI score, which was 75.8 points (close to the national average), while the Yellow River, the Huaihe River, and northwestern rivers scored a little more than 70 points, indicating a medium level of river happiness. Scores for the Liaohe River and the Haihe River lay between 60 and 70 points, indicating a medium-low level of river happiness.

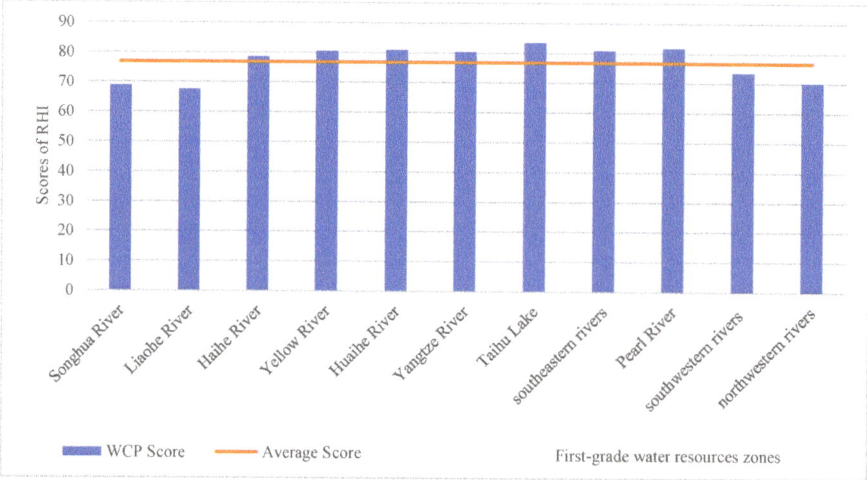

Figure 7. WCP evaluation results for China's first-grade water resource zones.

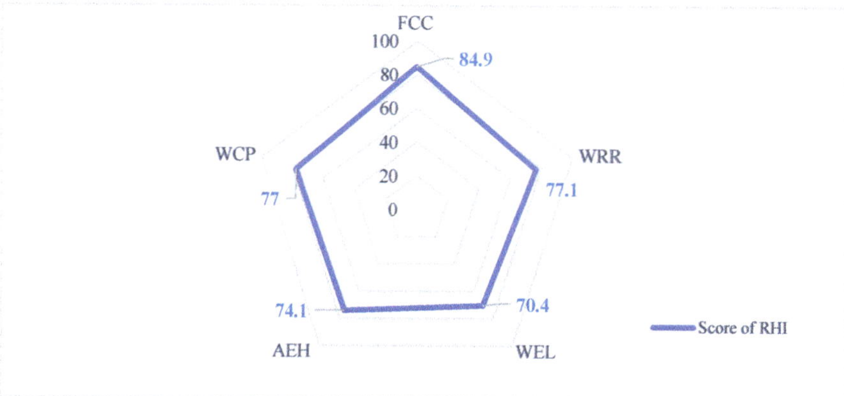

Figure 8. Evaluation results of the first-level indicators of the RHI in China.

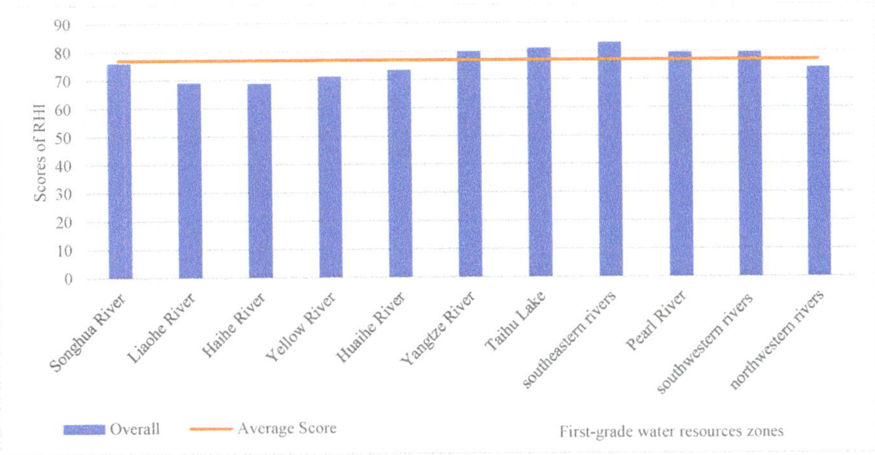

Figure 9. RHI evaluation results for China's first-grade water resource zones.

4.7. Evaluation of a Typical River Basin

The Yellow River, having nurtured and bred the Chinese civilization, is also an important economic zone in China, with the Huang-Huai-Hai Plain, the Fenwei Plain, and the Hetao Irrigation Area being major agricultural production areas, which contribute to around one third of the country's food and meat production. The Yellow River Basin, also known as the "Energy Basin", boasts abundant coal, oil, gas, and nonferrous metal resources, with coal reserves there accounting for more than half of the national total. It is an important base for energy, chemicals, raw materials, and basic industries in China.

The RHI scores 71.0 points in the Yellow River Basin, staying at a medium level and putting the river basin in eighth place among the 10 first-grade water resource zones nationwide. The evaluation results of the first-level indicators of the RHI are listed in Table 8. As shown in the table, the FCC score is the highest, reaching a good level; WCP is at a medium-high level; the AEH score is the lowest, at a poor level; and other indicators are at a medium-low level.

FCC. FCC scores 88.9 points in the Yellow River Basin, generally at a good level. Rate of Accepted Dikes (RAD) is 87.7%, the Rate of Accepted Reservoirs for medium-sized and large reservoirs (RAR), and the Rate of Accepted flood detention Basins (RAB) are 100% and RAR (for small reservoirs) is 98%, so RWA scores 94.7 points, reaching a near-excellent level. FMR (0.4 per million people) scores 91.9 points and ELR is 0.22%, which scores 85.3 points, both at a good level. DRC scores 73.7 points, at a medium level, indicating relatively weak resilience.

WRR. WRR scores 64.6 points in the Yellow River Basin, generally at a medium-low level. Water Supply Coverage (WSC) is 91.3% and the Rate of actual Irrigated Areas (RIA) is 79.9%, so WSR scores 86.3 points, reaching a good level; LSI scores 77.0 points, at a medium level; AWP is 620.2 m^3, scoring 44.8 points, and as Water resources Utilization Rate (WUR) exceeds 70% and GDP Output per cubic meter of Water use (GOW) is RMB 180.5/m^3, CSD scores 42.0, both at a poor level that is still a long way from the medium level.

WEL. WEL scores 66.2 points in the Yellow River Basin, generally at a medium-low level. To be specific, the groundwater exploitation coefficient is 1.04 and GPI scores 26.0 points, staying at a very poor level; WEI scores 69.8 points, standing at a medium-low level; QDS scores 74.5 points, at a medium level; the length of rivers with Class I~III water quality accounts for 80.3% and that of rivers with water quality inferior to Class V accounts for 9.2%, and the Rate of Eutrophic Lakes and reservoirs (REL) is 25%, so WQI scores 82.2 points, at a medium-high level.

Table 8. RHI evaluation results for the Yellow River Basin.

Criterion Layer	Indicator Layer		Benchmark Value	Status Quo Value	Score	Weight	Subtotal
FDS	FMR		0%	0.40%	91.9	0.30	88.9
	ELR		0%	0.22%	85.3	0.30	
	RWA	RAL	100%	87.5%	87.5	0.12	
		PAR	100%	Medium-sized and large reservoirs: 100%, small reservoirs: 98%	99.2	0.12	
		PAD	100	100.00	100.0	0.06	
	DRC		100	73.7	73.7	0.10	
RWR	AWP		10,000 m^3/per	620.21m^3/per	44.8	0.20	64.6
	WSR	WSC	100%	91.31%	91.3	0.17	
		PIA	100%	79.86%	79.9	0.13	
	CSD	WER	40%	70.76%	50.6	0.12	
		GOW	509 m^3	180.51 m^3	34.0	0.13	
	LSI	GDP	132,400 (RMB)	59,502.17 (RMB)	43.1	0.08	
		ENC	25%	26.58%	94.1	0.09	
		ALE	81	74.16	91.6	0.08	
LWE	WQI	RQI	The proportion of the length of river with Class I~III water quality ≥ 90%	The proportion of the length of rivers with Class I~III water quality: 80.3%, the proportion of the length of rivers with water quality inferior to Class V: 9.2%	87.1	0.18	66.2
		PEL	0%	25.00%	75.0	0.12	
	QDS		100%	74.50%	74.5	0.30	
	GPI		0.3	1.04	26.0	0.20	
	WEI		100	14.90	69.8	0.20	
HAE	REF		100%	25.00%	25.0	0.30	56.8
	NHR	SAR	100%	79.02%	79.0	0.125	
		LCI	-	0.61%	75.4	0.125	
	IBI		1	0.429	42.9	0.20	
	SWC		-	-	85.7	0.25	
PWC	CPI	HPC	10	9	90.0	0.15	80.6
		CDC	6	4.92	82.0	0.10	
	MCI		6	4.86	81.0	0.25	
	WLI		6	3.84	78.4	0.25	
	PAE	ARW	-	-		0.00	
		PER	100%	76.3%	76.3	0.25	

AEH. AEH scores 56.8 points in the Yellow River Basin, generally at a poor level. SWC scores 85.7 points, at a medium-high level, the highest among the second-level indicators, followed by NHR, which scores 77.2 points, at a medium level; IBI scores 42.9 points, at a poor level; and REF scores 25.0 points, at a very poor level.

WCP. WCP scores 80.6 points in the Yellow River Basin, generally at a medium-high level. CPI scores 86.8 points, reaching a good level; MCI scores 81.0 points, at a medium-high level; and PAE scores 76.3 points and WLI scores 78.4 points, both at a medium level.

The RHI evaluation results for the Yellow River Basin suggest that the main problems are in the following areas: first, inadequate post-disaster recovery capability is the main risk affecting water security in the Yangtze River Basin; second, an inherent shortage of water resources and a high rate of water resource exploitation and utilization remain the biggest restrictions on high-quality economic and social development; third, serious groundwater overdraft in some areas and heavy pollution of tributaries are major problems to be solved as early as possible in order to create a livable water environment in the river basin; fourth, the low rate of rivers with accepted ecological flows is a weak link to be addressed in order to maintain healthy water ecology; and fifth, the brand effect of water culture is yet to be enhanced and the water landscape impact is yet to be improved.

5. Discussion

5.1. Uncertainty of the Evaluation Method

Happiness is a subjective feeling for people, and rivers that make people feel happy should have common objective characteristics in terms of water security, water resources, water environment, water ecology, and water culture. Based on the five dimensions, the RHI is composed of altogether 30 indicators, including five first-level indicators, namely FCC, WRR, WEL, AEH, and WCP, each of which consist of four second-level indicators and corresponding third-level indicators. In the process of evaluation, the determination of weight coefficients is crucial. At present, research on evaluation indicators more often employs entropy weight, standard deviation, and Criteria Importance Through Intercriteria Correlation (CRITIC) methods, which can calculate indicator weights through mathematical methods based on original data, but render the scope of application limited, as they fully reflect the characteristics of selected data. Since the degree of flood control construction, water environment protection, utilization, and demand varies greatly from river to river due to the complexity of each river, selecting a universal indicator system that takes into account the characteristics of different rivers is vital to the evaluation of rivers.

5.2. Applicability of the Evaluation Method

Current evaluation of rivers mainly deals with the hydrological condition of natural attributes, ecosystems, and social functions of rivers, with focus on the health of rivers and different indicator systems established for different rivers. For instance, the Upper Mississippi River Restoration Program (UMRRP) assessed the ecosystem health of the Upper Mississippi River using seven categories of indicators, such as hydrology, sedimentation, water quality, land cover, aquatic vegetation, invertebrate and fish, and 25 specific monitoring indicators. The status and trend reports published indicate that most indicators remain relatively stable, with ecosystems being healthy in the north and relatively unhealthy in the south [28,29]. With a growing demand for quality of life, healthy evaluation of a river's happiness should adopt a people-centered approach that not only assesses the general social and economic functions of the river by systematically measuring its water security, water supply, and water environment services, but also scientifically examines its aquatic ecosystem quality and water culture prosperity in the light of the natural endowments and cultural background of the river basin [30]. Therefore, in the case of the Mississippi River, the RHI scores 80.1 points, at a medium-high level. To break it down, FCC scores 90+ points, which is at a good level; WRR and WEL score 80+ points, both at a medium-high level; and AEH and WCP score less than 80 points, both at a medium level. The evaluation results show the following characteristics of the Mississippi: first, the river basin is generally well managed and reaches a medium-high level in terms of water security, water resources, and water environment, indicating that the long-term systematic governance of the river basin produced desirable results and a high level of public awareness and engagement in water governance; second, low scores on the eco-hydrological process variation index and the longitudinal connectivity index imply the great impact of human activity on natural habitats and a high rate of water resource development; and third, historical and cultural inheritance and protection is inadequate in the river basin. From the perspective of the evaluation results, the RHI performs better in comprehensively describing the overall situation of a large river.

5.3. Policy Recommendations

River health, as a comprehensive concept, is increasingly embodied in domestic and foreign water resource management systems, but there are no specific rules for the implementation of river evaluation indicator systems, and the indicator systems selected for a single river may vary. In order to scientifically assess the status of rivers and lakes in different regions, an international guideline for river and lake evaluation should be introduced, which can take into account the differences and commonalities of river ecosystems and provide a unified, standardized technical system for evaluation. This requires a unified indicator system, under which proper adjustments can be made to indicators according to the situation of a specific river basin, but such adjustments should meet scientific cognition, assessment standards, and other requirements. Major rivers and lakes should be assessed on a regular basis. In response to underperforming indicators, such as RWA, GPI, and REF, the problem of unbalanced spatial distribution of water resources should be addressed as early as possible. For underperforming indicators, such as AWP and WUR, the principle of giving priority to saving water must be put into practice. For underperforming indicators such as WEI, IBI, and CPI, scientific sector-specific decisions should be made.

6. Conclusions

As social and economic development needs and anthropogenic threats grow, countries around the world are suffering from numerous river and lake problems, such as altered hydrological processes, damaged physical structures, polluted water, and declining aquatic biodiversity, and the evaluation of rivers and lakes is changing from purely water quality evaluation to a more comprehensive one. This paper gives a new definition of a happy river, which enriches the new connotation of water management. A happy river refers to a river that can maintain its own health, support high-quality economic and social development in the river basin, as well as reflect human–water harmony, thus give people a high sense of security, gain, and satisfaction. The RHI method is based on Maslow's hierarchy of needs theory and draws lessons from the World Happiness Report and SDG Report. Compared with the previous studies, the study insisted on a people-centered approach, and takes the river culture into consideration. This paper develops evaluation indicators, such as FCC, WRR, WEL, AEH, and WCP in terms of water resources, security, ecology, environment, and culture, and assesses the overall river happiness of major rivers and lakes nationwide by examining the situation of China's 10 first-grade water resource zones and the Taihu Lake Basin in 2019.

The analysis of the evaluation indicators suggests that China's RHI scores 77.1 points, which is at a medium level, with FCC at a near-good level and WRR, WEL, AEH, and WCP all at a medium level. Meanwhile, the RHI in first-grade water resource zones in southern China stands above the national average, in contrast to the situation in the north, mainly because WRR and WEL are lower in the north than in the south.

The analysis of the river basins evaluated supports that WEL for the Songhua River, the Liaohe River, and the Haihe River in the north is at a poor level, restricting the quality and stability of aquatic ecosystems; AEH for the Yangtze River, the Taihu Lake, and southeastern rivers in the south is at a below-average level, indicating the poor condition of important aquatic organisms, which is a major problem that requires attention in order to maintain healthy aquatic ecosystems in these regions. Problems in the Yellow River Basin, which involves 340 counties (county-level cities, districts, or banners) in 66 prefectures (prefecture-level cities, autonomous prefectures, or leagues) in nine provinces and autonomous regions, are more complex, with FCC reaching a good level, WCP at an above-average level, AEH at a poor level, and WRR and WEL at a below-average level. Based on the evaluation results, we therefore propose targeted basin governance measures: more attention should be paid to the intensive and economical use of water resources in northern China, and ecological flows of rivers must be effectively ensured in southern.

On the whole, happy rivers should be built in a way that seeks to maintain the health of rivers while pursuing greater benefits for the people by following the basic principle of human–water harmony on the premise of maintaining river health. With regard to the perspective of the future work, except the expert comprehensive evaluation method applied in the current study, the index weights can also be determined by the Analytic

Hierarchy Process (AHP), considering the characteristics of the river basin, social and economic conditions, or people's opinions. Secondly, at present the model may not include all aspects of a river, such as its hydrodynamics, transport, and species migration, so more relevant aspects can be supplemented and studied in subsequent work. Furthermore, building happy rivers and lakes is not only applicable to those in China, but also in the rest of the world. To this end, a guideline must be developed as soon as possible to provide technical support for the building of happy rivers and lakes. We will continue to select more representative world rivers to verify the accuracy and adaptability of the RHI evaluation method.

Author Contributions: Conceptualization, C.L.; methodology, C.L. and S.J.; formal analysis, C.L. and Q.J.; writing—original draft preparation, Q.J.; writing—review and editing, C.L. and S.J.; visualization, Q.J.; supervision, C.L.; All authors have read and agreed to the published version of the manuscript.

Funding: This research was funded by the National High Technology Research and Development Program of China (No.2021YFC3200205); and the Cooperation Project of Shandong Water Transfer Operation and Maintenance Center (No.37000000025002920210100001).

Institutional Review Board Statement: Not applicable.

Informed Consent Statement: Not applicable.

Data Availability Statement: The data that support the finding of this study are available from the corresponding author upon reasonable request.

Conflicts of Interest: The happy river research group of the China Institute of Water Resources and Hydropower Research is involved in relevant work.

References

1. GE, J.X. Rivers and Human Civilizations. *Folk. Stud.* **2021**, *6*. 5–13+158. (In Chinese)
2. Carpenter, D.; Schwartz, J.; Slate, L.; Sinha, S.; Brennan, K.; MacBroom, J. The Status of Urban Stream Restoration in the United States. In *Critical Transitions in Water and Environmental Resources Management*; American Society of Civil Engineers: Reston, VA, USA, 2004; pp. 1–13.
3. Palmer, M.A.; Bernhardt, E.S.; Allan, J.D.; Lake, P.S.; Alexander, G.; Brooks, S. Standards for Ecologically Successful River Restoration. *J. Appl. Ecol.* **2005**, *42*, 208–217. [CrossRef]
4. Denneman, W.D.; De Pree, A.; Reininga, G.A.O.; Van Der Braak, J. Environmental Aspects of the Restoration of River Ecosystems in the Netherlands. *Water Sci. Technol.* **1995**, *31*, 147–150. [CrossRef]
5. Chick, J.H.; Ickes, B.S.; Pegg, M.A.; Barko, V.A.; Hrabik, R.A.; Herzog, D.P. *Spatial Structure and Temporal Variation of Fish Communities in the Upper Mississippi River System*; Geological Survey La Crosse Wi Upper Midwest Environmental Sciences Center: La Crosse, WI, USA, 2005.
6. Gloss, S.; Lovich, J.E.; Melis, T.S. *The State of the Colorado River Ecosystem in Grand Canyon: A Report of the Grand Canyon Monitoring and Research Center 1991–2004*; US Department of the Interior, US Geological Survey: Washington, DC, USA, 2005; Volume 1282.
7. Zuo, Q.T.; Han, C.H.; Han, C.H. Study on the Theoretical Basis and Support System of River Governor System. *Yellow River* **2017**, *39*. 1–6+15 (In Chinese)
8. Peng, H.; Han, Q.; Cao, J.P. Study on Evaluation Index System of River Chief System and Lake Chief System for Taihu Basin. *China Water Resour.* **2019**, *06*. 11–15+5 (In Chinese)
9. Yu, X.B.; Tang, D.S. Comprehensive Implementation of River Chief System Effectiveness Evaluation in Jiangsu Province Based on AHP-EVM. *Yellow River* **2020**, *42*. 63–68+73 (In Chinese)
10. Wang, G.J.; Liu, Z.; Lang, M.X. Performance Evaluation of River-Lake Chief System and its Implications. *China Water Resour.* **2021**, *02*, 15–18. (In Chinese)
11. Speech at the Symposium on Ecological Protection and High-Quality Development of the Yellow River Basin. Available online: http://www.gov.cn/xinwen/2019-10/15/content_5440023.htm (accessed on 2 August 2022).
12. To Write a New Chapter in River Protection and Management in the New Era. Available online: http://www.mwr.gov.cn/xw/slyw/201912/t20191205_1373783.html (accessed on 2 August 2022).
13. Wu, E.N.; Yang, K.; Che, Y.; Yuan, W. Characterization of Rivers Health Status and its Assessment. *Adv. Water Sci.* **2005**, *16*, 602–608.
14. Scrimgeour, G.J.; Wicklum, D. Aquatic Ecosystem Health and Integrity: Problems and Potential Solutions. *J. N. Am. Benthol. Soc.* **1996**, *15*, 254–261. [CrossRef]
15. Wright, J.F.; Sutcliffe, D.W.; Furse, M.T. *Assessing the Biological Quality of Freshwaters: RIVPACS and Other Techniques*; Freshwater Biological Association: Ambleside, UK, 2000; pp. 1–24.

16. Smith, M.J.; Kay, W.R.; Edward, D.H.D.; Papas, P.J.; Richardson, K.S.J.; Simpson, J.C. AusRivAS: Using Macroinvertebrates to Assess Ecological Condition of Rivers in Western Australia. *Freshw. Biol.* **1999**, *41*, 269–282. [CrossRef]
17. Karr, J.R. Assessments of Biotic Integrity Using Fish Communities. *Fisheries* **1981**, *6*, 21–27. [CrossRef]
18. Ladson, A.R.; White, L.J.; Doolan, J.A.; Finlayson, B.L.; Hart, B.T.; Lake, P.S.; Tilleard, J.W. Development and Testing of an Index of Stream Condition for Waterway Management in Australia. *Freshw. Biol.* **1999**, *41*, 453–468. [CrossRef]
19. Raven, P.J.; Holmes, N.T.H.; Dawson, F.H.; Everard, M. Quality Assessment Using River Habitat Survey Data. *Aquat. Conserv. Mar. Freshw. Ecosyst.* **1998**, *8*, 477–499. [CrossRef]
20. Zuo, Q.T. Human-Water Harmony Theory: From Idea to Theory System. *Water Resour. Hydropower Eng.* **2009**, *40*, 25–30. (In Chinese)
21. Zuo, Q.T.; Zhang, Y.; Lin, P. Index System and Quantification Method for Human-Water Harmony. *J. Hydraul. Eng.* **2008**, *4*, 440–447. (In Chinese)
22. Maslow, A.H. *Motivation and Personality*, 3rd ed.; China Renmin University Press: Beijing, China, 2012. (In Chinese)
23. Helliwell, J.F.; Layard, R.; Sachs, J. World Happiness Report 2012. New York: UN Sustainable Development Solutions Network. Available online: https://worldhappiness.report/ed/2012/ (accessed on 2 August 2022).
24. Research Group of the Happy River. Analysis of the Connotation and Index System for the Happy River. *China Water Resour.* **2020**, *23*, 1–4. (In Chinese)
25. Liu, C.S.; Wang, J.H.; Jiang, Y.Z.; Qiu, Y.Q.; Zhang, H.T.; Guan, X.Y. River Happiness Index: The Evaluation System of River to Enriching the People. *J. China Inst. Water Resour. Hydropower Res.* **2021**, *19*, 441–448. (In Chinese)
26. China Institute of Water Resources and Hydropower Research. *China River Happiness Report 2020*; China Water Resources and Hydropower Press: Beijing, China, 2021. (In Chinese)
27. Transforming our World: The 2030 Agenda for Sustainable Development. Available online: https://sdgs.un.org/2030agenda (accessed on 2 August 2022).
28. Upper Mississippi River Restoration Program. Available online: https://www.mvr.usace.army.mil/Missions/Environmental-Stewardship/Upper-Mississippi-River-Restoration/ (accessed on 2 August 2022).
29. Johnson, B.L.; Hagerty, K.H. *Status and Trends of Selected Resources of the Upper Mississippi River System: A Synthesis Report of the Long Term Resource Monitoring Program*; US Geological Survey, Upper Midwest Environmental Sciences Center: La Crosse, WI, USA, 2008.
30. Pietrucha-Urbanik, K. Assessment Model Application of Water Supply System Management in Crisis Situations. *Global NEST J.* **2014**, *16*, 893–900.

MDPI
St. Alban-Anlage 66
4052 Basel
Switzerland
www.mdpi.com

Water Editorial Office
E-mail: water@mdpi.com
www.mdpi.com/journal/water

www.ingramcontent.com/pod-product-compliance
Lightning Source LLC
Chambersburg PA
CBHW061511210926
51850CB00008B/1879
9783036586397